Chemesthesis

Chemical Touch in Food and Eating

Chemesthesis

Chemical Touch in Food and Eating

EDITED BY

Shane T. McDonald PhD

Kalsec©, Inc.
Kalamazoo, MI

David A. Bolliet MS

Kalsec©, Inc.
Kalamazoo, MI

John E. Hayes PhD

Sensory Evaluation Center and Department of Food Science
College of Agricultural Sciences
The Pennsylvania State University
University Park, PA, USA

FOREWORD BY JOHN PRESCOTT

WILEY Blackwell

This edition first published 2016 © 2016 by John Wiley & Sons, Ltd.

Registered Office
John Wiley & Sons, Ltd, The Atrium, Southern Gate, Chichester, West Sussex, PO19 8SQ, UK

Editorial Offices
9600 Garsington Road, Oxford, OX4 2DQ, UK
The Atrium, Southern Gate, Chichester, West Sussex, PO19 8SQ, UK
111 River Street, Hoboken, NJ 07030-5774, USA

For details of our global editorial offices, for customer services and for information about how to apply for permission to reuse the copyright material in this book please see our website at www.wiley.com/wiley-blackwell.

Library of Congress Cataloging-in-Publication data applied for

ISBN: 9781118951736

A catalogue record for this book is available from the British Library.

Wiley also publishes its books in a variety of electronic formats. Some content that appears in print may not be available in electronic books.

Cover image: © Steve Allen (PhotoDisc)

Set in 9.5/12pt Meridien by SPi Global, Pondicherry, India

1 2016

Contents

List of contributors

Yeranddy A. Alpizar PhD
Department of Cellular and Molecular Medicine, Laboratory of Ion Channel Research and TRP Research Platform Leuven (TRPLe), Leuven, Belgium

David A. Bolliet MS
Senior Manager Analytical R&D, Kalsec®, Inc., Kalamazoo, MI, USA

Ali Bouzari PhD
Co-Founder, CSO, Pilot R+D, Healdsburg, CA, USA

Brian Byrne PhD
CEO, Natural Advantage LLC, Oakdale, LA, USA

Nadia Byrnes PhD
Postdoctoral Fellow, Department of Viticulture and Enology, University of California Davis, Davis, CA, USA

E. Carstens PhD
Department of Neurobiology, Physiology and Behavior, University of California Davis, Davis, CA, USA

Pamela Dalton PhD, MPH
Member, Monell Chemical Senses Center, Philadelphia, PA, USA

Barry G. Green PhD
Director and Fellow, The John B. Pierce Laboratory, Professor, Department of Surgery (Otolaryngology), Yale School of Medicine, New Haven, CT, USA

Howard Haley BS
Fellow, Kalsec®, Inc., Kalamazoo, MI, USA

John E. Hayes PhD
Sensory Evaluation Center and Department of Food Science, College of Agricultural Sciences, The Pennsylvania State University, University Park, PA, USA

Christopher R. Loss PhD, AOS
Director of Academic Research, The Culinary Institute of America, Hyde Park, NY, USA

Mary-Jon Ludy PhD, RD
Assistant Professor of Clinical Nutrition, Department of Public and Allied Health, Bowling Green State University, Bowling Green, OH, USA

Richard D. Mattes MPH, PhD, RD
Distinguished Professor, Department of Nutrition Science, Purdue University, West Lafayette, IN, USA

Shane T. McDonald PhD
Principal Flavor Chemist, Kalsec®, Inc., Kalamazoo, MI, USA

Steven Pringle PhD
Corporate VP Sales & Marketing, Aroma Chemical Services International, Gmbh, Höxter-Stahle, Germany

Cecil J. Saunders PhD
Department of Otorhinolaryngology, Perelman School of Medicine, University of Pennsylvania, Philadelphia, PA, USA

Christopher T. Simons PhD
Assistant Professor of Sensory Science, Department of Food Science and Technology, The Ohio State University, Columbus, OH, USA

Wayne L. Silver PhD
Professor, Department of Biology,
Wake Forest University,
Winston-Salem, NC, USA

Karel Talavera PhD
Department of Cellular and Molecular Medicine, Laboratory of Ion Channel Research and TRP Research Platform Leuven (TRPLe),
Leuven, Belgium

Thomas Voets PhD
Department of Cellular and Molecular Medicine, Laboratory of Ion Channel Research and TRP Research Platform Leuven (TRPLe),
Leuven, Belgium

Cindy Ward PhD
Sensation Research,
Maineville, OH, USA

Foreword

Unless my reading list has a major gap, "*Chemesthesis: Chemical Touch in Food and Eating*" is the first book to focus entirely on the topic of chemesthesis since the 1990s *Irritation* (Green *et al.*, 1990). At that time, the study of oral and nasal somatosensory stimuli was commonly termed the *common chemical sense* or *pungency* or *irritation* or the *trigeminal sense.* Since none of these terms were entirely satisfactory (see Chapter 1 in the present book), in their Preface the Editors of *Irritation* proposed the term *chemesthesis,* meaning a general chemical sensibility, and the name has stuck.

At the time *Irritation* (the proceedings of a conference hosted by Monell Chemical Senses Center) was published, chemesthesis was talked about as the "forgotten flavor sense". It was clear that stimulation of various branches of the trigeminal nerves was crucial to food flavors – chili, containing the powerful irritant capsaicin, is consumed by a substantial proportion of the world's population every day, for example. At the same time, we knew very little about *how* irritant stimuli contributed to food flavor, and it was this knowledge gap that sparked my own interest in this area. Having recently entered the chemical senses field within the context of a food research laboratory, as an experimental psychologist, I saw immediately that key questions regarding food "spiciness" had not yet been answered. Moreover, I began to see just how pervasive these chemesthetic qualities were in everyday foods. I prepared a slide at the time for talks on this topic by going to my own kitchen and selecting items for which flavor was characterized by irritation. I had no trouble assembling for the photograph a dozen such examples: onion, ginger, soda water, rum, vodka, mustard, pepper, vinegar, and so on.

I was also fortunate in being able at that time to visit "irritation central", namely the laboratory of Barry Green at Monell. Psychophysicists are rare enough in the chemical senses generally, so having a psychophysicist specializing on questions about chemesthesis was a key reason that the field was able to progress.[1] As an important synergy, Green's colleagues Bruce Bryant, also at Monell, Tom Finger, and Wayne Silver were at the same time addressing questions regarding the physiology and anatomy of what became known as chemesthesis.

Another major influence was the *Irritation* book itself.[2] Here were chapters by the key researchers in the area, across a range of disciplines, producing an effective "state of the art" document. My own copy still has tucked inside it a drinks napkin

---

[1] Other psychophysicists such as Harry Lawless, Dave Stevens, and Beverly Cowart were also making important contributions to understanding chemesthesis at this time.

[2] My copy has joint editor Russ Mason's name written in pen on the inside cover. Did he give it to me? If I stole it, I hope that he feels that it went to a good home.

(complete with wine stain) from my 1991 visit to Monell on which is sketched the design for a study to ask if the reason why chili *appears* to suppress other flavors (and actually *does* suppress sweetness) in foods is because it more effectively garners attention. I still haven't done this study, and the question still hasn't been properly answered. A review chapter on chemesthesis by Green and Lawless (1991) at around the same time also became an important resource for those of us in this area.

Irritation also alerted me to an entirely different field of chemesthesis, namely the study of nasal and eye irritation. Studies in this area have been important for the understanding of the impact of air pollution, industrial chemical irritants, and responses to indoor air quality. In contrast, there were also investigations of the potential of irritant compounds for pest control. Such studies were, incidentally, a good indication of just how broadly applicable research into the chemical senses could be.

Since the early 1990s, research on chemesthesis has moved on substantially, and this is reflected in many chapters of this current volume. Our understanding of the mechanisms of chemesthesis has advanced over the past decade or so through discoveries in molecular biology, particularly with regard to transduction processes. A number of transient receptor potential (TRP) nerve channels are known to be activated by capsaicin, menthol, and other irritant compounds (see Chapters 4, 5 and 14). These TRPs are also temperature sensitive, thus justifying the use of the term "hot food" for food that is spicy, as well as for food to which heat has been applied. Even prior to this, we knew that spiciness and temperature enhanced one another, and our understanding of this and other psychophysical behaviors of irritant compounds have also increasingly been characterized (see Chapter 6). Two interesting phenomena in particular – sensitization and desensitization – have received much attention. If you think spicy food gets hotter as you consume it, you are probably very often right (sensitization); if it leaves your mouth feeling insensitive once you finish eating, again that's probably an accurate description (desensitization).

Knowledge of chemesthetic psychophysics has had some important practical consequences. Because spicy foods have become increasingly popular in the West, properties such as sensitization and desensitization have presented particular challenges for product developers. How is it possible to evaluate the sensory properties of, or consumer responses to, different versions of a spicy food when the heat builds up over different samples? Food sensory evaluation specialists have increasingly had to develop techniques to adapt standard ways of testing to such products (see Chapter 10).

To evaluate the impact of chemesthesis, especially in the context of foods, also requires some understanding of the qualities that stimulation of the various TRPs evokes. It is clear that chemesthesis is not only about heat, and there has been much debate about chemesthetic qualities. Are feelings of *warmth* merely low intensity heat? Is *stinging* the same as *bite*? Without yet having a clear consensus on a chemesthetic lexicon or a complete understanding of the underlying mechanisms of qualitatively different sensations, we do recognize that sensations of cooling produced by compounds such as menthol (see Chapter 7) are included within the definition of chemesthesis, as are sensations of numbing or tingling produced by compounds such as eugenol or carbon dioxide, respectively (see Chapter 8).

Where *Irritation* was clearly academic in intent, in that it pulled together different disciplinary approaches to the study of oral somatosensory sensations, in this current volume we have an additional emphasis on applying what is known to the sciences of eating. A wide variety of plant species have been used historically to add pungent sensations to foods (see Chapters 3 and 13). Chili with its powerful pungent compound capsaicin is the "flag bearer" for chemesthesis in food, yet it is clear that, wherever possible, almost all food cultures have used some form of chemesthetic spices or herbs to enliven food flavor, whether it is pepper, mustard, cinnamon, clove, ginger, and so on. There is clearly something very appealing about the addition of irritation to food flavors. Even monkeys on the Japanese island of Koshima have been observed "spicing" their food with salt by dipping it in sea water (Kawai, 1962).

But why? The widespread affection for high intensity burn in foods remains a mystery. There is no conundrum involving ice cream or chocolate preferences. Chili, in contrast, contains compounds that, when they aren't being added to foods, are being sprayed into the faces of felons, students, and other less desirable members of society. Indeed, the effects in both cases are not entirely dissimilar: tearing, pain, facial flushing, and excessive salivation. In neither case does our body seem to be welcoming a dose of capsaicin. So liking chili presents a problem. In *Irritation*, Paul Rozin addressed this question from multiple perspectives (Rozin, 1990) but the issue remains alive today. In particular, what determines the person to person differences that we see in liking for spicy food? It is clear that we must learn to like oral irritation, but in addition variations in genetics, personality and sensitivity to sensory stimulation generally may all be important (see Chapter 2).

Many of those who refuse the pleasures of pungency in food feel that hot spices overly dominate other tastes or flavors, reducing the pleasure of the latter. With the exception of sweetness, which does seem to decrease in the presence of chili heat, this complaint has been difficult to demonstrate (Prescott and Stevenson, 1995). But there is clearly a need to determine how chemesthesis interacts with other food sensory properties (see Chapter 9), if for no other reason than it will assist in optimizing foods that appeal to consumers.

Understanding the determinants of food palatability and food choices is, of course, essential in ensuring that diets promote health. In this regard, the recent report that regular consumption of chili actually seems to increase longevity (Lv *et al.*, 2015) has sparked predictable public interest. It has been known for some time that chili is a valuable source of vitamins C and D, and may also have antimicrobial properties within foods. Along with other pungent spices, it also promotes salivation, aiding digestion, and it is increasingly evident that such spices exert an impact on appetite, energy intake, blood sugar regulation, and metabolism generally, including thermogenesis (see Chapter 12).

So, a generation after *Irritation*, comes *Chemesthesis: Chemical Touch in Food and Eating*, which is timely not just for the continued relevance of chemesthesis to understanding food flavor, but also to hopefully inspire more food scientists, psychophysicists, and neuroscientists to continue to address the important questions that remain unanswered.

John Prescott
TasteMatters Research & Consulting
August, 2015

References

Green, B.G. and Lawless, H.T. 1991. The psychophysics of somatosensory chemoreception in the nose and mouth, in *Smell and Taste in Health and Disease*, (eds. T.V. Getchell *et al.*). New York: Raven Press, pp. 235–253.

Green, B.G., Mason, J.R., and Kare, M.R. (eds.) 1990. *Irritation*, New York: Marcel Dekker.

Kawai, M. 1962. *On the newly-acquired behaviors of the natural troop of Japanese monkeys on Koshima Island*, Seventh Annual Meeting of the Society for Primate Researches, Inuyama, Japan.

Lv, J., Qi, L., Yu, C. *et al.* 2015. Consumption of spicy foods and total and cause specific mortality: population based cohort study. *British Medical Journal*, **351**, h3942.

Prescott, J. and Stevenson, R.J. 1995. The effects of oral chemical irritation on tastes and flavors in frequent and infrequent users of chili. *Physiology & Behavior*, **58** (6), 1117–1127.

Rozin, P. 1990. *Getting to like the burn of chili pepper. Biological, psychological, and cultural perspectives*, in *Irritation*, (eds. B.G. Green, J.R. Mason, and M.R. Kare), New York: Marcel Dekker, pp. 231–269.

Preface

This edited volume began with a symposium at the 2013 Institute of Food Technologists (IFT) meeting in Chicago, although our collective interest in the topic is much older. The three of us, a flavor chemist (Shane), an analytical chemist (David), and a biopsychologist (John), bring three distinct perspectives to the study of chemesthesis, and we have recruited additional content area experts to produce the first comprehensive book on chemesthesis since the original book coining the term was published in 1990. In the last quarter century, our understanding of both chemesthetic stimuli and the biology underlying these sensations has exploded.

Shane, a flavor chemist, first became interested in chemesthesis when he began working at Kalsec® in 2007. One of the earliest extracts sold by Kalsec® was chili pepper oleoresin, and the company had performed some of the early work on quantifying the capsaicinoid content in oleoresins via analytical instrumentation, as opposed to human sensory panels. Later, Shane published an article in the trade journal *Perfumer & Flavorist* on the differences between the pungent expression of several common spice extracts such as capsicum, ginger, black pepper, and mustard extracts. With this in mind, Shane started making blends of pungent spice extracts to customize a pungent expression. In doing so, he became interested in another spice with chemesthetic properties, Szechuan (or Sichuan) pepper. Shane and David ended up working with the Flavor and Extracts Manufacturing Association (FEMA) to obtain Generally Recognized as Safe (GRAS) status on Szechuan Pepper Extract, which made the FEMA GRAS List 26 as FEMA #4754.

David suggested that we take advantage of this GRAS status by proposing a symposium on chemesthesis for the 2013 IFT Annual Meeting. We wanted to explore the cultivation, physiology, and psychology of chemesthesis. This resulted in a symposium entitled "Chemesthesis, Capsicum, Szechuan: It's a Spicy World!" In this session, chaired by Shane and David, Shane spoke about the tingling and buzzing sensations from Szechuan pepper, jambu oleoresin, and carbonation. The session also included three other presenters: Michael Mazourek from Cornell University, and John Hayes and Nadia Byrnes from Penn State. Michael spoke about selective plant breeding, while Nadia, who at the time of the symposium was a PhD candidate in John's research group, spoke about the psychology of the enjoyment and intake of spicy food. John rounded out the session with a talk on the biology behind chemesthesis.

At IFT, Shane was then approached by David McDade of Wiley who asked if we would be interested in editing a book on chemesthesis. At first, Shane was hesitant, having never edited a book before. Further, he did not consider himself expert enough. However, after thinking about it a bit more, Shane decided such a book should have a logical format exploring the various attributes of chemesthesis, such as physiology, raw materials, sensory evaluation, instrumental analysis, and food science aspects, as well as in depth discussion of the various types of chemesthesis (heating, cooling, tingling, etc.). In other words, we were interested in exploring the theme of chemesthesis in greater depth, with a goal of providing an updated reference text to the field. Shane eventually agreed to edit the book, with the assistance of David, both as someone to bounce ideas off of and for his excellent editorial skills. We also reached out to John again to be a co-editor, prizing his knowledge of the basic research on chemesthesis as well as his strong passion for the subject.

With our editorial team in place, the three of us began the hard work of finding the best contributors for the various chapters, and working with them, as well as Wiley, over the course of a couple of years to eventually produce the book you are now holding.

As is only appropriate, the book opens with a foreword by John Prescott and an introduction and brief history of chemesthesis by Barry Green. The next chapter is by Drs. Pam Dalton of Monell Chemical Senses Center and Nadia Byrnes, now a postdoctoral scholar at UC Davis. In their chapter, they discuss the psychology of chemesthesis – why do some of us come to enjoy what is classically considered a painful, defensive sensation?

Most chemesthetic agents used in food understandably come from natural spices and herbs. These materials are covered in Chapter 3 by Howard Haley and Shane McDonald. In Chapter 4, Yeranddy Alpizar, Thomas Voets, and Karel Talavera Pérez, biophysicists at KU-Leuven in Belgium, review the structural aspects of Transient Receptor Potential (TRP) channels and their role in chemesthesis.

The anatomy and physiology of chemesthesis is covered by Wayne Silver and Cecil Saunders. The next three chapters cover diverse types of chemesthesis including pungency and heat by John Hayes, cooling by Steve Pringle, and tingling and numbing by Chris Simons.

In the context of food, chemesthesis does not operate by itself – it interacts with other senses. These interactions are covered in Chapter 9 by Brian Byrne. How do we measure chemesthesis behaviorally and the stimuli that give rise to these sensations? Cindy Ward describes human sensory analysis in Chapter 10, while David Bolliet presents a review on instrumental analytical techniques in Chapter 11.

Chemesthesis and health are covered by Rick Mattes and Mary-Jon Ludy in Chapter 12. Food Science and culinary science aspects of chemesthesis are reviewed in Chapter 13 by Chris Loss and Ali Bouzari. The final chapter is an overview of the topic and a brief look to the future by neurobiologist Earl Carstens.

We feel this book provides a comprehensive review of various aspects of chemesthesis, and we hope it successfully balances in-depth nuanced discussions with a wide breadth of work from different disciplines that all relate back to chemesthesis. It is the product of hundreds of hours of work by many of the most talented people in the field, and we sincerely thank the contributors for all their efforts, as well as David McDade and the rest of the Wiley staff for encouraging us to both start, and complete, this project.

Shane T. McDonald
David A. Bolliet
John E. Hayes

CHAPTER 1
Introduction: what is chemesthesis?

Barry G. Green
The John B. Pierce Laboratory, Department of Surgery (Otolaryngology), Yale School of Medicine, New Haven, CT, USA

1.1 A brief history

The coolness of peppermint, the warmth of cinnamon, the heat of chilis, the tingling of carbonated beverages, the sting from a bee, the itch from a mosquito bite, the pungency of sniffed ammonia, the pain from an inflamed joint – these diverse sensations all share a common basis in chemesthesis. Not limited to the nose and mouth but experienced throughout much of the body, chemesthesis might simply be described as the chemical sensitivity of the body that is *not* served by the senses of taste or smell. But such a definition would not convey either the neurobiological complexity or the varied and important functions of chemesthesis. These and the concept of chemesthesis can be better appreciated by first considering the venerable concept that it replaced: "the common chemical sense".

For much of the 20th century, researchers in the chemical senses and related fields considered the common chemical sense to be a third specialized chemosense in addition to taste and smell. The concept was proposed by the Harvard zoologist G.H. Parker (1912) to describe the chemical sensitivity of the integument of fish and amphibians, which had previously simply been referred to as "the chemical sense" or "the undifferentiated chemical sense". By cutting individual cranial nerves and observing behavioral responses to concentrated solutions of HCl, NaOH, NaCl, and quinine applied to the bodies and tails of two species of fish, Parker concluded the sensitivity to chemical irritants was a property of "ordinary spinal nerves" rather than of the gustatory and olfactory nerves. He further proposed that the common chemical sense was a sensory system in vertebrates "as distinct as smell or taste" (Parker, 1912, p. 221), though closer in sensitivity and function to taste than to smell. A few years later, Crozier (1916) performed experiments on frogs that he argued provided further support for a common chemical sense. Some decades later, in his book titled *The Chemical Senses*, Moncrieff (1944) lent further credence to the concept by describing the common chemical sense as a separate modality that functions in concert with taste and smell.

However, some researchers were unhappy with the concept and argued instead that the chemical sensitivity of the skin and mucus membranes was a property of the sense of pain. Among them was M.H. Jones (1954), who conducted a study of her own after complaining that "the 'common chemical sense' is accepted by some writers without much tangible evidence and summarily rejected by others without much better evidence" (Jones, 1954, p. 696). Jones found that application of cocaine to the mucosal surface of the lower lip in humans reduced sensitivity to mechanical pain as well as to chemical pain, and so concluded that both forms of stimulation were sensed by cutaneous nerve endings of the pain system. In support of this conclusion, Jones quoted from Carl Pfaffmann's (1951) chapter on the chemical senses in Stevens' *Handbook of Experimental Psychology* in which he wrote, "Pain and the common chemical sensitivity appear...to be mediated by the same nerve endings" (Pfaffmann, 1951, p. 1144). It is notable, however, that this quotation was taken from a section in the chapter with the heading "The Common Chemical Sense", in which Pfaffmann went on to say, "On the other hand, it is quite clear that such chemical sensitivity is distinct from touch, and in the mouth and nose, distinct from taste and smell" (p. 1145). Pfaffmann's use of the term and affirmation of a chemical sensitivity separate from taste and smell may have helped to sustain the concept of a common chemical sense despite the clear evidence of its relationship to pain.

Further sustaining the terminology (if not Parker's original concept) were papers by Keele and others (Armstrong *et al.*, 1953; Bleehen and Keele, 1977; Keele, 1962) on the chemical sensitivity of pain, in which the possibility of specific "chemo-nociceptors" was proposed. While this body of work demonstrated beyond a doubt that chemosensory irritation was mediated at least in part by receptors of the pain sense, it also implied that the common chemical sense was in fact a specialized chemical sensitivity within the pain sense. Indeed, Keele titled his 1962 paper "The common chemical sense and its receptors". Other work published around the same time on the neurophysiological and perceptual response to capsaicin, the spicy-hot constituent of chilis (Jancso *et al.*, 1968; Szolcsanyi, 1977; Szolcsanyi *et al.*, 1988; Szolcsanyi and Jancso-Gabor, 1973), further strengthened the connection between pain and chemical irritation by showing that sensitization or desensitization by capsaicin also affected the sensitivity to both mechanical pain and heat pain (Green, 1986; Szolcsanyi, 1977; Szolcsanyi, 1985; Szolcsanyi *et al.*, 1988). This work paralleled and supported Jones' earlier evidence that cocaine reduced the sensitivity to both mechanical and chemical pain. Thus, whether or not specialized chemonociceptors existed, the evidence was clear that chemical irritants also stimulate nonspecific (polymodal) nociceptors, and thus are not sensed exclusively by a chemosensitive sub-modality of pain.

At about the same time the chemical sensitivity of the temperature senses was being brought to light through studies which showed that menthol evokes its sensory cooling effect by direct stimulation of cold fibers and not merely by evaporative cooling (Green, 1985; Schafer *et al.*, 1986; Schafer *et al.*, 1989). Remarkably, the sensitivity of cold fibers to menthol had been demonstrated decades before in electrophysiological studies of the gustatory nerves (Dodt *et al.*, 1953; Hensel and Zotterman, 1951), but the earlier findings had not found their way into published discussions of the common chemical sense. Evidence that warm fibers could also be chemically stimulated was less clear (Foster and Ramage, 1981), although experiments showing that capsaicin-sensitive receptors played a role in thermoregulation, and that capsaicin increased the perceived temperature of warm or hot water sipped into the mouth, suggested that capsaicin could modulate the excitability of the warmth system (Green, 1986; Szolcsanyi and Jancso-Gabor, 1973).

It was at this stage of understanding that a symposium on "chemical irritation" was held at the Monell Chemical Senses Center in 1988. The symposium brought together leading researchers in diverse fields of study to present their latest findings and to discuss current understanding and future research directions. Dissatisfaction with the concept of the common chemical sense surfaced throughout the symposium and was a central topic in the closing discussion, but no agreement was reached on an alternative terminology. Not until the proceedings of the meeting were being edited was the term "chemesthesis" coined and offered in the preface of the published volume as an alternative concept (Green *et al.*, 1990). Defined as the chemical *sensibility* of the skin and mucus membranes rather than as a chemical sense, the term was intended to communicate what the collective evidence had by that time shown, namely that cutaneous chemical sensitivity is multimodal in nature and derives primarily from chemically-sensitive receptors of the senses of pain and temperature.

Because it is defined as a property of the somatosensory system, chemesthesis serves as a unifying concept that includes chemosensory stimulation throughout the body, not just within the nose and mouth, where research on chemosensory irritation had most often been focused. Indeed, with the exception of the work of Keele and his colleagues, virtually all prior data on chemosensory irritation in humans had come from studies of oral and nasal sensitivity. Reflecting this research emphasis, chemosensory scientists routinely described chemicals that evoked sensations other than taste or smell as "trigeminal stimuli", since the nasal mucosa and the anterior regions of the oral cavity are both innervated by the trigeminal nerve (CN V). Tasteless and largely odorless chemicals such as vanilloids and aldehydes were typically described as "trigeminal irritants", and taste and odor stimuli that in high concentrations also produced sensations such as burning, stinging, or tingling (e.g., salts, acids, alcohols) were said to have a "trigeminal" component or quality. This terminology is still in use today and is appropriate and even preferable when the stimulus is limited to areas innervated solely by the trigeminal nerve (Hummel, 2000; Just *et al.*, 2007; Prah and Benignus, 1984; Scheibe *et al.*, 2006). Nonetheless, reference to trigeminal sensitivity can also oversimplify the neurobiology of oral and nasal chemosensory irritation. Because somesthesis on the back of the tongue is served by the glossopharyngeal nerve (CN IX) (Nagy *et al.*, 1982; Yamada, 1965; Zotterman, 1935), and the vagus nerve (CN X) innervates the airways and esophagus, when stimuli are either swallowed or inhaled they can be sensed by at least one other nerve that contains somatosensory, and thus chemosensory, receptors.

1.2 What is its relevance today?

As is evident from the varied contents of the chapters in the present volume, in the quarter century since the concept of chemesthesis was introduced, our understanding of the perception and neurobiology of this sensibility have advanced dramatically.

Whereas a serious topic of debate at the 1988 symposium was whether "trigeminal" stimulation had qualitative as well as quantitative dimensions, the clear evidence that chemicals can evoke tactile and thermal sensations as well as many varieties of painful sensations (e.g., burn, sting, bite, tingle) has settled the debate emphatically (e.g., Dessirier *et al.*, 2000; Green, 1991; Klein *et al.*, 2011; Zanotto *et al.*, 2007). Most relevant to the concept have been the discoveries that chemicals in the sanshool family can stimulate mechanoreceptors as well as nociceptors (Albin and Simons, 2010; Bryant and Mezine, 1999; Lennertz *et al.*, 2010), making chemesthesis a property of all three primary somatosensory modalities of touch, temperature, and pain, and that thermoreceptive and nociceptive sensory neurons express members of the transient receptor potential (TRP) family of receptors that are sensitive to chemicals and pH (Caterina *et al.*, 1997; Gerhold and Bautista, 2009; Koltzenburg, 2004; Patapoutian *et al.*, 2003; Peier *et al.*, 2002; Stucky *et al.*, 2009).

In addition, the discovery of extra-oral T2R "bitter" taste receptors in the mammalian and human airways that appear to play protective roles against inhalation of dangerous chemicals via sensory (i.e., apnea triggered by trigeminal or vagal afferents) and non-sensory (e.g., in motile cilia of the lung) mechanisms (Finger *et al.*, 2003; Gulbransen *et al.*, 2008; Tizzano *et al.*, 2010; Tizzano *et al.*, 2011) has further broadened understanding of the neurobiological basis and function of chemesthesis. But more than just increasing the scope and importance of the concept, these discoveries point to the role of chemesthesis as one of the body's important defenses against biological and chemical threats in the environment. Within this broader scope, chemesthesis can be considered part of the immune system via the sensitivity of pain fibers to endogenous inflammatory mediators (Jancso-Gabor *et al.*, 1980; Rang *et al.*, 1991), which were originally studied in the skin by Keele and his colleagues (Armstrong *et al.*, 1953; Bleehen and Keele 1977; Keele, 1962). We now know too that sensitivity to inflammation and tissue damage throughout the body is mediated in part by the same classes of multimodal pain receptors that respond to capsaicin and many other exogenous irritants, for example, TRPV1 (Blackshaw, 2014) and TRPA1 (Dhaka *et al.*, 2009; Talavera *et al.*, 2009; Wang *et al.*, 2010; Willis *et al.*, 2011). Accordingly, it was recently proposed that chemesthesis be considered as the sensory component of what might be termed the body's "chemofensor complex" (Green, 2012), the array of chemical defense mechanisms that function both together and separately to protect and rid the body of harmful chemicals and bacterial agents.

It is interesting to consider that this modern view places chemesthesis on an equal footing with taste and smell, though in terms of Gibson's (1966) pioneering concept of shared functionality within a perceptual system rather than shared categorization as special senses. One could argue that within the domain of chemical defenses, chemesthesis has the broadest range of functions of these three chemosensory components, having both an exteroceptive sentinel function and an interoceptive function as a signal of tissue damage and/or infection. Running as it does against the theme of specialized sensory systems that has historically dominated research in sensory neuroscience, an understanding of chemesthesis has evolved more slowly than in the classical sense modalities, where workers have been able to focus on specific sensory mechanisms serving specific stimuli and functions. Yet the wide ranging research presented in this volume testifies to the growing emphasis on multidisciplinary and multisensory approaches to the study of human sensory perception, which has contributed significantly to the broader and deeper understanding of chemesthesis that has begun to emerge.

References

Albin, K.C. and Simons, C.T. 2010. Psychophysical evaluation of a sanshool derivative (alkylamide) and the elucidation of mechanisms subserving tingle. *PLoS One,* **5** (3), e9520 available from: PM:20209090.

Armstrong, D., Dry, R.M.L., Keele, C.A., and Markham, J.W. 1953. Observations on chemical excitants of cutaneous pain in man. *Journal of Physiology,* **120**, 326–351.

Bautista, D.M., Jordt, S.E., Nikai, T. *et al.*, 2006. TRPA1 mediates the inflammatory actions of environmental irritants and proalgesic agents. *Cell,* **124** (6), 1269–1282 available from: PM:16564016.

Blackshaw, L.A. 2014. Transient receptor potential cation channels in visceral sensory pathways. *British Journal of Pharmacology,* **171** (10), 2528–2536 available from: PM:24641218.

Bleehen, T. and Keele, C.A. 1977. Observations on the algogenic actions of adenosine compounds on the human blister base preparation. *Pain,* **3** (4), 367–377.

Bryant, B.P. and Mezine, I. 1999. Alkylamides that produce tingling paresthesia activate tactile and thermal trigeminal neurons. *Brain Research,* **842** (2), 452–460.

Caterina, M.J., Schumacher, M.A., Tominaga, M., Rosen, T.A., Levine, J.D., and Julius, D. 1997. The capsaicin receptor: a heat-activated ion channel in the pain pathway. *Nature,* **389**, 816–824.

Crozier, W.J. 1916. Regarding the existence of the "common chemical sense" in vertebrates. *The Journal of Comparative Neurology,* **26**, 1–8.

Dessirier, J.M., Simons, C.T., Carstens, M.I., O'Mahony, M., and Carstens, E. 2000. Psychophysical and neurobiological evidence that the oral sensation elicited by carbonated water is of chemogenic origin. *Chemical Senses,* **25** (3), 277–284 available from: PM:10866986.

Dhaka, A., Uzzell, V., Dubin, A.E. *et al.* 2009. TRPV1 is activated by both acidic and basic pH. *Journal of Neuroscience,* **29** (1), 153–158 available from: PM:19129393.

Dodt, E., Skouby, A.P., and Zotterman, Y. 1953. The effect of cholinergic substances on the discharge from thermal receptors. *Acta Physiologica Scandinavia,* **28**, 101–114.

Finger, T.E., Bottger, B., Hansen, A., Anderson, K.T., Alimohammadi, H., and Silver, W.L. 2003. Solitary chemoreceptor cells in the nasal cavity serve as sentinels of respiration. *Proceedings of the National Academy of Sciences, U.S.A,* **100** (15), 8981–8986 available from: PM:12857948.

Foster, R.W. and Ramage, A.G. 1981. The action of some chemical irritants on somatosensory receptors of the cat. *Neuropharmacology,* **20**, 191–198.

Gerhold, K.A. and Bautista, D.M. 2009. Molecular and cellular mechanisms of trigeminal chemosensation. *Annals of the New York Academy of Sciences,* **1170**, 184–189 available from: PM:19686135.

Gibson, J.J. 1966. *The Senses Considered as Perceptual Systems,* Westport, CT: Greenwood Press.

Green, B.G. 1985. Menthol modulates oral sensations of warmth and cold. *Physiology & Behavior,* **35**, 427–434.

Green, B.G. 1986. Sensory interactions between capsaicin and temperature. *Chemical Senses,* **11**, 371–382.

Green, B.G. 1991. Capsaicin cross-desensitization on the tongue: Psychophysical evidence that oral chemical irritation is mediated by more than one sensory pathway. *Chemical Senses,* **16**, 675–689.

Green, B.G. 2012. Chemesthesis and the chemical senses as components of a "chemofensor complex". *Chemical Senses,* **37** (3), 201–206 available from: PM:22210122.

Green, B.G., Mason, J.R., and Kare, M.R. 1990. *Chemical Senses, vol. 2: Irritation,* New York: Marcel Dekker, Inc.

Gulbransen, B., Silver, W., and Finger, T.E. 2008. Solitary chemoreceptor cell survival is independent of intact trigeminal innervation. *Journal of Comparative Neurology,* **508** (1), 62–71 available from: PM:18300260.

Hensel, H. and Zotterman, Y. 1951. The effect of menthol on thermoreceptors. *Acta Physiologica Scandinavia*, **24**, 27–34.

Hummel, T. 2000. Assessment of intranasal trigeminal function. *International Journal of Psychophysiology*, **36** (2), 147–155.

Jancso, G., Kiraly, E., and Jancso-Gabor, A. 1980. Chemosensitive pain fibres and inflammation. *International Journal of Tissue Reactions*, **2**, 57–66.

Jancso, N., Jancso-Gabor, A., and Szolcsanyi, J. 1968. The role of sensory nerve endings in neurogenic inflammation induced in human skin and in the eye and paw of the rat. *British Journal of Pharmacology and Chemotherapy*, **32**, 32–41.

Jones, M.H. 1954. A study of the "common chemical sense". *American Journal of Psychology*, **67**, 696–699.

Just, T., Pau, H.W., Steiner, S., and Hummel, T. 2007. Assessment of oral trigeminal sensitivity in humans. *European archives of oto-rhino-laryngology*, **264** (5), 545–551.

Keele, C.A. 1962. The common chemical sense and its receptors. *Archives Internationales de Pharmacodynamie et de Therapie*, **139**, 547–557.

Klein, A.H., Carstens, M.I., Zanotto, K.L., Sawyer, C.M., Ivanov, M., Cheung, S., and Carstens, E. 2011. Self- and cross-desensitization of oral irritation by menthol and cinnamaldehyde (CA) via peripheral interactions at trigeminal sensory neurons. *Chemical Senses*, **36** (2), 199–208.

Koltzenburg, M. 2004. The role of TRP channels in sensory neurons. *Novartis Foundation Symposium*, **260**, 206–213 available from: PM:15283452.

Lennertz, R.C., Tsunozaki, M., Bautista, D.M., and Stucky, C.L. 2010. Physiological basis of tingling paresthesia evoked by hydroxy-α-sanshool. *The Journal of Neuroscience*, **30** (12), 4353–4361.

Moncrieff, R.W. 1944. *The Chemical Senses*, London: Leonard Hill Ltd.

Nagy, J.I., Goedert, M., Hunt, S.P., and Bond, A. 1982. The nature of the substance P-containing nerve fibres in taste papillae of the rat tongue. *Neuroscience*, **7** (12), 3137–3151 available from: PM:6186943.

Parker, G.H. 1912. The relations of smell, taste, and the common chemical sense in vertebrates. *Journal of the Academy of Natural Sciences of Philadelphia*, **15**, 221–234.

Patapoutian, A., Peier, A.M., Story, G.M., and Viswanath, V. 2003. ThermoTRP channels and beyond: mechanisms of temperature sensation. *Nature Reviews Neuroscience*, **4** (7), 529–539 available from: PM:12838328.

Peier, A.M., Moqrich, A., Hergarden, A.C. *et al.* 2002. A TRP channel that senses cold stimuli and menthol. *Cell*, **108** (5), 705–715 available from: PM:11893340.

Pfaffmann, C. 1951, Taste and smell, in *Handbook of Experimental Psychology*, (ed. S.S. Stevens), New York: John Wiley & Sons, Inc., pp. 1143–1171.

Prah, J.D. and Benignus, V.A. 1984. Trigeminal sensitivity to contact chemical stimulation: a new method and some results. *Perception & Psychophysics*, **35** (1), 65–68.

Rang, H.P., Bevan, S., and Dray, A. 1991. Chemical activation of nociceptive peripheral neurones. *British Medical Bulletin*, **47**, 534–548.

Schafer, K., Braun, H.A., and Isenberg, C. 1986. Effect of menthol on cold receptor activity. Analysis of receptor processes. *Journal of General Physiology*, **88** (6), 757–776 available from: PM:3794639.

Schafer, K., Necker, R., and Braun, H.A. 1989. Analysis of avian cold receptor function. *Brain Research*, **501** (1), 66–72 available from: PM:2804699.

Scheibe, M., Zahnert, T., and Hummel, T. 2006. Topographical differences in the trigeminal sensitivity of the human nasal mucosa. *Neuroreport*, **17** (13), 1417–1420.

Stucky, C.L., Dubin, A.E., Jeske, N.A., Malin, S.A., McKemy, D.D., and Story, G.M. 2009. Roles of transient receptor potential channels in pain. *Brain Research Reviews*, **60** (1), 2–23 available from: PM:19203589.

Szolcsanyi, J. 1977. A pharmacological approach to elucidation of the role of different nerve fibres and receptor endings in mediation of pain. *Journal of Physiology (Paris)*, **73** (3), 251–259 available from: PM:926026.

Szolcsanyi, J. 1985, Sensory receptors and the antinociceptive effects of capsaicin, in *Tachykinin antagonists*, (eds. R. Hakanson and F. Sundler), Amsterdam: Elsevier Science Publishers B.V., pp. 45–54.

Szolcsanyi, J., Anton, F., Reeh, P.W., and Handwerker, H.O. 1988. Selective excitation by capsaicin of mechano-heat sensitive nociceptors in rat skin. *Brain Research*, **446**, 262–268.

Szolcsanyi, J. and Jancso-Gabor, A. 1973, Capsaicin and other pungent agents as pharmacological tools in studies on thermoregulation, in *The Pharmacology of Thermoregulation,* 1st edn (eds. E. Schonbaum and P. Lomax), Basel: Karger, pp. 331–338.

Talavera, K., Gees, M., Karashima, Y., Meseguer, V.M., Vanoirbeek, J.A., Damann, N., Everaerts, W., Benoit, M., Janssens, A., Vennekens, R., Viana, F., Nemery, B., Nilius, B., and Voets, T. 2009. Nicotine activates the chemosensory cation channel TRPA1. *Nature Neuroscience,* **12** (10), 1293–1299 available from: PM:19749751.

Tizzano, M., Cristofoletti, M., Sbarbati, A., and Finger, T.E. 2011. Expression of taste receptors in solitary chemosensory cells of rodent airways. *BMC Pulmonary Medicine,* **11**, 3 available from: PM:21232137.

Tizzano, M., Gulbransen, B.D., Vandenbeuch, A., Clapp, T.R., Herman, J.P., Sibhatu, H.M., Churchill, M.E., Silver, W.L., Kinnamon, S.C., and Finger, T.E. 2010. Nasal chemosensory cells use bitter taste signaling to detect irritants and bacterial signals. *Proceedings of the National Academy of Sciences, U.S.A.,* **107** (7), 3210–3215 available from: PM:20133764.

Wang, Y.Y., Chang, R.B., and Liman, E.R. 2010. TRPA1 is a component of the nociceptive response to CO_2. *Journal of Neuroscience,* **30** (39), 12958–12963 available from: PM:20881114.

Willis, D.N., Liu, B., Ha, M.A., Jordt, S.E., and Morris, J.B. 2011. Menthol attenuates respiratory irritation responses to multiple cigarette smoke irritants. *FASEB Journal,* **25** (12), 4434–4444 available from: PM:21903934.

Yamada, K. 1965. The glossophyaryngeal nerve response to taste and thermal stimuli in the rat, rabbit and cat. *Kumamoto Medical Journal,* **18** (2), 106–108 available from: PM:0005825888.

Zanotto, K.L., Merrill, A.W., Carstens, M.I., and Carstens, E. 2007. Neurons in superficial trigeminal subnucleus caudalis responsive to oral cooling, menthol, and other irritant stimuli. *Journal of Neurophysiology,* **97** (2), 966–978 available from: PM:17151223.

Zotterman, Y. 1935. Action potentials in the glossopharyngeal nerve and in the chorda tympani. *Skandinavica Archive fur Physiologie,* **72**, 73–77.

CHAPTER 2

Psychology of chemesthesis – why would anyone want to be in pain?

Pamela Dalton[1] and Nadia Byrnes[2]
[1] *Monell Chemical Senses Center, Philadelphia, PA, USA*
[2] *Department of Viticulture and Enology, University of California Davis, Davis, CA, USA*

> The chile, it seems to me, is one of the few foods that has its own goddess.
>
> *Diana Kennedy, cookbook author*

> All I ask of food is that it doesn't harm me.
>
> *Michael Palin*

2.1 Introduction and background

The desire for spices is not a new fascination, in fact, it has been suggested that humans' desire for spices fueled the Age of Discovery and altered the course of history (Le Couteur and Burreson, 2004). While piperine, the compound responsible for the pungency of peppercorns, was already well loved in Europe at the time that Christopher Columbus introduced chili peppers to Europe, capsaicin did not catch hold as quickly. Shortly after its introduction, however, the chili pepper spread quickly around other parts the world and in less than 50 years had been incorporated into local cuisines across the globe. In the centuries since then, the desire for this pungent compound has not diminished. A recent Mintel report from June 2014 showed that nearly 75% of Americans are interested in trying spicy peppers, chilis, and spices in restaurant dishes (Fajardo, 2014). This study also showed that across the United States, restaurant patrons are demanding cuisines and foods that contain chemesthetic compounds.

Warming, cooling, tingling, stinging, and burning are sensations that can occur when pungent chemical compounds present in our foods stimulate the free nerve endings of the trigeminal nerve in the oral cavity, a "sensibility" known as "chemesthesis" (see Chapter 1). Compounds found in a number of foods, including the herbs and spices listed in Table 2.1, elicit chemesthetic sensations. The somatosensory system responsible for chemesthetic sensations is an innate part of the mammalian "pain" system. At low concentrations, these compounds may only

Table 2.1 Overview of herbs and spices and the chemesthetic compounds responsible for irritant qualities.

Spice/herb	Compounds
Clove	Eugenol
Cinnamon	Cinnamaldehyde
Ginger (fresh)	Gingerols
Ginger (dried)	Shogaols
Ginger (cooked)	Zingerone
Black and white pepper	Piperine
Oregano	Carvacrol
Szechuan peppers (Huajiao)	Sanshools
Mint	Menthol
Horseradish, mustard, wasabi	Allyl isothiocyanate
Chili peppers	Capsaicin

be perceived as a "chemical feel" or tingle; however, as concentrations increase, a "warming" or "tingling" sensation can give way to stinging and frank burning. Despite this, a considerable percentage of the population worldwide avidly consumes foods containing pungent spices, raising the question, "why would anyone choose to be in pain?"

The sensations of pungency elicited by these compounds can vary considerably in the area of stimulation, the quality of the sensation, and the time course over which the sensation waxes and wanes (e.g., Bennett and Hayes, 2012; Bryant and Mezine, 1999; Cicerale *et al.*, 2009; Cliff and Heymann, 1992; Mcdonald *et al.*, 2010). For example, menthol elicits cooling and tingling, while cinnamaldehyde elicits warming and burning. The slow but increasing burn of chili peppers (capsaicin) differs considerably from the rapid onset and offset of the pungency of horseradish (allyl isothiocyanate). Nevertheless, the tendency to enjoy or avoid these pungent sensations appears to vary across the population. Despite the wide variety of spice compounds capable of eliciting chemesthetic sensations, consumption of capsaicin, the pungent compound in chili peppers, is the most ubiquitous with some estimates suggesting approximately a quarter of the world's population consumes capsaicin on a daily basis (Rozin, 1990b). Hence, most studies on the liking or preference for chemesthetic agents have focused on factors related to capsaicin consumption and thus necessarily comprise the bulk of the research presented here.

This chapter will explore the factors underlying the variability in response to chemesthetic sensations from food and the mechanisms by which such sensations can shift from aversive to appetitive. The content is divided into three sections. The first addresses inter-individual biological differences, which may account for differences in perceived intensity of sensation on an individual's first encounter with capsaicin or other chemesthetic agents. These biological differences include genetic and phenotypic variation in taste-bud morphology and the receptor that capsaicin activates. The second section addresses social mechanisms by which an individual may come to enjoy the sensation that capsaicin elicits, even if their first experience with capsaicin-containing foods is aversive. The final section covers personality

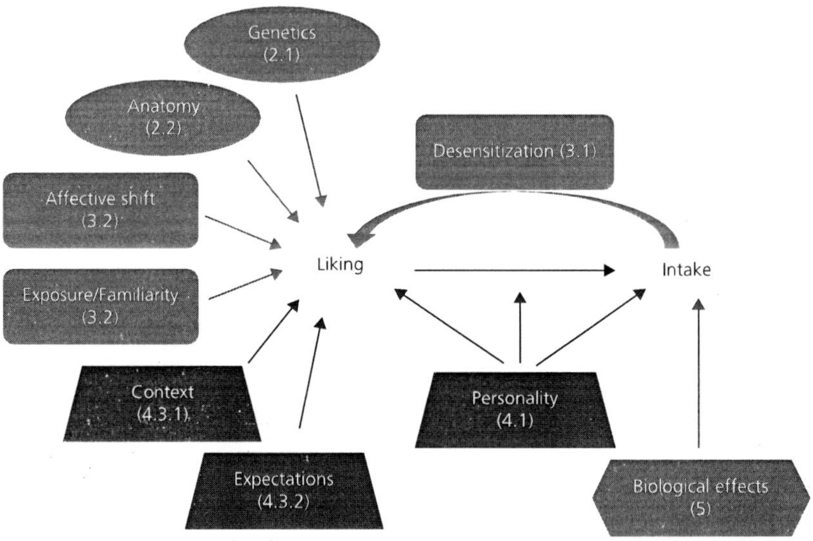

Fig. 2.1 Relationships between variables associated with liking and intake of spicy foods.

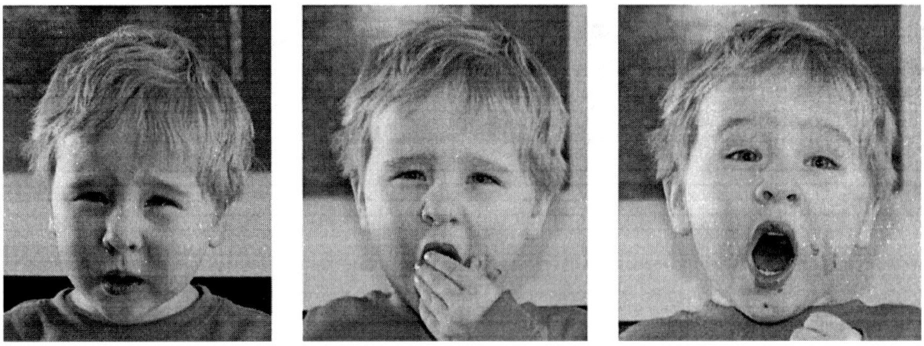

Fig. 2.2 Reaction of a naïve user (American child) to the first encounter with cinnamon-flavored candy.

traits that have been linked to food adventurousness and the liking of spicy foods as well as the degree to which cognitive factors such as expectations or context can determine preference or acceptability. Figure 2.1 illustrates the relationship between the variables explored in this chapter.

2.1.1 Individual variation in hedonic response

A wide range of hedonic responses to capsaicin has been reported, from individuals disliking any irritation to those individuals that simply cannot get enough pungency (Prescott and Stevenson, 1995a; Rozin and Schiller, 1980; Tepper *et al.*, 2004). Some individuals report even enjoying piquancy when it is isolated from food or beverages. It is generally assumed that an individual's first encounter with capsaicin is averse, given the response noted in young children and pets (Fig. 2.2), causing one to wonder why anyone would repeatedly consume something that is irritating.

However, there are numerous examples of foods that are initially aversive, yet which individuals learn to like, such as alcohol, coffee, and tobacco (Rozin and Schiller, 1980). For these foods, there are post-ingestive or social effects that influence liking and consumption (Rozin and Schiller, 1980), such as the energizing effects of caffeine, which may overcome the aversive bitterness of coffee. Post-ingestive effects of capsaicin consumption have been reported (Ludy and Mattes, 2011; Rozin and Schiller, 1980), and it has been posited that these effects may be a factor in the consumption of capsaicin-containing foods (Rozin and Schiller, 1980), however, until recently there was little evidence to support this hypothesis (but also see Chapter 12).

2.2 Physiological differences: maybe they can't feel the burn?

The question posed at the beginning of this chapter "Why would anyone want to be in pain?" makes the assumption that the individual consuming these irritating compounds can actually feel the irritation. There are well-established differences in the sensitivity of individuals to the pungency of capsaicin and the overall liking of the irritation sensation produced by capsaicin in foods (Lawless *et al.*, 1985; Prescott and Stevenson, 1995b; Stevenson and Prescott, 1994; Stevenson and Yeomans, 1993; Yoshioka *et al.*, 2001). Reasons proposed to explain the preference for consuming foods that elicit oral irritation include physiological differences such as genetic variation (Hayes *et al.*, 2013), oral anatomy (Miller and Reedy, 1990), and taste phenotype (Duffy, 2007; Duffy and Bartoshuk, 2000), all of which could possibly alter sensitivity to capsaicin. This section provides a brief overview of the biological variations that may result in individual differences in perceived intensity of capsaicin and capsaicin-containing foods.

2.2.1 Genetics: variability in sensation and diet

Genetic variation has previously been shown to explain differences in oral sensation and dietary choices (for a review see Hayes *et al.*, 2013). For example, variation in the *TAS2R38* gene associates with differences in bitterness perception and intake of vegetables. Most commonly, there are two haplotypes, or collections of alleles, of the *TAS2R38* gene that occur, the PAV haplotype and the AVI haplotype. These haplotypes are named for the amino acids that are present at three specific locations in the amino acid sequence. At position 49, a proline (P) or alanine (A) is present, at position 262, an alanine (A) or valine (V) is present, and at position 296, a valine (V) or isoleucine (I) is present. For further information, see Kim *et al.* (2003). Individuals carrying at least one copy of the PAV haplotype, or PAV carriers, tend to report the intensity of bitter compounds higher than AVI/AVI carriers (also known as AVI homozygotes), and report lower consumption of vegetables (Duffy *et al.*, 2010; Sacerdote *et al.*, 2007). Duffy and colleagues showed that *TAS2R38* also associates with alcohol intake, with individuals with PAV homozygotes consuming less alcohol than those with PAV heterozygotes (one copy of PAV haplotype and one copy of AVI haplotype), who consumed less alcohol than those with AVI homozygotes.

In addition to genetic variation accounting for differences in taste sensations, variation in levels of salivary protein content have been associated with perception and liking of astringent foods (Dinnella *et al.*, 2011; Dinnella *et al.*, 2010; Horne *et al.*, 2002). Individuals that experience higher levels of salivary protein depletion after stimulation with phenolic stimuli, or high responding (HR) individuals, show higher perceived levels and lower liking of astringent stimuli than their low responding (LR) counterparts. A recent study suggests that differences in salivary protein levels, and, thus, astringency perception, may be genetically determined (Törnwall *et al.*, 2012).

As with the perception of the chemesthetic sensation astringency, it has been suggested that the variability in the response to capsaicin is due to polymorphisms in the TRPV1 capsaicin receptor (Park *et al.*, 2007; Snitker *et al.*, 2009). Recently, Törnwall and colleagues presented evidence of a common genetic mechanism responsible for the liking of various types of oral pungency (Törnwall *et al.*, 2012). Between 18 and 58% of the variation in hedonic responses to oral pungency were explained by genetics: however, as this was a twin study, no specific genetic mechanism was identified. These values fall within the range of heritability previously reported for sweet and sour preferences (Keskitalo *et al.*, 2007).

While genotypic variation may account for some of the differences observed in perception and liking of various oral sensations, it is critical to note that phenotypic variation also plays a role in these perceptual differences. Variability in responses to 6-n-propylthiouracil (PROP) are explained for the most part by polymorphisms of the *TAS2R38* gene (Kim *et al.*, 2003), with carriers of the PAV allele tending to show a higher perceived intensity of suprathreshold PROP solutions than carriers of the AVI allele. Originally, the term "supertaster" was used to describe these individuals that perceived high intensity from PROP. However, work from Hayes and colleagues (Hayes *et al.*, 2008) showed that the *TAS2R38* genotype does not account for all observed variability in PROP perception and that other factors may play a role in determining PROP bitterness.

With relevance to understanding the phenotypic variation in chemesthetic sensations, researchers have shown that "supertasters", at least based on their response to PROP, have a broad heightened response to a wide range of chemosensory stimuli (Bajec and Pickering, 2008; Bartoshuk *et al.*, 1994; Hayes *et al.*, 2008; Hayes and Duffy, 2007; Pickering and Robert, 2006; Pickering *et al.*, 2004; Tepper and Nurse, 1998). Supertasters, who may perceive more intense irritation from capsaicin and other chemesthetic stimuli, may also show increased perception of side tastes in some chemesthetic compounds. It has been reported that individuals with higher taste responsivity report increased intensity of a bitter side taste from capsaicin, piperine, and zingerone on the posterior tongue (Green and Hayes, 2004). Perceiving bitterness in addition to the oral irritation of these stimuli may make these chemesthetic compounds more aversive to supertasters than to medium or non-tasters.

2.2.2 Anatomy: oral phenotypes and sensation

One of the proposed reasons for differences in sensory intensity and acuity is variation in oral anatomy. Fungiform papillae (FP) are one of the three types of papillae on the tongue that house taste buds. FP are located all over the tongue, but are most dense near the tip of the tongue. Since taste buds are housed in FP, counting

FP on the tip of someone's tongue can be used as a rough estimate of overall taste-bud density, and presumably, a measure of overall taste function (Miller and Reedy, 1990). Studies show that individuals with higher FP density perceive more intensity from bitter, sweet, and salty tastes (Miller and Reedy, 1990). A relationship has also been shown between FP density and PROP supertasting, suggesting that PROP supertasters have more FP (Bartoshuk *et al.*, 1994; Essick *et al.*, 2003; Miller and Reedy, 1990), although not all studies support this relationship (Feeney and Hayes, 2014; Fischer *et al.*, 2013; Garneau *et al.*, 2014).

It has also been suggested that PROP supertasters perceive more burn from capsaicin than PROP non-tasters (Karrer and Bartoshuk, 1991; Karrer *et al.*, 1992). The hypothesis linking burn perception and FP density arose from the understanding that nociceptive fibers are located in taste buds, located in FP. A higher density of FP in supertasters (Bartoshuk *et al.*, 1994; Miller and Reedy, 1990) would lead to a greater density of nociceptive fibers with which to perceive the burn of capsaicin. However, evidence to support the association is inconsistent. Work from Tepper and Nurse (1998) and unpublished work from Karrer and colleagues (1992) showed that PROP tasters had a higher density of FP and were more sensitive to capsaicin. In contrast, Prescott and Swain-Campbell tested the relationship between PROP supertasting and perceived capsaicin intensity and found no association, whether PROP supertasters were in a group separate from or combined with medium tasters (2000). Similarly, Törnwall and colleagues showed no association between PROP taster status and responses to oral pungency (2012). The inconsistency in findings indicate that while PROP is a reliable predictor of taste sensitivity, this compound may not be a reliable predictor of individual differences in response to chemesthetic stimuli.

2.3 Effects of exposure on chemesthetic response (social)

While biological variation may play a role in determining baseline sensitivity to the oral sensations elicited by chemesthetic agents, these genetic and phenotypic differences do not account for the fact that individuals can actually learn to enjoy the sensation that capsaicin produces, irrespective of initial experience. Individual differences in the liking of the sensation elicited by capsaicin have been proposed to arise primarily from prior experiences and familiarity with capsaicin and capsaicin-containing foods (Ludy and Mattes, 2012; Stevenson and Yeomans, 1995). This portion of the chapter is devoted to exploring how exposure and familiarity with capsaicin-containing foods might result in increased liking of capsaicin.

2.3.1 Desensitization

Frequent users of spicy foods often rate the burn of capsaicin as less intense and more pleasant than infrequent users of spicy foods (see Fig. 2.3), and it has been suggested that the reported liking of spicy foods is merely the effect of reduced sensitivity to the burning sensation via desensitization (Lawless *et al.*, 1985; Prescott and Stevenson, 1995a; Stevenson and Yeomans, 1995).

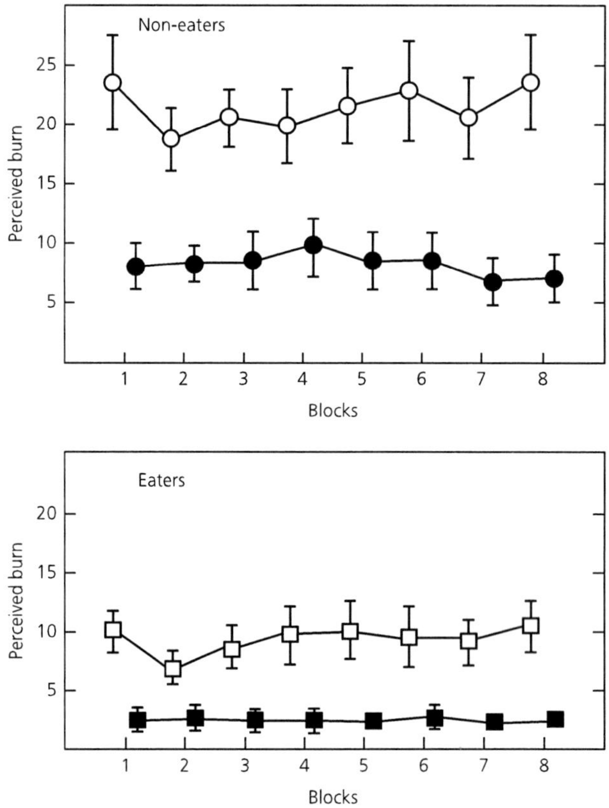

Fig. 2.3 Magnitude estimation (mean ratings, plus or minus 1 standard error of the mean) relative to a sound standard, of the intensity of burn of capsaicin solutions by chili eaters and chili non-eaters. Open symbols show the ratings just after expectoration of a capsaicin rinse solution and closed symbols show the rating after four intervening judgments on qualities of other solutions (Source: Lawless *et al.*, 1985. Reproduced with permission from Oxford University Press).

In other words, individuals increase their consumption of capsaicin-containing foods not because they enjoy the burning sensation, but because they no longer perceive it, as a result of their prior consumption. Both acute and chronic desensitization to capsaicin are well-established phenomena following exposure (Cowart, 1987; Green *et al.*, 1993; Karrer and Bartoshuk, 1991; Lawless *et al.*, 1985; Logue and Smith, 1986; Rozin *et al.*, 1981; Stevenson and Prescott, 1994). Rozin and Schiller examined desensitization and liking in two populations with varying levels of capsaicin consumption – American adults and adults and children from a rural Mexican village (1980). They hypothesized that the higher consumption of capsaicin containing foods among the Mexicans would lead to: (1) higher detection thresholds for capsaicin among the Mexicans, (2) positive correlations between detection thresholds and degree of exposure, and (3) higher detection thresholds among individuals who report liking capsaicin than among those who are neutral to it or dislike it. In contrast to their expectations, only a small, non-significant difference in threshold for capsaicin was apparent between the Mexican and American groups. Additionally, there was no difference in thresholds between American chili likers and dislikers. When examining detection thresholds, preference levels, and tolerance thresholds they observed an astonishing consistency in the degree of correlation between these variables among both Mexicans and Americans. Chili preference and tolerance levels correlated with detection thresholds in the range of from 0.20 to 0.39, while preference and tolerance levels correlations were between 0.8 and 0.9. These results suggest that decreased sensitivity (i.e., increased detection threshold) does not predict increased liking of capsaicin-containing foods. Rozin and Schiller also explored whether threshold increased with exposure, and hence with age, but no significant effects were seen (1980). Collectively, the work examining the effect of desensitization on liking suggests that while there may be a small desensitization effect from eating chili pepper in moderate amounts (Rozin *et al.*, 1981), the effect is slight and is unlikely to play an important role in determining the liking of capsaicin.

2.3.2 Affective shift: "learning to like"

Rather than increased liking resulting from decreased perceived intensity, it has been proposed that spicy food likers actually enjoy the burning sensation that comes from eating capsaicin (Rozin and Schiller, 1980; Stevenson and Yeomans, 1993). Indeed, in the surveys conducted by Rozin and Schiller, Mexican subjects did not seem to like the flavor of chili peppers when the pungency was removed (Rozin and Schiller, 1980). Stevenson and Yeomans (1993) observed a similar effect, in that pleasantness ratings were higher for likers than non-likers, even when the samples being compared were rated as having the same perceived burning intensity. Proposed mechanisms for this affective shift include the physiological consequences of ingestion, such as increased salivary flow, which may help in digestion of the starch-heavy diets that are common in areas where capsaicin is regularly consumed, such as in traditional Mexican cuisine (Rozin and Schiller, 1980). Other mechanisms include the association of capsaicin with satiating, or otherwise pleasant foods, which would lead to conditioned liking for the burning sensation (Rozin and Schiller, 1980; Rozin and Vollmecke, 1986).

2.3.2.1 Culture and familiarity

Yet another proposed mechanism for the observed shift in liking of capsaicin, and by far the most extensively studied, is the hypothesis of "mere exposure". In the late 1960s, Zajonc suggested "mere exposure of the individual to a stimulus enhances his attitude toward it" (1968). The effects of mere exposure have been observed in rodents with various food stimuli, with repeated exposure to saccharin (Carroll *et al.*, 1975; Domjan, 1972, 1976; Domjan and Gillan, 1976; Mitchell *et al.*, 1977; Nachman, 1959), casein (Domjan and Bowman, 1974), milk (Williams, 1968), garlic flavor (Capretta and Rawls, 1974), sucrose octa-acetate (Warren and Pfaffmann, 1959), and coffee and vinegar (Siegel, 1974) increasing rodents' liking of the initially unfavorable stimuli.

The "exposure hypothesis" has shown significant effects in adult humans using a variety of stimuli such as musical passages (Mull, 1957), human faces (Zajonc,

1968), paintings (Maslow, 1937), and foods. Under controlled conditions, repeated taste exposures and modeling behaviors have been shown to increase preference and acceptance of various foods in infants (Sullivan and Birch, 1994), children (Horne *et al.*, 2004; Lakkakula *et al.*, 2010; Sullivan and Birch, 1990; Wardle *et al.*, 2003; Wardle *et al.*, 2003), and adults (Pliner, 1982). In examining factors that are important to determining children's liking of foods, Birch observed that familiarity was one of two key dimensions (Birch, 1979a, 1979b). Birch and Marlin later showed, in a six-week exposure period to various types of cheeses, that preference for the cheeses in children was a clear function of exposure (Birch and Marlin, 1982). The effect of exposure on acceptance and preference of foods is not limited to children. Similar exposure effects have also been observed in adults (e.g., Crandall, 1985; Pliner, 1982), with liking increasing between initial and final exposure of the stimuli (Stein *et al.*, 2003).

While there is an extensive body of literature that supports the affective shifts from "mere exposure", data on the number of exposures necessary to increase liking or preference are inconsistent. In young children one exposure may be sufficient, while in school-aged children and adults up to 15 exposures may be necessary. Additionally, the type of food presented is important in determining how many exposures will be necessary to induce a liking or preference alteration (Costa *et al.*, 2014; Horne *et al.*, 2004; Liem and De Graaf, 2004; Sullivan and Birch, 1994; Wardle *et al.*, 2003; Wardle *et al.*, 2003). For example, Costa and colleagues (Costa *et al.*, 2014) recently determined that even while strong positive correlations ($r = 0.99$) were observed between exposure and acceptance of goat's milk yogurts, rapid, repeated exposure over six days did not suffice to generate a significant change in acceptance of the product.

It is important to note that the literature presented here highlights the effects of exposure by consuming foods, though evidence also suggests that it might not be necessary to actually consume the novel food to see the effects of mere exposure on liking (1980). Work by Birch showed stable preference enhancements in preschoolers for vegetables chosen by peers (Birch, 1980). It is possible that even though a child is not consuming chili themselves at a young age, their exposure to adults and peers consuming chili might influence their liking of chili (Rozin and Vollmecke, 1986). Rozin reported that the shift from disliking to liking in many chili-eating cultures occurs between the ages of five and nine years old (Rozin, 1990b; Rozin and Schiller, 1980), while these children begin receiving capsaicin-containing foods around three to five years of age (Rozin and Schiller, 1980). Considering this, and the convincing evidence for mere exposure affecting increased acceptance and liking for foods, it would not be unexpected that the same effects exist between exposure to and liking of capsaicin. Indeed, Stevenson and Yeomans showed that under controlled conditions, repeated exposure to capsaicin enhanced ratings of burn pleasantness and that this shift was not due to sensory adaptation (Stevenson and Yeomans, 1995).

We can also ask to what degree the preference for pungency is due to early exposure to taste and chemesthetic compounds present in amniotic fluid or breast milk. While the degree to which capsaicin is found in either amniotic fluid or excreted into breast milk is unknown, rodent and human studies have shown that the pungent compound in black pepper (piperine) is excreted into breast milk shortly after ingestion (Khachik *et al.*, 1997). In humans, food flavors transmitted this way in experimental studies have been shown to lead to increased preference and consumption for those flavors at the time of weaning (Beauchamp and Mennella, 2009; Mennella *et al.*, 2001). Thus, it is intriguing to suggest that the acceptability of chemesthetic compounds paired with food flavors may arise from prenatal experience to the cultural diet.

The increase in acceptance and liking of a stimulus that happens as a result of exposure is thought to be due to the dissipation of neophobia regarding the new stimulus (Hill, 1978; Rozin, 1990a). In a review of animal literature, Hill suggests that neophobia protects animals by limiting their interaction with unfamiliar substances until there is evidence that the substance is not dangerous (Hill, 1978). While this reduction in neophobia may result in liking for some individuals, Rozin and Schiller hypothesized that the enjoyment that some individuals derive from consuming capsaicin has to do with the fact that the body perceives capsaicin as dangerous (Rozin and Schiller, 1980). The researchers suggested that some people enjoy the thrill that comes from the disparity between bodily responses that the stimulus is harming the body (i.e., burning sensation in the mouth, watering eyes, and running nose when consuming capsaicin) and the cognitive realization that the stimulus is neither dangerous nor life threatening. Rozin and Schiller speculated that there is enjoyment that comes from experiencing a constrained risk such as this, similar to the same type of enjoyment that comes from riding roller-coasters or gambling. Rozin and Schiller termed these types of activities "benignly masochistic" (Rozin and Schiller, 1980). In research with Americans and Mexicans who reported enjoying spicy foods, a number of individuals showed preferred levels of spice that were equal to their maximum tolerable level of spice (Rozin and Schiller, 1980). Spicy food dislikers overall showed a larger distance between preferred level of spice and maximum tolerated level of spice than the reported spicy food likers.

2.4 Cognitive factors underlying chemesthetic response: state versus trait

This section provides an overview of a variety of personality instruments and traits that have been used in the exploration of food choice motives including the Sensation Seeking Scale (Zuckerman *et al.*, 1964), Arnett's Inventory of Sensation Seeking (Arnett, 1994), the Novelty Seeking subscale (Cloninger, 1987), Extraversion (from Eysenck's EPQ: Eysenck and Eysenck, 1978), the Sensitivity to Punishment and Sensitivity to Reward Questionnaire (Torrubia *et al.*, 2001), Private Body Consciousness (Miller *et al.*, 1981), Food Neophobia (Pliner, 1982), and the Food Involvement Scale (Bell and Marshall, 2003). A brief overview of the personality traits measured by these scales will be given, followed by a review of work linking the liking and consumption of spicy foods, specifically capsaicin-containing foods, to traits measured by these scales.

2.4.1 Personality traits

2.4.1.1 Sensation seeking

Aside from food neophobia, finickiness, and pickiness, one of the best-studied personality traits in the food choice literature is sensation seeking. Sensation seeking was first defined as the "need for varied, novel, and complex sensations and experiences" (Zuckerman *et al.*, 1964). The first version of Zuckerman's Sensation Seeking Scale to measure this trait was published in 1964, and since then the scale has evolved to the current version: Zuckerman's Sensation Seeking Scale-V (SSS-V) (Zuckerman, 2007; Zuckerman *et al.*, 1964). In the time between the first and most recent versions of the scale, four factors emerged: thrill and adventure seeking (TAS), experience seeking (ES), disinhibition (DIS), and boredom susceptibility (BS; Zuckerman, 1996). The TAS subscale consists of items that show the desire to engage in physical activities, such as mountain climbing or skydiving, which provide unusual experiences and sensations. The items on the ES subscale show the desire to seek new sensations and experiences through the mind, such as music, art, and travel. These sensations are often sought through a generally nonconforming lifestyle and was informally called the "hippie factor" in the 1970s. The DIS subscale consists of items that indicate the desire to seek sensations through other people, or through a "hedonic lifestyle" (Zuckerman, 2007). Activities of this lifestyle include drinking to disinhibit, attending wild parties, and seeking sexual variety. The BS scale represents an aversion to monotony and a desire to avoid or break away from monotonous conditions.

2.4.1.1.1 Biological basis for sensation seeking

As with a number of other personality theorists, Zuckerman suggested that there might be a biological basis for the differences observed in sensation seeking (Zuckerman, 2007). Twin studies have shown that the heritability of sensation seeking, as measured with SSS-V, is around 0.58 (Fulker *et al.*, 1980; Hur and Bouchard Jr, 1997). These estimates are high compared with heritability measures of other personality measures, which fall between 0.30 and 0.50 (Bouchard, 1994; Loehlin, 1992). In 1995, Zuckerman proposed a biological model that illustrates interactions between the three behavioral mechanisms that are assumed to underlie sensation seeking: arousal, inhibition, and approach systems (Zuckerman, 1995). Dopamine functioning has been associated with approach behaviors, serotonin has been associated with inhibition behaviors, and norepinephrine has been associated with arousal behaviors (Berridge and Stalnaker, 2002). Thus, sensation seeking is supposedly associated with strong dopamine reactivity, and weak serotonin and norepinephrine reactivity. Importantly, reactivity refers to the sensitivity of receptor cells. It was also suggested by Zuckerman that enzymes such as monoamine oxidase (MAO) and dopamine beta-hydroxylase (DBH) may also affect reactivity (Zuckerman, 2007).

2.4.2 New forms of sensation seeking scales

While Zuckerman's Sensation Seeking Scale is well validated and has been used in a number of studies, there are a number of criticisms of the wording and format of the scales (Arnett, 1994; Haynes *et al.*, 2000). The original Sensation Seeking Scales contained language that is unfamiliar to younger generations. These terms include "hippies," "swingers," and "jet-setters." The forced choice response format has also been criticized because some individuals may not feel like either answer option is representative (Arnett, 1994). Additionally, there are a number of items on Zuckerman's scale that include strenuous physical activities, which may create an age bias, and thus age-related differences in responses should be interpreted with caution. To address these issues, a number of different scales have been created, some that are based on the original construct of sensation seeking as defined by Zuckerman, others that redefine the trait of sensation seeking, and still others that assess a related dimension to sensation seeking. The following sections will present an overview of some of these scales.

2.4.2.1 Arnett's Inventory of Sensation Seeking (AISS)

The original definition of sensation seeking included novelty and complexity as important characteristics of stimuli that would be sought out by high sensation seekers, but it did not include intensity. Arnett emphasized the importance of the intensity of the stimuli in his conceptualization of sensation seeking and divided his scale into two subscales, the novelty seeking (NS) and intensity seeking (IS) subscales (Arnett, 1994) each with ten questions. In addition to this reconceptualization, Arnett gave more emphasis to the effect that environment might have on personality and removed questions that were age and gender biased.

Although the AISS and SSS-V have different theoretical foundations, they are designed to measure the same personality construct; however, reported correlations between the scales are between 0.41 and 0.71 (Andrew and Cronin, 1997; Arnett, 1994; Carretero Dios and Salinas Martínez De Lecea, 2008; Ferrando and Chico, 2001; Haynes *et al.*, 2000; Zuckerman, 2007). Two studies specifically addressed the degree of equivalence of the scales and found that in younger individuals (undergraduate students, mean age = 22.61 years) the scales measure the same construct (Ferrando and Chico, 2001). However, in older individuals, differences between the scales are apparent (Carretero Dios and Salinas Martínez De Lecea, 2008).

2.4.2.2 Novelty Seeking (NS)

Like Zuckerman (Zuckerman, 2002), Cloninger believed that novelty seeking, a closely related trait to sensation seeking, was a major personality factor, rather than a facet of a different factor, as it is in the Big Five personality traits (Cloninger, 1987; Cloninger *et al.*, 1993). Cloninger's Tridimensional Personality Questionnaire (TPQ) (Cloninger, 1987) is made up of three factors, novelty seeking (NS), harm avoidance (HA), and reward dependence (RD). Zuckerman and Cloninger disagree over the level of similarity between the constructs of NS and SS, with Zuckerman (Zuckerman, 1988) suggesting that NS is practically identical to trait SS, while Cloninger argued that SS is a more factorially complex construct that is possibly a combination of all three dimensions of this personality model, NS, HA, and RD (Cloninger, 1987). Some studies assessing the relatedness of NS to SSS-V show strong correlations between the measures ranging from 0.34 to 0.55 (Mccourt *et al.*, 1993; Zuckerman and Cloninger, 1996). Correlations between the total scale scores support Zuckerman's argument that NS and SS are closely related. However, the evidence that NS and HA are related to different parts of SS support Cloninger's claim.

2.4.2.3 Sensitivity to Punishment and Sensitivity to Reward Questionnaire (SPSRQ)

The Sensitivity to Punishment and Sensitivity to Reward Questionnaire (SPSRQ) is an operationalization of Jeffrey Gray's personality theory, which is not associated with a specified scale. While there have been a number of attempts to measure Gray's personality dimensions, the Sensitivity to Punishment and Sensitivity to Reward scales are regarded as the best measure of Gray's Behavioral Inhibition System and Behavioral Approach System, respectively (Caseras *et al.*, 2003).

The SPSRQ is made up of two scales, the Sensitivity to Punishment scale (SP), which measures behavioral inhibition under specific conditions of punishment or threat, and the Sensitivity to Reward scale (SR), which measures approach behaviors to conditioned and unconditioned stimuli that indicate rewards, such as money, sexual partners, praise, and social status (Caseras *et al.*, 2003; Torrubia *et al.*, 2001). These two scales represent distinct systems such that an individual's personality is a combination of sensitivity to punishment and reward (someone does not necessarily have to be low SP if they are high SR). The English language version of the scale was developed by O'Connor and colleagues (O'Connor *et al.*, 2004). The final version of the SPSRQ is made up of two scales, each with 24 questions.

In the initial study publishing the SPSRQ, Torrubia and colleagues conducted extensive analysis to explore the relation of the SP and SR scales to the array of scales that measure either impulsivity, anxiety, or dimensions closely related to these traits, including Zuckerman's SSS-V and Cloninger's TPQ (Torrubia *et al.*, 2001). Comparing SP to Zuckerman's and Cloninger's measures, SP showed significantly negative correlations with SSS-V total score and subscale scores, while a significant positive relationship was observed between SP and Cloninger's HA dimension (Torrubia *et al.*, 2001). Comparing the SR subscale to these measures, SR and NS showed significant positive, although weak, associations, while SR and Zuckerman's SSS-V showed significant positive associations with SSS-V total score and subscale scores, ranging in strength from weak to moderate (Torrubia *et al.*, 2001).

Measures of impulsivity have been associated with SS, NS, and SR (Torrubia *et al.*, 2001; Zuckerman, 2007). While impulsivity may be linked with each of these constructs, and could play a role in the liking of spicy foods, trait impulsivity is a multidimensional concept that is not wholly measured by any of these constructs (Evenden, 1999; Eysenck and Eysenck, 1978; Gerbing *et al.*, 1987; Parker and Bagby, 1997; Pickering and Gray, 1999; Torrubia *et al.*, 2001).

2.4.2.4 Private Body Consciousness (PBC)

Another personality theory that has been linked to the consumption of spicy foods is the Private Body Consciousness (PBC). Private Body Consciousness refers to degrees of self-awareness and self-consciousness that are only observable to the individual (Miller *et al.*, 1981). These state changes include changes in heart rate,

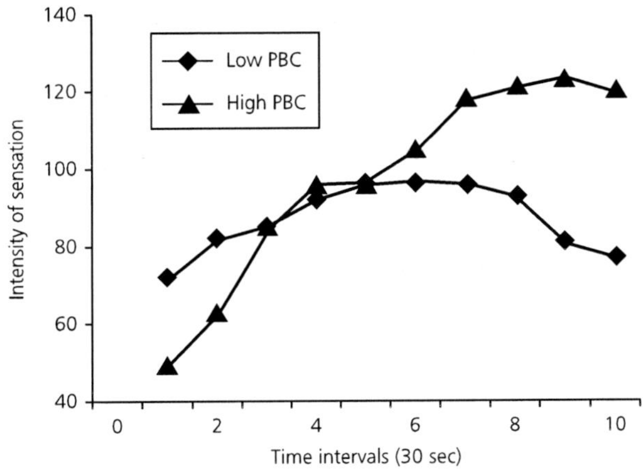

Fig. 2.4 Periodic mean normalized estimates of the intensity of the oral irritation produced by capsaicin over 5 min by subjects (n = 10/gp) classified as having high or low private body consciousness (Source: Adapted from Stevens, 1990).

hunger pains, and body temperature. Individuals are asked to rate how well five statements characterize them using a five-point Likert scale (0 – extremely uncharacteristic to 4 – extremely characteristic). Individuals with high PBC have reportedly been able to detect and identify differences in sensory properties of foods compared with low PBC individuals as a result of their increased sensitivity to sensory stimuli, as seen in Fig. 2.4 (Jaeger *et al.*, 1998; Miller *et al.*, 1981; Stevens, 1990; Ueland, 2001). PBC has also been linked with sensitivity to sensations caused by spicy foods such that high PBC individuals rate the perceived burning from capsaicin as more intense than low PBC individuals (Ferguson and Ahles, 1998; Martin *et al.*, 1991; Stevens, 1990), however not all studies support these findings (Byrnes and Hayes, 2013).

2.4.2.5 Food Neophobia and Food Involvement Scale

Two other scales that have been associated with food preference and consumption are the Food Neophobia Scale (FNS; Pliner and Hobden, 1992) and the Food Involvement Scale (FIS; Bell and Marshall, 2003). Pliner and Hobden developed the FNS as a measure of an individual's reluctance or willingness to eat new foods. There is a hypothetical evolutionary significance for the development of food neophobia, as discussed earlier and it has been proposed that food neophobia is considered a personality trait (Pliner and Hobden, 1992). While a number of manipulations, such as the overall degree of novelty in the eating situation, have been shown to influence the degree of food neophobia in humans and other animals, a fairly stable propensity to approach or avoid new foods has been suggested (Pliner and Hobden, 1992). This stability, along with the hypothetical evolutionary significance of the trait, motivated researchers to explore any hereditary component of food neophobia (Knaapila *et al.*, 2007). Findings from this study suggest that roughly two thirds of the variance in food neophobia can be attributed to genetic effects with 69% in Finnish families and 67% in British families (Knaapila *et al.*, 2007).

The scale consists of ten statements to which the respondent replies, using a seven-point scale, how strongly they agree or disagree with (1 = "strongly disagree" to 7 = "strongly agree"). The range of possible scores is 10 to 70, with higher scores representing high neophobia, thus lower levels of approach behavior towards new foods. Alpha coefficients for the FNS are sufficiently high ($r = 0.88$; Pliner and Hobden, 1992) as are test–retest reliabilities at two to four weeks, and 12 weeks (2 to 4 weeks: $r = 0.91$, $p < 0.01$, 12 weeks: $r = 0.82$, $p < 0.01$; Pliner and Hobden, 1992).

One would expect the trait of food neophobia to be related to Zuckerman's trait Sensation Seeking, since those who are high Sensation Seekers should be more likely to choose novel foods to increase their arousal level versus low Sensation Seekers. However, results in the literature do not fully support this notion. Significant correlations have been reported between scores on the ES subscale of Zuckerman's SSS-V (Zuckerman and Neeb, 1979), to the number of unfamiliar foods an individual is willing to taste (Otis, 1984), and to measures of neophobia ($r = -0.46$, $p < 0.05$; Pliner and Hobden, 1992). However, in a different study examining the relationship between the FNS and the SSS-V, no main effect of sensation seeking was seen on the number of novel foods chosen, though a significant interaction effect with level of arousal state was observed (Pliner and Melo, 1997). In the low arousal state, individuals with high trait SS tried significantly more novel foods than their low SS counterparts. In overall adventurousness, individuals who seek out experiences that are new and/or exciting tend to be less neophobic than less adventurous individuals (Terasaki and Imada, 1988). Without considering food choice motives, scores on the FNS significantly predicted intake of spices in a group of Dutch undergraduate students (Eertmans *et al.*, 2005).

The FIS, developed by Bell and Marshall (2003), measures the importance of food in a person's life by asking about the extent to which an individual enjoys talking and thinking about food and engages in food-related activities at all five stages of the food's "lifecycle" (acquisition, preparation, cooking, eating, and disposal; Goody, 1982). Factor analysis showed two subscales that exist in the FIS, the set and disposal (S&D) construct and the preparation and eating (P&E) construct (Marshall and Bell, 2004). Individuals with high FIS scores were shown to make finer discriminations between food items in sensory evaluations and hedonic ratings (Bell and Marshall, 2003). There have been no associations of the FIS to liking of capsaicin-containing foods but given the relationships of this scale with the Food Neophobia Scale and the VARSEEK scale (Bell and Marshall, 2003; Marshall and Bell, 2004), it is possible that individuals with high FIS scores are more adventurous with foods, and thus more willing to try capsaicin-containing foods even if they do not enjoy their first encounter.

2.4.3 Personality and food choice

Food neophobia has been used extensively in studying food choice, though little work has been done examining the relationship between food neophobia and the liking of spicy foods. The Food Involvement Scale has also seen limited application in the field of food choice. In one study the relationships between personality traits, a range of food choice motives, and intake of a variety of foods were assessed in a group of Dutch undergraduate students (Eertmans *et al.*, 2005). Scores on the FNS significantly predicted intake of spices when not taking into account food choice motives (negative correlation; Eertmans *et al.*, 2005). Eertmans and colleagues (2005) observed that food involvement showed a positive effect on the intake of spices, when food choice motives were not taken into account. Importantly, this study only specified that "spices" did not include salt or pepper, thus it is possible that this relationship may be driven by spices other than those containing capsaicin.

There have been several attempts to associate sensation-seeking traits with the liking for pungent sensations in foods and beverages. Early work linking personality traits with foods focused on the relationship between food preferences and the oral–passive versus oral–sadistic Freudian personality dimension (Wolowitz, 1964). Later work suggested that sensation seekers tend to prefer bold foods, such as foods that are crunchy, sour, and spicy over sweet, soft, and bland foods (Kish and Donnenwerth, 1972; Brown *et al.*, 1974). Rozin and Schiller's 1980 study (Rozin and Schiller, 1980), which suggests a relationship between preference for chili peppers and sensation seeking, is one of the most well-known studies linking personality and a liking for spicy food and is often treated as the definitive work that firmly establishes the relationship between these variables. In this study, eating chili peppers was linked with a number of "benignly masochistic" and thrill-seeking activities, such as riding rollercoasters, gambling, and consumption of initially aversive substances, such as alcohol and coffee in American college students (1980). A common thread among these experiences is that they are initially aversive yet certain individuals learn to enjoy them, perhaps due to appreciation that the initial perceived risk is harmless and thus any riskiness can be considered a "constrained risk" (Rozin and Schiller, 1980). However, while Rozin convincingly argued that sensation seeking and the liking of spicy foods are linked, it is important to note that the data were not collected using validated measures of personality, but rather, the link to sensation seeking was inferred based on a theorized common "constrained risk" across the activities, such as riding rollercoasters, gambling, and eating spicy foods.

A number of studies report weak, yet significant correlations between measures of sensation seeking and the liking of, or preference for, spicy and chili-containing foods (Brown *et al.*, 1974; Logue and Smith, 1986; Terasaki and Imada, 1988). However, more recent work shows strong positive correlations between the liking and intake of spicy foods and the personality measures Sensation Seeking (AISS) and Sensitivity to Reward (SPSRQ; Byrnes and Hayes, 2013). Later work showed that the effects of Sensation Seeking and Sensitivity to Reward may act on intake of spicy foods through different pathways (Byrnes and Hayes, 2015). Additionally, it was observed that the effects of these personality traits were different between men and women, indicating that there may be biological differences related to these personality measures that influence liking and intake differently. Interestingly, measures of risk taking and impulsivity were not associated with liking of spicy foods.

While Impulsivity, Sensation Seeking, Sensitivity to Reward, and Novelty Seeking are related (see Section 2.4.1), the exact orientation of these scales to one another is not clear. It is clear, however, that these scales do not measure the same underlying construct, perhaps accounting for the differences seen in the association of these scales with the liking and intake of spicy foods. It is likely that Sensation Seeking and

Sensitivity to Reward tap a similar dimension regarding the intake of spicy foods, however the mechanism by which these traits relate to intake may differ. It is possible that for individuals scoring high in Sensation Seeking, the effect on intake acts through liking, perhaps reflecting an intrinsically pleasant response evoked by capsaicin. In individuals scoring high in Sensitivity to Reward or Risk Taking, it is possible that the effect of personality on the intake of spicy foods reflects extrinsic motivation for consumption, such as social praise or the association with traits such as strength or daring as reported by Rozin and Schiller (1980).

2.4.4 Cognitive factors underlying chemesthetic response: states

Inter-individual variation in the acceptability and liking of chemesthetic sensations can also result from cognitive and emotional factors that influence perception. Such influences arise from what are generally considered to be "top-down" factors, which comprise features that the individual brings from their personal experience, that is, pre-existing knowledge or general beliefs or expectations (Solso, 2001). How these non-sensory factors can be influential in determining the liking and acceptability of chemesthetic sensations in foods and beverages will now be described.

2.4.4.1 Context

The context in which a sensation is experienced can be instrumental in determining its acceptability (Zellner *et al.*, 2007). For example, the pleasantness of the oral tingling sensation from mint compounds such as those used in toothpastes or mouthwashes may acquire its acceptability through flavor-sensation conditioning with continued use. However, such a sensation can be considered undesirable when experienced with other flavors or in other products. Similarly, pungent sensations experienced in unfamiliar products or with unfamiliar flavors are likely to be deemed unpleasant.

2.4.4.2 Expectations

The degree to which expectations can bias the hedonic evaluation and acceptability of an irritant stimulus has been demonstrated in studies using odorous chemicals that also stimulate the nasal and ocular trigeminal system. Dalton (1999) exposed individuals to a mildly pungent concentration of methyl salicylate (wintergreen) under three different information conditions: no information, positive health information, and negative health information. Individuals given the negative health information rated the sensations evoked as far more irritating and intense than those given the positive health information. Indeed, many of the individuals in the positive information condition reported feeling "refreshed" and "invigorated" following the exposure.

Expectations have also been shown to influence the rated bitterness of coffee (Olson and Dover, 1976) and pomegranate juice (Cardello and Sawyer, 1992) and the pungency of olive oil (Caporale *et al.*, 2006). In all cases, the expectations provided by the experimenters were assimilated into the ratings of the actual products and thereby shifted the hedonic response in the hypothesized direction. In a similar study, health-related information about a bittersweet beverage increased a behavioral measure of acceptability, but not hedonic ratings (Stein *et al.*, 2003). Thus, there is reason to suspect that expectations about the pungency of a food or beverage can shift its perceptual characteristics and acceptability.

2.5 Benefits of liking

The basis for the widespread consumption of chemesthetic agents such as chili peppers in the diet have been explored via adaptive and evolutionary mechanisms. These explanations have ranged from their benefits in food preservation, masking food spoilage and inducing perspiration and therefore an evaporative cooling effect in tropical environments. Unfortunately, there is little evidence to support these contentions (Rozin, 1990b).

However, there is evidence to support both nutritional and sensory benefits from consumption of chemesthetic agents (see Chapter 12). As suggested earlier, the stimulation from chemesthetic agents has been linked to increased salivary flow and gastric juice secretion, and hence aids in digestion. In addition, studies have shown that several types of pungent spices (ginger, piperine, capsaicin) stimulate bile flow, which leads to enhanced fat digestion and absorption, but without fat accumulation due to their effects on inducing thermogenesis and enhanced lipid oxidation (Prakash and Srinivasan, 2012). Anti-inflammatory and hypocholesterolaemic effects have also been identified (Liang *et al.*, 2013; Pal *et al.*, 2009; Srinivasan, 2013). Thus, in addition to the sensory pleasures certain individuals obtain from the consumption of chemesthetic agents, there may be post-ingestive consequences, which also provide reinforcement for their consumption.

2.6 Summary

Despite widespread consumption and numerous studies, no clear consensus has emerged to explain humans' liking for oral irritation from spicy foods. Genetic influences on oral sensitivity may account for individual differences at baseline, while personality differences or cognitive factors may account for the willingness to explore and ultimately incorporate novel food sensations, including initially aversive ones. Finally, post-ingestive benefits for some may also contribute to continued liking and consumption. To be sure, it is likely that more than one factor plays a role in the acquisition and maintenance of preference for spicy oral sensations and it is abundantly clear that pungent spices will continue to elicit highly valued and appetitive sensations in the diets of many cultures, worldwide.

References

Andrew, M. and Cronin, C. 1997. Two measures of sensation seeking as predictors of alcohol use among high school males. *Personality and Individual Differences*, **22**, 393–401.

Arnett, J. 1994. Sensation seeking : a new conceptualization and a new scale. *Personality and Individual Differences*, **16**, 289–296.

Bajec, M.R. and Pickering, G.J. 2008. Thermal taste, PROP responsiveness, and perception of oral sensations. *Physiology & Behavior*, **95**, 581–590.

Bartoshuk, L.M., Duffy, V.B., and Miller, I.J. 1994. PTC/PROP tasting: anatomy, psychophysics, and sex effects. *Physiology & Behavior*, **56**, 1165–1171.

Beauchamp, G.K. and Mennella, J.A. 2009. Early flavor learning and its impact on later feeding behavior. *Journal of Pediatric Gastroenterology and Nutrition*, **48**, S25–S30.

Bell, R. and Marshall, D.W. 2003. The construct of food involvement in behavioral research: scale development and validation. *Appetite*, **40**, 235–244.

Bennett, S.M. and Hayes, J.E. 2012. Differences in the chemesthetic subqualities of capsaicin, ibuprofen, and olive oil. *Chemical Senses*, **37**, 471–478.

Berridge, C.W. and Stalnaker, T.A. 2002. Relationship between low-dose amphetamine-induced arousal and extracellular norepinephrine and dopamine levels within prefrontal cortex. *Synapse*, **46**, 140–149.

Birch, L.L. 1979a. Dimensions of preschool children's food preferences. *Journal of Nutrition Education*, **11**, 77–80.

Birch, L.L. 1979b. Preschool children's food preferences and consumption patterns. *Journal of Nutrition Education*, **11**, 189–192.

Birch, L.L. 1980. Effects of peer models' food choices and eating behaviors on preschoolers' food preferences. *Child Development*, **51**, 489–496.

Birch, L.L. and Marlin, D.W. 1982. I don't like it; I never tried it: effects of exposure on two-year-old children's food preferences. *Appetite*, **3**, 353–360.

Bouchard, T.J. 1994. Genes, environment, and personality. *Science*, **264**, 1700–1701.

Brown, L.T., Ruder, V.G., Ruder, J.H., and Young, S.D. 1974. Stimulation seeking and the Change Seeker Index. *Journal of Consulting and Clinical Psychology*, **42**, 311.

Bryant, B.P. and Mezine, I. 1999. Alkylamides that produce tingling paresthesia activate tactile and thermal trigeminal neurons. *Brain Research*, **842**, 452–460.

Byrnes, N.K. and Hayes, J.E. 2013. Personality factors predict spicy food liking and intake. *Food Quality and Preference*, **28**, 213–221.

Byrnes, N.K. and Hayes, J.E. 2015. Gender differences in the influence of personality traits on spicy food liking and intake. *Food Quality and Preference*, **42**, 12–19.

Caporale, G., Policastro, S., Carlucci, A., and Monteleone, E. 2006. Consumer expectations for sensory properties in virgin olive oils. *Food Quality and Preference*, **17**, 116–125.

Capretta, P.J. and Rawls, L.H. 1974. Establishment of a flavor preference in rats: importance of nursing and weaning experience. *Journal of Comparative and Physiological Psychology*, **86**, 670–673.

Cardello, A.V. and Sawyer, F.M. 1992. Effects of disconfirmed consumer expectations on food acceptability. *Journal of Sensory Studies*, **7**, 253–277.

Carretero Dios, H. and Salinas Martínez De Lecea, J.M. 2008. Using a structural equation model to assess the equivalence between assessment instruments: the dimension of sensation seeking as measured by Zuckerman' s SSS-V and Arnett's AISS. *International Journal of Clinical and Health Psychology*, **8**, 219–232.

Carroll, M.E., Dinc, H.I., Levy, C.J., and Smith, J.C. 1975. Demonstrations of neophobia and enhanced neophobia in the albino rat. *Journal of Comparative and Physiological Psychology*, **89**, 457–467.

Caseras, X., Avila, C., and Torrubia, R. 2003. The measurement of individual differences in Behavioural Inhibition and Behavioural Activation Systems: a comparison of personality scales. *Personality and Individual Differences*, **34**, 999–1013.

Cicerale, S., Breslin, P.A., Beauchamp, G.K., and Keast, R.S. 2009. Sensory characterization of the irritant properties of oleocanthal, a natural anti-inflammatory agent in extra virgin olive oils. *Chemical Senses*, **34**, 333–339.

Cliff, M. and Heymann, H. 1992. Descriptive analysis of oral pungency. *Journal of Sensory Studies*, **7**, 279–290.

Cloninger, C.R. 1987. A systematic method for clinical description and classification of personality variants. *A proposal. Archives of General Psychiatry*, **44**, 573–588.

Cloninger, C.R., Svrakic, D.M., and Przybeck, T.R. 1993. A psychobiological model of temperament and character. *Archives of General Psychiatry*, **50**, 975–990.

Costa, M., Balthazar, C., Franco, R., *et al.* 2014. Changes on expected taste perception of probiotic and conventional yogurts made from goat milk after rapidly repeated exposure. *Journal of Dairy Science*, **97**, 2610–2618.

Cowart, B.J. 1987. Oral chemical irritation: does it reduce perceived taste intensity? *Chemical Senses*, **12**, 467–479.

Crandall, C.S. 1985. The liking of foods as a result of exposure: Eating doughnuts in Alaska. *The Journal of Social Psychology*, **125**, 187–194.

Dalton, P. 1999. Cognitive influences on health symptoms from acute chemical exposure. *Health Psychology*, **18**, 579–590.

Dinnella, C., Recchia, A., Tuorila, H., and Monteleone, E. 2011. Individual astringency responsiveness affects the acceptance of phenol-rich foods. *Appetite*, **56**, 633–642.

Dinnella, C., Recchia, A., Vincenzi, S., Tuorila, H., and Monteleone, E. 2010. Temporary modification of salivary protein profile and individual responses to repeated phenolic astringent stimuli. *Chemical Senses*, **35**, 75–85.

Domjan, M. 1972. CS preexposure in taste-aversion learning: Effects of deprivation and preexposure duration. *Learning and Motivation*, **3**, 389–402.

Domjan, M. 1976. Determinants of the enhancement of flavored-water intake by prior exposure. *Journal of Experimental Psychology: Animal Behavior Processes*, **2**, 17–27.

Domjan, M. and Bowman, T.G. 1974. Learned safety and the CS-US delay gradient in taste-aversion learning. *Learning and Motivation*, **5**, 409–423.

Domjan, M. and Gillan, D. 1976. Role of novelty in the aversion for increasingly concentrated saccharin solutions. *Physiology & Behavior*, **16**, 537–542.

Duffy, V.B. 2007. Variation in oral sensation: implications for diet and health. *Current Opinion in Gastroenterology*, **23**, 171–177.

Duffy, V.B. and Bartoshuk, L.M. 2000. Food acceptance and genetic variation in taste. *Journal of the American Dietetic Association*, **100**, 647–655.

Duffy, V.B., Hayes, J.E., Davidson, A.C., *et al.* 2010. Vegetable intake in college-aged adults is explained by oral sensory phenotypes and *TAS2R38* genotype. *Chemosensory Perception*, **3**, 137–148.

Eertmans, A., Victoir, A., Vansant, G., and Van Den Bergh, O. 2005. Food-related personality traits, food choice motives and food intake: Mediator and moderator relationships. *Food Quality and Preference*, **16**, 714–726.

Essick, G.K., Chopra, A., Guest, S., and Mcglone, F. 2003. Lingual tactile acuity, taste perception, and the density and diameter of fungiform papillae in female subjects. *Physiology & Behavior*, **80**, 289–302.

Evenden, J.L. 1999. Varieties of impulsivity. *Psychopharmacology*, **146**, 348–361.

Eysenck, S.B.E. and Eysenck, H.J. 1978. Impulsiveness and venturesomeness: their position in a dimensional system of personality description. *Psychological Reports*, **43**, 1247–1255.

Fajardo, K. 2014. Innovation on the Menu: Flavor Trends, report, Mintel Group Ltd.

Feeney, E.L. and Hayes, J.E. 2014. Regional differences in suprathreshold intensity for bitter and umami stimuli. *Chemosensory Perception*, **7**, 147–157.

Ferguson, R.J. and Ahles, T.A. 1998. Private body consciousness, anxiety and pain symptom reports of chronic pain patients. *Behavoiur Research and Therapy*, **36**, 527–535.

Ferrando, P.J. and Chico, E. 2001. The construct of sensation seeking as measured by Zuckerman's SSS-V and Arnett's AISS: a structural equation model. *Personality and Individual Differences*, **31**, 1121–1133.

Fischer, M.E., Cruickshanks, K.J., Schubert, C.R., *et al.* 2013. Factors related to fungiform papillae density: the beaver dam offspring study. *Chemical Senses*, **38**, 669–677.

Fulker, D.W., Eysenck, S.B., and Zuckerman, M. 1980. A genetic and environmental analysis of sensation seeking. *Journal of Research in Personality*, **14**, 261–281.

Garneau, N.L., Nuessle, T.M., Sloan, M.M., *et al.* 2014. Crowdsourcing taste research: genetic and phenotypic predictors of bitter taste perception as a model. *Frontiers in Integrative Neuroscience*, May 27;8:33, DOI 10.3389/fnint.2014.00033. eCollection 2014.

Gerbing, D.W., Ahadi, S.A., and Patton, J.H. 1987. Toward a conceptualization of impulsivity: Components across the behavioral and self-report domains. *Multivariate Behavioral Research*, **22**, 357–379.

Green, B.G. and Hayes, J.E. 2004. Individual differences in perception of bitterness from capsaicin, piperine and zingerone. *Chemical Senses*, **29**, 53–60.

Green, B.G., Shaffer, G.S., and Gilmore, M.M. 1993. Derivation and evaluation of a semantic scale of oral sensation magnitude with apparent ratio properties. *Chemical Senses*, **18**, 683–704.

Goody, J. 1982. *Cooking, Cuisine and Class: A Study in Comparative Sociology*. Cambridge: Cambridge University Press.

Hayes, J.E., Allen, A.L., and Bennett, S.M. 2013. Direct comparison of the generalized visual analog scale (gVAS) and general labeled magnitude scale (gLMS). *Food Quality and Preference*, **28**, 36–44.

Hayes, J.E., Bartoshuk, L.M., Kidd, J.R., and Duffy, V.B. 2008. Supertasting and PROP bitterness depends on more than the TAS2R38 gene. *Chemical Senses*, **33**, 255–265.

Hayes, J.E. and Duffy, V.B. 2007. Revisiting sugar–fat mixtures: sweetness and creaminess vary with phenotypic markers of oral sensation. *Chemical Senses*, **32**, 225–236.

Hayes, J.E., Feeney, E.L., and Allen, A.L. 2013. Do polymorphisms in chemosensory genes matter for human ingestive behavior? *Food Quality and Preference*, **30**, 202–216.

Haynes, C.A., Miles, J.N.V., and Clements, K. 2000. A confirmatory factor analysis of two models of sensation seeking. *Personality and Individual Differences*, **29**, 823–839.

Hill, W.F. 1978. Effects of mere exposure on preferences in nonhuman mammals. *Psychological Bulletin*, **85**, 1177–1198.

Horne, J., Hayes, J., and Lawless, H.T. 2002. Turbidity as a measure of salivary protein reactions with astringent substances. *Chemical Senses*, **27**, 653–659.

Horne, P., Tapper, K., Lowe, C., *et al*. 2004. Increasing children's fruit and vegetable consumption: a peer-modelling and rewards-based intervention. *European Journal of Clinical Nutrition*, **58**, 1649–1660.

Hur, Y.-M. and Bouchard, T.J., Jr. 1997. The genetic correlation between impulsivity and sensation seeking traits. *Behavior Genetics*, **27**, 455–463.

Jaeger, S.R., Andani, Z., Wakeling, I.N., and Macfie, H.J.H. 1998. Consumer preferences for fresh and aged apples: a cross-cultural comparison. *Food Quality and Preference*, **9**, 355–366.

Karrer, T. and Bartoshuk, L. 1991. Capsaicin desensitization and recovery on the human tongue. *Physiology & Behavior*, **49**, 757–764.

Karrer, T., Bartoshuk, L., Conner, E., *et al*. 1992. *PROP status and its relationship to the perceived burn intensity of capsaicin at different tongue loci*, Abstracts, Fourteenth Annual Meeting of the Association for Chemoreception Sciences (AChemS XIV), Oxford: IRL Press at Oxford University Press.

Keskitalo, K., Knaapila, A., Kallela, M., *et al*. 2007. Sweet taste preferences are partly genetically determined: identification of a trait locus on chromosome 16. *The American Journal of Clinical Nutrition*, **86**, 55–63.

Khachik, F., Spangler, C.J., Smith, J.C., *et al*. 1997. Identification, quantification, and relative concentrations of carotenoids and their metabolites in human milk and serum. *Analytical Chemistry*, **69**, 1873–1881.

Kim, U.-K., Jorgenson, E., Coon, H., *et al*. 2003. Positional cloning of the human quantitative trait locus underlying taste sensitivity to phenylthiocarbamide. *Science*, **299**, 1221–1225.

Kish, G.B. and Donnenwerth, G.V. 1972. Sex differences in the correlates of stimulus seeking. *Journal of Consulting and Clinical Psychology*, **38**, 42–49.

Knaapila, A., Tuorila, H., Silventoinen, K., *et al*. 2007. Food neophobia shows heritable variation in humans. *Physiology & Behavior*, **91**, 573–578.

Lakkakula, A., Geaghan, J., Zanovec, M., Pierce, S., and Tuuri, G. 2010. Repeated taste exposure increases liking for vegetables by low-income elementary school children. *Appetite*, **55**, 226–231.

Lawless, H., Rozin, P., and Shenker, J. 1985. Effects of oral capsaicin on gustatory, olfactory and irritant sensations and flavor identification in humans who regularly or rarely consume chili pepper. *Chemical Senses*, **10**, 579–589.

Le Couteur, P. and Burreson, J. 2004. *Napoleon's Buttons: How 17 Molecules Changed History*, London: Penguin Books.

Liang, Y.T., Tian, X.-Y., Chen, J.N., *et al*. 2013. Capsaicinoids lower plasma cholesterol and improve endothelial function in hamsters. *European Journal of Nutrition*, **52**, 379–388.

Liem, D.G. and De Graaf, C. 2004. Sweet and sour preferences in young children and adults: role of repeated exposure. *Physiology & Behavior,* **83**, 421–429.

Loehlin, J.C. 1992. *Genes and environment in personality development,* Sage Publications, Inc.

Logue, A. and Smith, M.E. 1986. Predictors of food preferences in adult humans. *Appetite,* **7**, 109–125.

Ludy, M.J. and Mattes, R.D. 2011. Noxious stimuli sensitivity in regular spicy food users and non-users: Comparison of visual analog and general labeled magnitude scaling. *Chemosensory Perception,* **4**, 123–133.

Ludy, M.J. and Mattes, R.D. 2012. Comparison of sensory, physiological, personality, and cultural attributes in regular spicy food users and non-users. *Appetite,* **58**, 19–27.

Marshall, D. and Bell, R. 2004. Relating the food involvement scale to demographic variables, food choice and other constructs. *Food Quality and Preference,* **15**, 871–879.

Martin, J.B., Ahles, T.A., and Jeffery, R. 1991. The role of private body consciousness and anxiety in the report of somatic symptoms during magnetic resonance imaging. *Journal of Behavior Therapy and Experimental Psychiatry,* **22**, 3–7.

Maslow, A. 1937. The influence of familiarization on preference. *Journal of Experimental Psychology,* **21**, 162–180.

Mccourt, W.F., Gurrera, R.J. and Cutter, H.S. 1993. Sensation seeking and novelty seeking. Are they the same? *Journal of Nervous and Mental Disease,* **181**, 309–312.

Mcdonald, S., Barrett, P., and Bond, L. 2010. What kind of hot is it? *Perfumer & Flavorist,* **35**, 32–39.

Mennella, J.A., Jagnow, C.P., and Beauchamp, G.K. 2001. Prenatal and postnatal flavor learning by human infants. *Pediatrics,* **107**, E88.

Miller, I.J., Jr and Reedy, F.E., Jr. 1990. Variations in human taste bud density and taste intensity perception. *Physiology & Behavior,* **47**, 1213–1219.

Miller, L.C., Murphy, R., and Buss, A.H. 1981. Consciousness of body: private and public. *Journal of Personality and Social Psychology,* **41**, 397–406.

Mitchell, D., Scott, D.W., and Mitchell, L.K. 1977. Attenuated and enhanced neophobia in the taste-aversion "delay of reinforcement" effect. *Animal Learning & Behavior,* **5**, 99–102.

Mull, H.K. 1957. The effect of repetition upon the enjoyment of modern music. *The Journal of Psychology,* **43**, 155–162.

Nachman, M. 1959. The inheritance of saccharin preference. *Journal of Comparative and Physiological Psychology,* **52**, 451–457.

O'Connor, R.M., Colder, C.R., and Hawk, L.W. 2004. Confirmatory factor analysis of the Sensitivity to Punishment and Sensitivity to Reward Questionnaire. *Personality and Individual Differences,* **37**, 985–1002.

Olson, J.C. and Dover, P. 1976. Effects of expectation creation and disconfirmation on belief elements of cognitive structure. *Advances in Consumer Research,* **3**, 168–175.

Otis, L.P. 1984. Factors influencing the willingness to taste unusual foods. *Psychological Reports,* **54**, 739–745.

Pal, M., Angaru, S., Kodimuthali, A., and Dhingra, N. 2009. Vanilloid receptor antagonists: emerging class of novel anti-inflammatory agents for pain management. *Current Pharmaceutical Design,* **15**, 1008–1026.

Park, J.J., Lee, J., Kim, M.A., *et al.* 2007. Induction of total insensitivity to capsaicin and hypersensitivity to garlic extract in human by decreased expression of TRPV1. *Neuroscience Letters,* **411**, 87–91.

Parker, J.D. and Bagby, R.M. 1997. Impulsivity in adults: a critical review of measurement approaches, in *Impulsivity: Theory, Assessment, and Treatment,* (eds. C.D. Webster and M.A. Jackson), New York: Guildford Press, pp. 142–155.

Pickering, A.D. and Gray, J.A. 1999. The neuroscience of personality. *Handbook of Personality: Theory and Research,* **2**, 277–299.

Pickering, G.J. and Robert, G. 2006. Perception of mouthfeel sensations elicited by red wine are associated with sensitivity to 6-n-propylthiouracil. *Journal of Sensory Studies,* **21**, 249–265.

Pickering, G.J., Simunkova, K., and Dibattista, D. 2004. Intensity of taste and astringency sensations elicited by red wines is associated with sensitivity to PROP (6-n-propylthiouracil). *Food Quality and Preference,* **15**, 147–154.

Pliner, P. 1982. The effects of mere exposure on liking for edible substances. *Appetite*, **3**, 283–290.
Pliner, P. and Hobden, K. 1992. Development of a scale to measure the trait of food neophobia in humans. *Appetite*, **19**, 105–120.
Pliner, P. and Melo, N. 1997. Food neophobia in humans: effects of manipulated arousal and individual differences in sensation seeking. *Physiology & Behavior*, **61**, 331–335.
Prakash, U.N. and Srinivasan, K. 2012. Fat digestion and absorption in spice-pretreated rats. *Journal of the Science of Food and Agriculture*, **92**, 503–510.
Prescott, J. and Stevenson, R. J. 1995a. Effects of oral chemical irritation on tastes and flavors in frequent and infrequent users of chili. *Physiology & Behavior*, **58**, 1117–1127.
Prescott, J. and Stevenson, R.J. 1995b. Pungency in food perception and preference. *Food Reviews International*, **11**, 665–698.
Prescott, J. and Swain-Campbell, N. 2000. Responses to repeated oral irritation by capsaicin, cinnamaldehyde and ethanol in prop tasters and non-tasters. *Chemical Senses*, **25**, 239–246.
Rozin, P. 1990a. Acquisition of stable food preferences. *Nutrition Reviews*, **48**, 106–113.
Rozin, P. 1990b. Getting to like the burn of chili pepper: biological, psychological, and cultural perspectives, in *Chemical Senses, Vol. 2: Irritation*, (eds. B.G. Green, F.R. Mason, and M.R. Kare), New York: Dekker.
Rozin, P., Mark, M., and Schiller, D. 1981. The role of desensitization to capsaicin in chili pepper ingestion and preference. *Chemical Senses*, **6**, 23–31.
Rozin, P. and Schiller, D. 1980. The nature and acquisition of a preference for chili pepper by humans. *Motivation and Emotion*, **4**, 77–101.
Rozin, P. and Vollmecke, T.A. 1986. Food likes and dislikes. *Annual Review of Nutrition*, **6**, 433–456.
Sacerdote, C., Guarrera, S., Smith, G.D., *et al.* 2007. Lactase persistence and bitter taste response: instrumental variables and mendelian randomization in epidemiologic studies of dietary factors and cancer risk. *American Journal of Epidemiology*, **166**, 576–581.
Siegel, S. 1974. Flavor preexposure and "learned safety". *Journal of Comparative and Physiological Psychology*, **87**, 1073.
Snitker, S., Fujishima, Y., Shen, H., *et al.* 2009. Effects of novel capsinoid treatment on fatness and energy metabolism in humans: possible pharmacogenetic implications. *American Journal of Clinical Nutrition*, **89**, 45–50.
Srinivasan, K. 2013. Dietary spices as beneficial modulators of lipid profile in conditions of metabolic disorders and diseases. *Food & Function*, **4**, 503–521.
Solso, R.L. 2001. *Cognitive Psychology*, 6th edn, Boston: Allyn and Bacon, Inc., chap. 4.
Stein, L.J., Nagai, H., Nakagawa, M., and Beauchamp, G.K. 2003. Effects of repeated exposure and health-related information on hedonic evaluation and acceptance of a bitter beverage. *Appetite*, **40**, 119–129.
Stevens, D. A. 1990. Personality variables in the perception of oral irritation and flavor. in *Chemical Senses, Vol. 2: Irritation*, (eds. B.G. Green, F.R. Mason, and M.R. Kare), New York: Marcel Dekker.
Stevenson, R.J. and Prescott, J. 1994. The effects of prior experience with capsaicin on ratings of its burn. *Chemical Senses*, **19**, 651–656.
Stevenson, R.J. and Yeomans, M.R. 1993. Differences in ratings of intensity and pleasantness for the capsaicin burn between chilli likers and non-likers; implications for liking development. *Chemical Senses*, **18**, 471–482.
Stevenson, R.J. and Yeomans, M.R. 1995. Does exposure enhance liking for the chilli burn? *Appetite*, **24**, 107–120.
Sullivan, S.A. and Birch, L.L. 1990. Pass the sugar, pass the salt: Experience dictates preference. *Developmental Psychology*, **26**, 546–551.
Sullivan, S.A. and Birch, L.L. 1994. Infant dietary experience and acceptance of solid foods. *Pediatrics*, **93**, 271–277.
Tepper, B.J., Keller, K.L., and Ullrich, N.V. 2004. Genetic variation in taste and preferences for bitter and pungent foods: implications for chronic disease risk. *Challenges in Taste Chemistry and Biology*, **867**, 60–74.
Tepper, B.J. and Nurse, R.J. 1998. PROP taster status is related to fat perception and preferences. *Annals of the New York Academy of Sciences*, **855**, 802–804.

Terasaki, M. and Imada, S. 1988. Sensation Seeking and Food Preferences. *Personality and Individual Differences*, **9**, 87–93.

Törnwall, O., Silventoinen, K., Kaprio, J., and Tuorila, H. 2012. Why do some like it hot? Genetic and environmental contributions to the pleasantness of oral pungency. *Physiology & Behavior*, **107**, 381–389.

Törnwall, O., Silventoinen, K., Keskitalo-Vuokko, K., *et al.* 2012b. Genetic contribution to sour taste preference. *Appetite*, **58**, 687–694.

Torrubia, R., Avila, C., Molto, J., and Caseras, X. 2001. The Senstivity to Punishment and Sensitivity to Reward Questionnaire (SPSRQ) as a measure of Gray's anxiety and impulsivity dimensions. *Personality and Individual Differences*, **31**, 837–862.

Ueland, Ø. 2001. Private body consciousness, in: *Food, People and Society*, New York: Springer.

Wardle, J., Cooke, L.J., Gibson, E.L., *et al.* 2003. Increasing children's acceptance of vegetables; a randomized trial of parent-led exposure. *Appetite*, **40**, 155–162.

Wardle, J., Herrera, M., Cooke, L., and Gibson, E.L. 2003. Modifying children's food preferences: the effects of exposure and reward on acceptance of an unfamiliar vegetable. *European Journal of Clinical Nutrition*, **57**, 341–348.

Warren, R.P. and Pfaffmann, C. 1959. Early experience and taste aversion. *Journal of Comparative and Physiological Psychology*, **52**, 263–266.

Williams, R.A. 1968. Effects of repeated food deprivations and repeated feeding tests on feeding behavior. *Journal of Comparative and Physiological Psychology*, **65**, 222–226.

Wolowitz, H.M. 1964. Food preferences as an index or orality. *The Journal of Abnormal and Social Psychology*, **69**, 650.

Yoshioka, M., Doucet, E., Drapeau, V., Dionne, I., and Tremblay, A. 2001. Combined effects of red pepper and caffeine consumption on 24 h energy balance in subjects given free access to foods. *British Journal of Nutrition*, **85**, 203–211.

Zajonc, R.B. 1968. Attitudinal effects of mere exposure. *Journal of Personality and Social Psychology*, **9**, 1–27.

Zellner, D.A., Saito, S., and Gonzalez, J. 2007. The effect of stress on men's food selection. *Appetite*, **49**, 696–699.

Zuckerman, M. 1988. Sensation seeking and behavior disorders. *Archives of General Psychiatry*, **45**, 502–503.

Zuckerman, M. 1995. Good and bad humors: Biochemical bases of personality and its disorders. *Psychological Science*, **6**, 325–332.

Zuckerman, M. 1996. The psychobiological model for impulsive unsocialized sensation seeking: a comparative approach. *Neuropsychobiology*, **34**, 125–129.

Zuckerman, M. 2002. Zuckerman-Kuhlman Personality Questionnaire (ZKPQ): an alternative five-factorial model, in *Big Five Assessment*, (eds. B. De Raad and M. Perugini), Seattle: Hogrefe & Huber, pp. 377–396.

Zuckerman, M. 2007. *Sensation Seeking and Risk*, Washington, DC: American Psychological Association.

Zuckerman, M. and Cloninger, C.R. 1996. Relationships between Cloninger's, Zuckerman's, and Eysenck's dimensions of personality. *Personality and Individual Differences*, **21**, 283–285.

Zuckerman, M., Kolin, E.A., Price, L., and Zoob, I. 1964. Development of a sensation-seeking scale. *Journal of Consulting Psychology*, **28**, 477–482.

Zuckerman, M. and Neeb, M. 1979. Sensation seeking and psychopathology. *Psychiatry Research*, **1**, 255–264.

CHAPTER 3

Spice and herb extracts with chemesthetic effects

Howard Haley and Shane T. McDonald
Kalsec®, Inc., Kalamazoo, MI, USA

3.1 Why plants have chemesthetic properties

Many plant constituents and extracts contain chemesthetic agents (sensates), which includes herbs (leafy material from temperate climes) such as mint; spices (any part of certain tropical plants) such as ginger, cinnamon, and cloves; as well as fruits such as chili (capsicum). But why do plants contain these compounds? Plants lack a nervous system, and cannot move to escape insects and vertebrates that want to eat them or their fruits, and are subject to organisms that cause disease. Plants utilize a variety of defenses, such as needles, tough skin, and production of a host of toxic compounds (Roach, 2008). Here, we discuss one class of such compounds, volatile and non-volatile plant compounds that evoke chemesthetic effects.

Capsicum fruits with high capsaicinoid concentrations inhibit fungi that are able to penetrate the fruit when insects have pierced the skin (Roach, 2008). Also, chilis use capsaicinoids for "directed deterrence". Birds, which are effective dispersers of capsicum species seeds, are not deterred by pungency, as avian TRPV1 does not respond to capsaicin. However, mammals, which either eat seeds or are ineffective dispersers of seeds, are thwarted from eating the fruits by the fruit's pungency (Tewksbury and Nabhan, 2001). Plants also are subject to disease and damage caused by bacteria, fungi, insects, protists, and vertebrates. Plants often produce terpenes in their volatile oils, which are effective antimicrobials. Mint plants produce terpenes such as menthol, which deter microbial and insect pathogens (Freeman and Beattie, 2008).

In response, organisms that feed on plants developed the ability to sense these potentially dangerous compounds via "chemofensor" senses, including chemesthesis (Green, 2011). At some point, humans began to enjoy the chemesthetic effects of certain plant materials and started to use them in foods (see Chapters 2 and 6).

This chapter reviews the most important commercial plant materials that are utilized in food for their chemesthetic effects. The production of spice extracts of these materials is then discussed, as well as basic quality control methodology.

3.2 Hot pungent spices: capsicum species

Capsicum is a member of the *Solanaceae*, or nightshade family. This diverse, primarily tropical, family also includes potatoes (*Solanum tuberosum*), eggplants (*S. melongena*), tomatoes (*Lycopersicon esculentum*), and tobacco (*Nicotiana tabacum*) (Purseglove *et al.*, 1981a).

Capsicum species originated in South America, where they have been part of the human diet since 7500 BC. There is evidence of its cultivation by Native Americans as early at 5200–3400 BC. It has since been spread globally, starting with Christopher Columbus. There are about 20 wild species, but only five are of commercial interest: *Capsicum annuum* (sweet bell, jalapeno, paprika, serrano, and lots more), *C. baccatum* (aji), *C. pubescens* (rocoto), *C. frutescens* (tabasco, malagueta), and *C. chinense* (habanero and scotch bonnet) (Raju *et al.*, 2010; Bosland, 1996).

The interest in this genus is the presence (in most varieties) of the pungent vanillamide compounds called capsaicinoids, which results in a "hot" or pungent sensation (see Chapter 6). There are a variety of capsaicinoids, the most important being capsaicin (69%), dihydrocapsaicin (22%), and lesser quantities of nordihydrocapsaicin (7%), homocapsaicin (1%), and homodihydrocapsaicin (1%), plus other minor capsaicinoids (Raju *et al.*, 2010). Pungency is traditionally measured in Scoville Heat Units (SHU), described later in this chapter. SHU for the capsaicinoids are capsaicin (16.1 million), dihydrocapsaicin (16.1 million), nordihydrocapsaicin (9.3 million), homocapsaicin (6.9 million), and homodihydrocapsaicin (8.1 million) (Todd *et al.*, 1977). The capsaicinoids are non-volatile and can only be captured in solvent extracts, oleoresins for example, and not in the essential oils distillates. Capsicum species actually have little essential oil. Much of the pungency lies with the pericarp of the fruit, including the dissepiment, the white inner partitions (Attokaran, 2011a–h). Indeed, removal of the seeds and pericarp substantially reduces the pungency, such that deseeded jalapeños may be referred to as capones ("castrated ones").

Although native to South America, capsicums are now grown around the world. They are annuals or short-lived perennials, small shrubs, with pungency and fruit size and shape varying considerably between species and cultivars. Some chilis, such as bell peppers, have no pungency at all, whereas others, for example, the habanero, are intensely hot. They are naturally pollinated by bees or ants, and can be self-pollinated or cross-pollinated. They grow in tropical areas from sea level up to 2000 m. They are killed by frost, and the pungent varieties usually require longer times to mature (Purseglove *et al.*, 1981a).

What influences the pungency of the pepper? Generally, fruits are most pungent when harvested at full maturity. The biggest factors that determine pungency are species and cultivar. These can be manipulated by selected breeding to increase pungency (Purseglove *et al.*, 1981a). As we go to print, the hottest chili pepper on record is the "Carolina Reaper" with a reported pungency of over 1.5 million SHU (Guinness World Records, 2014). It is a hybrid of two *Capsaicin chinense* varieties: Pakistani Naga and a red habanero type (peppersource.eu., 2014).

Research to manipulate the pungency of chilis has resulted in a traditionally hot habanero variety that processes little heat, called the "habanada" (Stewart *et al.*, 2007). Other work has manipulated varieties so that the minor capsaicinoids became predominant in the chili. One variety of pepper was elevated in octanoyl vanillamide. This capsaicinoid was notable in that in sensory tests, the pungency peaked 30 seconds after ingestion, the same as capsaicin. However, the pungency of octanoyl vanillamide diminished more quickly than capsaicin. This could benefit consumers who do not want a lingering heat. Another variety was elevated in the typically minor capsaicinoid myristal vanillamide, which did not attain peak pungency until 4 minutes had elapsed (Mazourek, 2013). So, with modern breeding techniques, the future may include plant derived extracts with various levels of pungency and pungency time/intensity profiles, which could help product developers finely tailor the pungency profiles of their finished products.

3.3 Other hot pungent spices

3.3.1 Cinnamon and cassia

Cinnamon trees are part of the *Lauraceae* family, or laurel family, as are avocados and sassafras. Included in the *Cinnamomum* genus are "true" cinnamon (*Cinnamomum verum*, formerly known as *C. zeylanicum*) and related plants termed "cassia" (e.g., *C. burmannii, C. cassia,* and *C. loureirii*). True cinnamon favors subtropical well-watered areas such as Sri Lanka and the Seychelles. The bushes are raised either from seedling or vegetatively. The tree is not allowed to grow tall, but is cut back every few years (i.e., "coppiced") to yield multiple shoots. The inner bark can start being harvested from the shoots after 3–4 years. The bark is allowed to ferment for 24 hours to aid in removing the outer bark, which has an acidic flavor. Yields start to decline after ten years. Cassia can be found in Indonesia, China, Vietnam, and other locations. It is grown and harvested much like true cinnamon, except for the longer intervals between initial and subsequent coppicings, around ten years, and that the outer bark need not be fermented (Rosegarten, 1973).

The inner bark of cinnamon is formed into rounded quills, and can be sold as is, but is usually powdered. The most important flavor constituent and principal pungent compound is *trans*-cinnamaldehyde. This aldehyde can vary from 30% to over 90% of the oil. Factors affecting the *trans*-cinnamaldehyde content include the age of the bush, type of bark harvested (inner bark or twigs, outer bark, etc.), variety, growing conditions, and place of origin. Fermentation can greatly increase the *trans*-cinnamaldehyde content, possibly due to release of the aldehyde from glycosides and other conjugates (Purseglove *et al.*, 1981b).

Unlike capsaicin, *trans*-cinnamaldehyde is volatile and can be perceived in the essential oil as well as the oleoresin. This also means, however, that the odor and pungency are linked, making it difficult for the flavor chemist who wants to separate the effect of pungency from the aroma. The essential oil content of true cinnamon bark is about 1%, consisting of 75% *trans*-cinnamaldehyde and perhaps 2% eugenol. Cassia bark has from 0.5 to 5% essential oils, with 87% *trans*-cinnamaldehyde and 1% eugenol. The oil from the leaves of cinnamon, however, is over 70% eugenol. Cassia leaf oil is about 70% *trans*-cinnamaldehyde (Vernin *et al.*, 1994; Attokaran, 2011a–c).

3.3.2 Black and white pepper

This spice is from *Piper nigrum*, a climbing vine of the very large genus *Piper*, which includes some other pungent spices of lesser commercial importance, such as *P. longum*, or long pepper and *P. officinarum*, or Javanese pepper.

Piper nigrum prefers wet tropical climates. They are usually propagated vegetatively via cuttings that are placed in the soil and allowed to grow. As a vine, they require supports such as trees or wooden structures. Lateral branches are encouraged as they bear the fruit. Pruning discourages the growth of non-productive branches. The plant must be cultivated as it is subject to a host of microbial diseases, weeds, and insect pests. After two years the plant is allowed to form flowering spikes, and six months later the drupes, about 60 to a spike, can be harvested. Black pepper is picked when the fruit is somewhat unripe and green. The berries are put into a pile in the sun to ferment and to start the transformation from green to black. The berries are then spread out and allowed to dry for several days, whereupon they shrivel and turn black. Ground black pepper has both white and black particles in it: the outer cortical is black while the interior pith is white.

White pepper comes from the same berries, but the berries are allowed to soak for days to initiate a "retting" process. This makes it easier to remove the dark pericarp, retaining the white interior (Purseglove *et al.*, 1986d). Besides a milder flavor and lighter color, white pepper often has a "barnyard" note from indole and *para*-cresol (Steinhaus and Schieberle, 2005). A type of white pepper can also be made by mechanical abasing to remove the pericarp, yielding a white product with an aroma almost identical to black pepper.

Pepper can be used in peppercorn or powder form, or as an oleoresin. The essential oil is used for flavor and aroma, but is not pungent. The major pungent principle in black and white pepper is piperine, a vanillamide. A number of other pungent and tingling compounds are found in the oleoresin, but most of the pungency is from piperine (Dawic *et al.*, 2012). Dried black peppers average 2–4% piperine. A typical black pepper oleoresin has 40% piperine and 20% essential oil. The piperine can be crystallized out of the oleoresin and sold to provide pungency without aroma (Attokaran *et al.*, 2011d).

Most of the work carried out to improve pepper is to improve yields and to make it more disease and pest resistant. The yield of piperine depends on the cultivar, climate, temperature, and where it is grown.

3.3.3 Ginger

Ginger (*Zingiber officinale* Rosc) is a member of the *Zingiberaceae* family. Besides ginger, this family includes turmeric (*Curcuma longa* or *C. domestica*), cardamom (*Elettaria cardamomum*), and galangal (*Alpinia galanga*). The portion of the ginger plant utilized for spice is the rhizome, which is a bulbous subterranean portion of the stem.

Ginger is believed to originate in Southeast Asia, and now is cultivated in India, Asia, Indonesia, parts of Africa, and famously in Jamaica. Ginger is usually grown in moist tropical climates. The plants are perennial, but for spice purposes are grown as an annual and propagated vegetatively from cuttings of the rhizome. To obtain good yields, the fields must be heavily mulched and fertilized. The rhizomes are usually ready to harvest after seven months. If harvested later, the rhizomes become fibrous and are better suited to the production of ginger powder (Purseglove *et al.*, 1981c).

As with capsicum and black pepper, the pungent principles of ginger are non-volatile vanillamides. Fresh ginger is characterized by the pungency of gingerols. There are a series of gingerols, with the most common being the homologous series 6-ginerol, 8-gingerol, and 10-gingerol, having increasing lengths of the hydrocarbon tails. With heating, the gingerols can undergo a dehydration reaction to form the shogaol counterparts, (again, the most common being 6-, 8-, and 10-shogaol) (McHale *et al.*, 1989). They are generally more pungent than their gingerol counterparts (about 160 000 SHU for shogaols versus 80 000 SHU for gingerols). When fresh ginger is dehydrated, it can undergo retroaldol reactions to form paradols, the most simple being zingerone. Zingerone, at about 50 000 SHU, is less pungent than gingerols, while some other paradols have pungencies of up to 100 000 SHU (Richard, 1991). Ginger processing can affect the type of pungent sensates as well as total pungency, and the analysis of these compounds can indicate relative quality. Ginger is available as powder, oleoresin, fresh ginger, and candied ginger. The essential oil is not pungent.

3.4 Nasal heat spices

These spices are noted for a nasal burn rather than the oral burn typical of capsicum.

3.4.1 Mustard

Mustard is a much-branched annual herbaceous plant belonging to the *Brassicaceae* (formerly *Cruciferae*) family. This large family contains many important and frequently pungent vegetables, such as cabbage (*B. oleracea*), broccoli (*B. oleracea* var. *botrytis*), cultivated radish (*Raphanus sativus* L.), wasabi (*Wasabia japonica*), horseradish (*Armoracia rusticana*), and watercress (*Nasturtium officinale*) (United States Department of Agriculture, 2015). Mustard is derived from several species, such as black or brown mustard (*B. nigra*) and white mustard (*Sinapis Alba*), and Indian mustard (*B. juncea*). Originally native to Europe, mustard is widely grown in temperate to tropical areas all over the world, such as in India, Canada, the United States, South America and Africa (Reineccius, 1994). The mustard plant is propagated via the seed. The mustard seed is used for spices and the essential oil.

Freshly ground mustard seed has very little aroma, but contains the glucosinolate sinigrin. When the mustard flour comes in contact with water, the glucosinolate decomposes under the action of myrosin and the pungent volatile compound allyl isothiocyanate is formed. Mustard can be used as whole seeds, ground into powder, used as part of prepared mustard, or as the volatile oil distilled from the seeds (Lewis, 1994).

3.4.2 Horseradish

Armoracia rusticana is a hardy perennial herb originally from Eastern Europe, but now grown around the world, especially in cooler climates. The plant is propagated via root cuttings (Rosegarten, 1973). The root is the part used for the pungent spice (Reineccius, 1994). As with mustard, it contains sinigrin, which can be converted into allyl isothiocyanate upon disruption of the root tissue. The disturbed root as well as the horseradish oil also contains significant amounts of phenethyl isothiocyanate from the action of glucosinolate on gluconasturtiin. The amounts of the pungent compounds depends on the cultivar as well as growing conditions (Tomsone *et al.*, 2013). The root has little aroma, but grated or cut the aroma is very sharp and burning like mustard oil, which disappears very quickly. As a result, horseradish is often served as a prepared condiment.

3.4.3 Wasabi

Wasabia japonica is another member of the mustard family. Originally from Japan, it is now also cultivated in New Zealand and the northwest coast of the United States as well as the Blue Ridge mountains of Tennessee and North Carolina. It can be propagated by seed, but is usually grown from root cuttings. It prefers growing next to mountain streams in cool climates in the shade. While all parts of the plant can be used for food, the characteristic burning flavor is concentrated in the root or rhizome. It is difficult to grow, and a grated form loses flavor within 20 minutes of grating. Wasabi is often substituted by horseradish (realwasabi, 2005).

Like horseradish, wasabi oil is high in allyl isothiocyanate. However, wasabi oil lacks phenethyl isothiocyanate, which is a major constituent of horseradish oil. The lack of this compound and the presence of a number of minor isothiocyanates are believed to be responsible for the difference in flavor between authentic wasabi and horseradish (Saltana *et al.*, 2003).

3.5 Cooling spices

3.5.1 Mint

As with the *Capsicum* species, the mint family is grown and bred for its chemesthetic agents (menthol and its derivatives) as well as for aroma and other characteristics. As discussed earlier, capsicums contain chemesthetic agents that are non-volatile and do not directly modify the aroma of the spices. Menthol compounds, however, have minty aromas/flavors as well as cooling effects. Thus, for natural extracts of mint, cooling and mint aroma/flavor go hand-in-hand.

Mint (*Mentha* genus) is part of the *Lamiaceae* (or *Labiatae*) family, a large family containing many economically important plant herb species, such as *Ocimum* (basil), *Origanum* (oregano), *Rosmarinus* (rosemary), *Salvia* (sage), and *Thymus* (thyme) (Boelens, 1991). While these other *Lamiaceae* genera are important herbs with essential oils that contain chemesthetic compounds, only the *Mentha* genus is cultivated specifically for its chemesthetic properties.

The principal cooling sensate in *Mentha* is the terpene menthol. The two species most associated with menthol are peppermint (*M. piperita*) and cornmint (*M. arvensis*). Peppermint is grown mainly in the United States for its flavor, for which the cooling chemesthetic effect is part of that effect. Cornmint is produced in India as well as China. This species is considered to be inferior in flavor compared with peppermint, but has a higher menthol content. Thus, cornmint is usually grown for the production of menthol, which also creates the byproduct dementholized mint oil (Boelens, 1991).

Peppermint plants are usually propagated by cutting their runners. Typically, the plants are harvested in nearly full bloom, and allowed to partially wilt to remove some of the moisture. The partially dried leaves are subjected to hydrodistillation, by running steam through the leaf mass to capture the essential oil. In the case of cornmint, menthol, which is a solid at room temperature, can be crystallized out of chilled essential oil (Lawrence, 2008).

There is much variation in the essential oil content of peppermint, depending on the cultivar, growing region, and growing conditions. A typical peppermint leaf yields 0.3–0.4% volatile oil and has 29–48% menthol in addition to other compounds with lesser cooling abilities, such as menthyl acetate (3–10%) and menthone (20–31%). Cornmint oil yields about 1–2% volatile oil of which menthol is 70–95% of the composition, along with menthone (10–20%), and other ingredients (Leung and Foster, 1996b).

3.5.2 Eucalyptus

Eucalyptus is a genus of trees in the *Myrtaceae* family, which also includes cloves (*Syzygium aromaticum*). Of about 1000 species of eucalyptus trees, only about ten varieties are used for extraction, including *E. globulus*. The Blue Mallee variety is the only one that can be mechanically harvested and produces oil with a high eucalyptol content. The essential oil is steam distilled from the fresh or partially dried leaves. Eucalyptus oil contains 70–85% eucalyptol (also referred to as 1, 8-cineole). The product is often sold as eucalyptus 80/85 oil, referring to the eucalyptol percentage (Lebermuth, 2014).

Eucalyptol has a cooling effect as well as a camphoraceous aroma (Burdock, 2010). Eucalyptol has a somewhat different cooling effect than menthol, by stimulating cooling receptors but not noxious cold (Takaishi *et al.*, 2012). As a result, eucalyptol works as a natural analgesic (Takaishi *et al.*, 2012).

3.6 Numbing spices

3.6.1 Cloves

Cloves are the dried unopened flowers of the evergreen tree *Syzygium aromaticum*, which, as mentioned previously, is part of the *Myrtaceae* family. Cloves originated in the Moluccas (or Maluku Islands), formerly referred to as the Spice Islands, which are now a part of Indonesia. The plant is also grown in Malaysia, India, Sri Lanka, and Tanzania. The trees are propagated by seeds, and can be harvested in from six to eight years for up to 60 years (Lewis, 1984).

The clove bud is used for the spices. For the oleoresin or essential oil, the buds, leaves, and twigs can be utilized. The clove bud is considered to have the best flavor, with yields of about 15–18% oil, and contains 60–90% eugenol, the principal chemesthetic agent. Clove leaf oil is considered inferior for flavor, yielding 2–3% oil, which contains 82–88% eugenol. The twigs yield 4–6% oil, containing 90–95% eugenol. The stem and leaf oils are often blended with the bud oil to reduce the price, or they can be used as a source of eugenol (Leung and Foster, 1996a).

Eugenol is an analgesic, (Park *et al.*, 2011), and is often used in dental applications. It has been shown to activate receptors associated with trigeminal stimulation (Saunders *et al.*, 2011). One such non-medical application of cloves is kretek clove cigarettes, which are very popular in Indonesia. Kreteks contain tobacco as well as 15–40% dried clove buds. They were believed to be developed to sooth the throat and lungs of asthma sufferers. The analgesic properties of eugenol numbs the nerve endings in the mouth and throat, lessening the harshness of smoking (Polzin *et al.*, 2007).

3.6.2 Wintergreen

The wintergreen plant (*Gaultheria procumbens*) is part of the *Ericaceae* family, which includes heath, heather, and blueberries. Wintergreen is native to North America, and produces edible red berries. The essential oil is distilled from the leaves. The plant is cut off near the base and the leaves are allowed to dry. The leaves are then steeped in hot water for at least 24 hours to ferment, a process that helps to release the oil. The fermentation also hydrolyzes the gaultherin, a glycoside, to release methyl salicylate, the compound that gives it the characteristic aroma. The leaves can then be steam distilled to obtain the essential oil (Lebermuth, 2014). Methyl salicylate is also an analgesic, finding numerous uses as a topical pain reliever besides its application as an aroma/flavor compound.

3.7 Tingling spices

3.7.1 Jambu

Spilanthes oleracea (also referred to as *S. acmella* or *Acmella oleracea*) is an herbaceous short plant with yellow flowers in the *Compositae* (or *Asteraceae*) family, which includes asters, daisies, and sunflowers. It is indigenous to northern Brazil, and the leaves and stems are used in cooking in the state of Para (hence another name, paracress). The *Spilanthes* genus is wide-ranging over tropical and sub-tropical climates (Costa *et al.*, 2013). It can be found in damp as well as in cultivated areas. The flowers and leaves have a pungent sensation accompanied by tingling and numbing, and is used in a number of medicinal applications (Ramsewak *et al.*, 1999). It is referred to as the "toothache plant" for its use to relieve the pain in toothaches and other oral pain situations. The active amide that is responsible for the chemesthetic effect is spilanthol, which can be obtained from various parts of the plant, with the highest concentration found in the flower (1%) (Burdock, 2010). The tingling and numbing effect can be combined with capsicums for a unique effect similar to that found in Szechuan pepper (Katzner, 2002).

3.7.2 Szechuan pepper

This spice can be obtained from a number of members of the *Zanthoxylum* (or *Xanthoxylum*) genus, a member of the *Rutaceae*, or citrus family. Common names include Chinese prickly ash, Sichuan pepper, toothache tree, hua jiao (Chinese) and sancho (Japanese). The plants are ash-like treelets or shrubs protected by sharp thorns. The male and female plants have to be in close proximity in order to fertilize.

Propagation is by seeds as well as by stem or root cuttings. The genus is wide ranging, including eastern and southeast Asia, North and South America, and Africa. Species used as spices are native to semi-arid plains in the Szechuan province of China. The pericarp is used for the spice (Mehdi *et al.*, 2013). FEMA GRAS 26 lists the following species for use in extracting oleoresin: *Z. bungeanum*, *Z. piperitum*, *Z. simulans*, and *Z. schinifolium* (Marnett *et al.*, 2013). Two popular varieties, red hua jiao and green hua jiao, have different aromas but have the same principal compound responsible for its tingling effect: the alkylamide α-hydroxysanshool. This compound is similar in structure and effect to the spilanthol in jambu and is found in the oleoresin but not in the essential oil (Lim, 2012). Like spilanthol, α-hydroxysanshool can be blended with capsicum to modify the perception of both in an effect called "má là", where "má" refers to the tingling of the Szechuan pepper and the "là" refers to the heat pungency of capsicum (McDonald, 2013).

3.8 Spice and herb extracts

Herbs can be sold and used fresh, but the perishability and seasonal character of fresh herbs limit their use to mainly home and culinary applications. Spices tend be dried for use, and can be sold whole in some instances, such as black pepper or cloves. However, processers may prefer ground spices and herbs for their uniformity and ease of addition into powdered spice blends. Microbial and pest contamination can be quite a concern, so spices often have to be sterilized by ethylene oxide, irradiation, steam, or some other treatments.

3.8.1 Extracts

Spice extracts are often used in industry. The two most common extracts are essential oils and oleoresins.

3.8.1.1 Essential oils

Essential oils are volatile, aromatic liquids produced by plants (Attokaran, 2011f). The essential oils are removed from the spice matrix in a process referred to as steam distillation or hydrodistillation. The herb or spice is partially or completely dried, and then ground to facilitate the steam intermixing with the plant material. Coarse grinding is preferred in order to prevent the powdered spice from packing too tightly. In steam distillation, the spice is placed in a bed in an extraction vessel, and steam generated from an external source is allowed to flow up through the spice bed. The heat and pressure from the steam ruptures the cells and allows the oils to intermix with the steam. The most volatile components form an azeotrope with the steam, which flows out of the reactor through a condenser. The water and essential oil flow into a collector, where the essential oil, which is not miscible with the water, either floats to the surface or, less often, sinks to the bottom. The oil can then be separated from the water. Non-volatile components such as plant matrix, waxes, pigments, and non-volatile sensates remain in the pot. Generally, essential oils are not exposed to temperatures greater than 100 °C and as such do not suffer much thermal damage. Vacuum can be employed to reduce the boiling point.

Hydrodistillation is similar, but the plant material is mixed with water in the pot, and the water is allowed to boil in order to form the steam.

The essential oils of cloves, cinnamon/cassia, and mint will contain the volatile chemesthetic agent eugenol, *trans*-cinnamaldehyde, and menthol, respectively. However, the chemesthetic agent of black pepper, piperine, is not volatile. Therefore, the essential oil of black pepper will possess the aroma of black pepper, but will completely lack the pungent "bite". In order to recover that, solvent extraction must be employed to create an oleoresin.

3.8.1.2 Oleoresins

Oleoresins are the products of the extraction of spices and herbs using an organic solvent, followed by the removal of the said solvent. For a solvent extract, the spice/herb is usually partially or completely dried to inhibit microbial decomposition and to aid in releasing the oils. Residual water in the plant material may be extracted by the solvent, which would affect the polarity of the solvent resulting in undesirable constituents being extracted. The product is finely ground to allow intimate mixing of the solvent and spice. However, if the spice is ground too fine, channeling can appear so that incomplete extraction of the material will occur.

A number of solvents can be used for the production of oleoresins. According to the United States Code of Federal Regulate 21CFR 173, suitable solvents for spice oleoresins include hexane, methanol, acetone, and methylene chloride, with limits to residual solvents listed (United States Food and Drug Administration, 2014). Other solvents include aliphatic hydrocarbons (propane, butane, and hexane), alcohols such as methanol, ethanol, and isopropyl alcohol, and carbonyl compounds such as acetone and methyl acetates (Bernardo-Gil, 1997). Halogen compounds such as methylene chloride are excellent solvents, but are now utilized less often due to adverse consumer perceptions.

If the solvent is not to be removed following extraction, then one of the few solvents that can be used is ethyl alcohol, resulting in a finished product called a tincture. Tinctures are of low strength because of the level of alcohol, hence it is more practical to remove the solvent to obtain a more concentrated extract (Lewis, 1984). The target molecules to be extracted can be influenced by the polarity and selectivity of the solvent. The boiling point should be low to allow removal of the solvent without loss of the desired flavor compounds. Other factors include: being relatively unreactive, stable to oxygen and light, low viscosity, safe to use, commercially available at a reasonable cost, and of course, legal to use.

The simplest extraction is a batch percolation at ambient temperature. The prepared spice or herb is placed in a vertical vessel with a false bottom, where the solvent is sprayed on the material and collected in the bottom. To obtain more of the extract, a batch countercurrent system can be employed using multiple vessels.

More common are continuous extractors where the material is added to segments in a continuous chain. Fresh solvent is sprayed on the discharge end to extract the most depleted material. The partially enriched solvent (now called the miscella) is sprayed onto the next segment. This is repeated for typically 6–12 segments. In this way, the most enriched solvent comes into contact with the freshest material.

Once the miscella has been collected and removed from the extraction vessel, the residual plant matrix material is termed "spent" and is not useable for human food consumption without further processing. If the solvent was water or ethanol, the extract can be used as is. However, if other organic solvents are employed, the solvent must be removed to below the residual solvent levels stipulated by law. Allowable solvents and solvent residues in oleoresins have been summarized by the International Organization of the Flavor Industry (IOFI, 2014). This is typically done by careful distillation to remove the solvent but not stripping out the essential oil. The finished product is called an oleoresin (Attokaran, 2011g).

An alternative process of solvent extraction is termed two-stage extraction. This process is fairly common in India. First, the prepared spice or herb is steamed distilled to remove the essential oil. Then, the residue is solvent extracted in the same way as described for the oleoresins. After desolventization, sufficient essential oil as is desired for flavor is added back to the oleoresin. The advantage of this method is that the essential oil is not in contact with the organic solvent. Therefore, the oil can be sold as is, or recombined with the oleoresin. There is also no loss of volatiles during desolventization (Attokaran, 2011h).

3.8.1.3 Alternative extraction processes

As an alternative to organic solvents, high pressure can also be used to extract the flavoring substances from spices and herbs. The materials are dried and then ground. High pressure, typically about 200 000 kPa, which generates a temperature of about 90–120 °C, is employed for up to one minute to exude the oleoresin from the spices. This process works best with spice seeds, which require no addition of oil. Plants with little or no essential oil, such as capsicum, require the addition of a suitable vegetable oil, such as canola oil, as a solvent. The press is designed to allow counter-current extraction to increase the extraction efficiency, but does require additional oil. The advantage of the press extract is that there is no residual organic solvent, which some consumers may object to. Also, no desolventization step is required that could result in the loss of volatile substances. This process is detailed in two US patents (Todd, 1998, 1999).

Carbon dioxide (CO_2) liquefies under pressure. While some extractions occur with liquid carbon dioxide, more common in industry is extraction using supercritical carbon dioxide. As CO_2 approaches the critical pressure (72.74 atm, or 7.37 MPa) and temperature (30.97 °C), the vapor phase becomes denser and the liquid phase becomes less dense. Above the critical point, a new phase occurs with properties of both a vapor and a liquid. Supercritical and near-critical CO_2 exhibit useful properties, such as low viscosity and surface tension, but what makes them particularly interesting is their density-dependent solvent power (Pereda *et al.*, 2007). Changes in the pressure can affect the ability of the liquid to extract volatile and certain non-volatile constituents in spices and herbs. This process can produce a wide array of extracts, ranging from those similar to essential oils (which are steam or hydrodistilled), to products similar to oleoresins (which are extracted with organic solvents), containing esters, heavy terpenes, waxes, resins, and many pigments (Bernardo-Gil, 1997).

Generally, the solubility of a spice extract in supercritical CO_2 increases with pressure or density and decreases with temperature up to a limiting "cross-over" pressure, above which the solubility increases with both increasing temperature and pressure. This phenomenon allows for the CO_2 to be recycled by lowering the temperature or increasing the pressure, or both, in the separators. By manipulating pressure and temperature, good selectivity and even some fractionation can be accomplished.

In a batch operation, ground dried spices (less than 10% moisture) are added to baskets that are placed in the extraction chamber, then supercritical CO_2 is added and the pressure and temperature are manipulated in a typical operating range of 100 000–5 000 000 kPa and 40–80 °C for 2–4 hours. Then the CO_2 flows into a separation chamber where it is depressurized and heated to convert it into gaseous CO_2, which has little solvent ability. The liquid extract comes out of solution and the CO_2 reverts to a vapor, which can be captured and reused. Semi-continuous operation can be achieved using multiple extraction chambers and separation chambers. Fractionation can be accomplished either by step pressure/temperature combinations, step depressurizing, or using co-solvents (Martinez and Vance, 2007).

Supercritical CO_2 is essentially non-polar; it is a poor choice for the extraction of polar compounds. This is often an advantage in extracting spices, as the polar sugars and other materials are left behind. If more polarity is desired, a food-grade polar co-solvent, such as ethanol, water, or ethyl acetate, can be added.

Chemesthetic herbs and spices can be extracted using supercritical CO_2. Extraction of red chili peppers produces a light layer, which contains most of the capsaicin, and a heavy layer, which contains most of the pigmentation and triglycerides. With ginger, the light fraction contains the essential oil, which has the aroma but not the pungency, and a heavy fraction containing the gingerols and shogaols. Likewise, black pepper can be separated into a fraction with a high essential oil content, and another fraction with up to 60% piperine.

3.9 Regulatory control of spices and herb extracts with chemesthetic properties

The American Spice Trade Association (ASTA) lists a number of methods for the quality control of spices and herbs (ASTA, 2015). Microbial specifications tend to be of lesser importance for essential oils and oleoresins as the processes that produce oleoresins and essential oils tend to destroy most bacteria. Carrot pellets contaminated with *Salmonella* exhibited a 6-log reduction after solvent extraction (Kalsec©, 2015). Pests and parts of pests that were in the spices are also removed during processing.

Determination of the volatile oil can be quite important in oleoresins of spices and herbs with volatile chemesthetic agents, such as cloves, mustard, and cassia. Specific gravity and refractive index can be compared with standards to see if the extracts are authentic. For example, ASTA method 17.0 employs indirect titration to determine *trans*-cinnamaldehyde in cinnamon and cassia oil. Gas chromatography–mass spectrometry (GC–MS) can also be used to positively identify and quantify volatile chemesthetic agents. Spices such as capsicum do not have any essential oils and the chemesthetic agents, such as capsaicin, are non-volatile. While there are gas chromatographic methods for analyzing capsaicinoids, high-pressure liquid chromatography (HPLC) is more common (ASTA method 21.3). Piperine in black pepper oleoresin can be determined using ASTA method 12.1. Sensory methods for determining pungency can also be used (ASTA method 21.2).

As varieties of capsicum are used to provide color (e.g., paprika), problems with illegal dyes such as Sudan Red have led to HPLC methods to detect such dyes in capsicums (ASTA methods 28.0 and 29.0).

3.10 Advantages of spices, essential oils, and oleoresins

Spices can contribute particulates that are important in specific applications where the visual effect is desired, such as ground cinnamon on a cinnamon roll. The volatile species in spices are naturally encapsulated and may work better in some high-temperature applications. However, as compared with essential oils or oleoresins, ground or whole spices have microbial and pest concerns and are often less stable for color and flavor due to exposure to oxygen, they can be bulky, are not very water or oil soluble, and may take some time to express their flavor in applications. The chemesthetic agent concentrations can be variable depending on seasonal, regional, or even batch to batch variation.

Essential oils are concentrated liquids with the aroma of the spices. If the spice sensates are volatile, the chemesthetic agents will be concentrated in the essential oil. Essential oils lack most pigmentation, and due to the lack of waxes and other insoluble compounds, tend to have low viscosity. The volatile constituents may be lost during thermal processing.

Oleoresins from solvent or expeller extractions are more concentrated than the spices themselves. They contain the complete flavor profile of the spices, and often also include pigmentation, anti-oxidant activity, and viscous materials such as waxes. They are usually standardized to a specific potent unit, such as capsaicinoids or a volatile oil. Oleoresins are liquid oil-soluble products, but can be converted into powders via plating or made water dispersible by the use of emulsifiers, surfactants, or by partitioning. They are more stable than the spices, and tend to be more heat resistant than the essential oils.

Capsicum oleoresin is often decolorized and deodorized to a dosing product that imparts pungency without flavor or color.

Oleoresins can be fairly viscous and pigmented, and in some cases will separate into the essential oil and non-volatile layers, such as oleoresin black pepper. Some of the most volatile constituents of oleoresins may be lost when the oleoresin is desolventized.

Owing to the selectivity of supercritical CO_2 extraction, oleoresins may contain as much of the constituents of the spices as traditional solvent extraction, or as little as what is found in an essential oil. They do not have organic solvent residues either, unless an organic solvent was used as a co-solvent. Extractions can be achieved with fairly low temperatures of around 40–60 °C so that thermal damage to the flavor is minimal.

A summary of the spice constituents extracted by various techniques is depicted in Table 3.1 (Bernardo-Gil, 1997).

Table 3.1 Summary of spice constituents extracted by various techniques. Source: Bernardo-Gil, 1997. Reproduced with permission.

Composition	Extraction method*				
	Hydrodistillation	Organic solvents	CO_2 (8 MPa)†	CO_2 (10 MPa)†	CO_2 (30 MPa)†
Essential oils	●	●	●	●	●
Esters and heavy terpenes	○	●	◐	●	●
Fatty acids	○	●	○	●	●
Triglycerides	○	●	○	●	●
Waxes	○	●	○	◐	●
Resins	○	●	○	○	●
Pigments	○	◐	○	○	◐

*● Total extraction; ◐ partial extraction; ○ no extraction.
†Extraction temperature = 40 °C.

References

American Spice Trade Association. 2015Download analytical methods by method, http://www.astaspice.org/food-safety/astas-analytical-methods-manual/download-analytical-methods-by-method/ (accessed 25 September 2015).

Attokaran, M. 2011a. Cassia, in *Natural Food Flavors and Colorants*, Blackwell Publishing and the Institute of Food Technologists, chap. 34.

Attokaran, M. 2011b. Cinnamon leaf, in *Natural Food Flavors and Colorants*, Blackwell Publishing and the Institute of Food Technologists, chap. 35.

Attokaran, M. 2011c. Cinnamon, in *Natural Food Flavors and Colorants*, Blackwell publishing and the Institute of Food Technologists, chap. 31.

Attokaran, M. 2011d. Black pepper, in *Natural Food Flavors and Colorants*, Blackwell Publishing and the Institute of Food Technologists, chap. 35.

Attokaran, M. 2011e. Capsicum, in *Natural Food Flavors and Colorants*, Blackwell Publishing and the Institute of Food Technologists, chap. 25.

Attokaran, M. 2011f. Essential oils, in *Natural Food Flavors and Colorants*, Blackwell Publishing and the Institute of Food Technologists, chap. 4.

Attokaran, M. 2011g. Solvent extraction, in *Natural Food Flavors and Colorants*, Blackwell Publishing and the Institute of Food Technologists, chap. 8.

Attokaran, M. 2011h. Spices, in *Natural Food Flavors and Colorants*, Blackwell Publishing and the Institute of Food Technologists, chap. 3.

Bernardo-Gil, G. 1997. Extraction and isolation methods. 5.0 in Topics in Food Chemistry, Technical Institute of Portugal (Lisbon), http://web.ist.utl.pt/ist11061/fidel/fidel3topics.html (accessed 24 September 2015).

Boelens, M. 1991. Spices and condiments II, in *Volatiles Compounds in Foods and Beverages*, (ed. H. Maarse), Dekker, chap. 13.

Bosland, P. 1996. Capsicum: Innovative uses for an ancient crop, in *Progress in New Crops*, (ed. J. Janick), Arlington, VA: ASHS Press, pp. 479–487.

Burdock, G. 2010. *Fenarolli's Handbook of Flavor Ingredients*, 6th edn, Boca Raton, FL: CRC Press.

Costa, S., Arumugam, D., Gariepy, Y., Rocha, S., and Raghavan, V. 2013. Spilanthol extraction using microwave: calibration curve for gas chromatography. *Chemical Engineering Transactions*, **35**, 1783–1788.

Dawic, C., Henze, A., Frank, O., *et al.* 2012. Structural and sensory characterization of key pungent and tingling compounds from black pepper (*Piper nigrum* L.). *Journal of Agricultural and Food Chemistry*, **60**, 2884–2895.

US Food & Drug Administration. 2014. Secondary direct food additives permitted in food for human consumption. *Code of Federal Regulations*, 21CFR 173.

Freeman, B.C. and Beattie, G.A. 2008. An overview of plant defenses against pathogens and herbivores. *The Plant Health Instructor*, DOI: 10.1094/PHI-I-2008-0226-01.

Green, B. 2011, Chemesthesis and the chemical senses as components of a "chemofensor complex". *Chemical Senses*, **37**, 201–206.

Guinness World Records. 2015. *Hottest Chili*, http://www.guinnessworldrecords.com/world-records/1/hottest-chili (accessed 23 September 2015).

IOF. 2014. IOFI IL 1515 Enclosure. IOFI Review of Extraction and Carrier Solvents for Flavors, http://www.femaflavor.org/iofi/information-letters/updated-list-residual-and-carrier-solvents-food-1514 (accessed 1 March 2014).

Kalsec©, Inc. 2015. Kalsec© validates kill step of Salmonella in its extraction process, http://www.kalsec.com/files/misc/Press%20Release%20-%20Kalsec%20Validates%20Kill%20Step%20of%20Salmonella%20in%20its%20Extraction%20Process%201_28_15.pdf (accessed 24 September 2015).

Katzner. G. 2002. Spice pages: paracress (*Spilanthes acmella/oleracea*, Para cress). http://gernot-katzers-spice-pages.com/engl/Spil_acm.html (accessed 23 September 2015).

Lawrence, B. 2008. *Peppermint Oil*, Carol Stream, IL: Allured Publishing Company.

LebermuthCompany. 2014. June newsletter, 2014. *Wintergreen*, p. 3.

Leung, A., and Foster, S. 1996a. Cloves, in *Encyclopedia of Common Natural Ingredients Used in Food, Drugs, and Cosmetics*, 2nd edn, New York: John Wiley & Sons, Inc., pp. 174–177.

Leung, A., and Foster, S. 1996b. Mints, in *Encyclopedia of Common Natural Ingredients Used in Food, Drugs, and Cosmetics*, 2nd edn, New York: John Wiley & Sons, Inc., pp. 368–372.

Lewis, Y. 1984. *Spices and Herbs for the Food Industry*, Orpington, UK: Food Trade Press.

Lim, T. 2012. *Zanthoxylum simulans*, in *Edible Medicinal and Non-medicinal Plants: Vol. 4, Fruits*. Springer Science Business Media, pp. 904–911.

Marnett, L., Cohen, S., Fukushima, S., *et al.* 2013. GRAS Flavoring Substances 26. *Food Technology*, **8**, 39–56.

Martinez, J., and Vance, S. 2007. Supercritical extraction plants. Equipment, process, and costs, in *Supercritical Fluid Extraction of Nutraceuticals and Bioactive Compounds* (ed. J. Martinez), Boca Raton, FL: CRC Press, chap. 2.

Mazourek, M. 2013. Enhancing quality capsicum through plant breeding. Presented on July 16, 2013 at the Institute of Food Technologists Annual Meeting, Chicago, IL.

McDonald, S. 2013. Application of tingling sensates in food. Presented on July 16, 2013 at the Institute of Food Technologists Annual Meeting, ChicagoIL.

McHale, D., Laurie, W., and Sheridan, J. 1989. Transformation of the pungent principles in extracts of ginger. *Flavour & Fragrance Journal*, **4**, 9–15.

Medhi, K., Deka, M., and Bhau, B. 2013. The genus *Zanthoxylum* – a stockpile of biological and ethnomedicinal properties. 2:697 DOI: 10.4172/scientificreports.697.

Park, S., Sim, Y., Lee, J., *et al.* 2011. The analgesic effects and the mechanisms of orally administered eugenol. *Archives of Pharmacal Research*, **34** (3), 501–507.

Peppersource.eu. 2014. *Carolina Reaper*, http://www.pepperseeds.eu/carolina-reaper.html (accessed 23 September 2015)

Pereda, S., Bottini, S., and Brignole, E. 2007. Fundamentals of supercritical fluid technology, in *Supercritical Fluid Extraction of Nutraceuticals and Bioactive Compounds* (ed. J. Martinez), Boca Raton, FL: CRC Press, chap. 1.

Polzin, G., Stanfill, S., Brown, C., Ashley, D., and Watson, C. 2007. Determination of eugenol, anethole, and coumarin in the mainstream cigarette smoke in Indonesian clove cigarettes. *Food and Chemical Toxicology*, **45**, 1948–1953.

Purseglove, J, Brown, E., Green, C., and Robbins, S. 1981a. Chillies: *Capsicum* spp., in *Spices*, vol. 1, Longman, chap. 7.

Purseglove, J, Brown, E., Green, C., and Robbins, S. 1981b. Cinnamon and cassia, in *Spices*, vol. 1, Longman, chap. 3.

Purseglove, J, Brown, E., Green, C., and Robbins, S. 1981c. Ginger, in *Spices*, vol. 2, Longman, chap. 8.

Purseglove, J, Brown, E., Green, C., and Robbins, S. 1981d. Pepper, in *Spices*, vol. 1, Longman, chap. 2.

Raju, P.S., Chauhan, O. P., and Bawa, A. S. 2010. Chili flavor, in *Handbook of Fruit and Vegetable Flavors* (ed. Y.H. Hui), Chichester: John Wiley & Sons, Ltd, chap. 10.

Ramsewak, R., Erickson, A., and Nair, M. 1999. Bioactive *N*-isobutylamides from the flower buds *of Spilanthes acmella*. *Phytochemistry*, **51**, 729–732.

Real wasabi LLC. 2005. Wasabi is quite picky about its growing conditions, http://www.realwasabi.com/cultivation/index.asp (accessed 23 September 2015).

Reineccius, G., 1994. *Source Book of Flavors*, 2nd edn, New York: Chapman and Hall, chap. 7, pp. 176–364.

Richard, H. 1991. Spices and condiments I, in *Volatiles Compounds in Foods and Beverages* (ed. H. Maarse), Dekker, chap. 12.

Roach, J. 2008. Fungus puts the heat in chili peppers, study says. 8/11/2008. National Geographic News, http://news.nationalgeographic.com/news/2008/08/080811-chili-bugs.html (accessed 3 October 2015).

Rosegarten, F. 1973. *The Handbook of Spices, a New Revised Edition*, New York: Jove Publications.

Saltana, T., Savage, G., McNeil, D., Porter, G., and Clark, B. 2003. Comparison of flavour compounds in wasabi and horseradish. WFL Publisher. *Science and Technology*, http://world-food.net/comparison-of-flavour-compounds-in-wasabi-and-horseradish/ (accessed 24 September 2015).

Saunders, C., Winston, Y., Patel, T., Muday, J., and Silver, W. 2013. Dissecting the role of TRPV1 in detecting multiple trigeminal irritants in three behavioral assays for sensory irritation. [v1; ref status: indexed http://f1000r.es/p8]F1000Research 21013, 2:74, DOI:10.12688/f1000research.2-74.v1 (accessed 23 September 2015).

Steinhaus, M., and SchieberleP., 2005. Characterization of odorants causing an atypical aroma in white pepper powder (*Piper nigrum* L.) based on quantitative measurements and orthonasal breakthrough thresholds. *Journal of Agricultural and Food Chemistry*, **53** (15), 6049–6055.

Stewart, C., Mazourek, M., Stellari, G., O'Connell, M., and Jahn, M. 2007. Genetic control of pungency in *C. chinense* via the *Pun1* locus. *Journal of Experimental Botany*, **58** (5), 979–991.

Takaishi, M., Fujita, F., Uchida, K., *et al.* 2012. 1,8-Cineole, a TRPM8 agonist, is a novel natural antagonist of human TRPA1. Molecular Pain, 8, 8, http://www.molecularpain.com/content/8/1/86, (accessed 23 September 2015).

Tewksbury, J. and Nabhan, G. 2001. Directed deterrence by capsaicin in chillies. *Nature*, **412**, 403.

Todd, G. 1998. High temperature countercurrent solvent extraction of capsicum solids. US patent 5 ,773,075.

Todd, G. 1999. High temperature extraction of spices and herbs. US Patent 5,985,352.

Todd, P., Bensinger, M., and Biftu, T. 1977. Determination of pungency due to capsicum by gas-liquid chromatography. *Journal of Food Science*, **42**, 660–665.

Tomsone, L., Kruma, Z., Galoburda, R., and Talou, T. 2013. Composition of volatile compounds of horseradish roots (*Armoracea rustican* L.) depending on genotype. *Proceedings of the Latvia University Agriculture*, **29**, 324.

United State Department of Agriculture (USDA). 2015. Natural resources conservation service. Plants database. Classification for kingdom *Plantae* down to family *Brassicaceae*. https://plants.usda.gov/java/ClassificationServlet?source=profile&symbol=Brassicaceae&display=31 (accessed 23 September 2015).

Vernin, G., Vernin, E., Metzger, J., Pujol, L., and Parkanyi, C. 1994. GC/MS Analysis of cinnamon and cassia oils: A comparative study, in *Spices, Herbs, and Edible Fungi* (*Developments in Food Science*), vol. 34 (ed. G. Charalambous), Amsterdam: Elsevier, pp. 411–425.

CHAPTER 4

Molecular mechanisms underlying the role of TRP channels in chemesthesis

Yeranddy A. Alpizar, Thomas Voets, and Karel Talavera

Department of Cellular and Molecular Medicine, Laboratory of Ion Channel Research and TRP Research Platform Leuven (TRPLe), Leuven, Belgium

4.1 Introduction

The chemical senses allow organisms to sample the chemical environment and to respond with appropriate behaviors towards potentially harmful or beneficial substances. Chemosensation is usually associated with the senses of taste and olfaction, which serve to detect and identify chemical compounds dissolved in the saliva or that are airborne, respectively. However, chemosensation also occurs at the skin, mucosa, and viscera through interactions of compounds with specialized receptors expressed in sensory nerves and epithelial cells. These processes, collectively referred to as chemesthesis, largely rely on the activation of various members of the transient receptor potential (TRP) family of cation channels. For instance, TRPV1 is known to be activated by pungent compounds present in chilis and by extracellular acidosis, TRPM8 is activated by menthol and other cooling compounds, and TRPA1 stands out for its strong sensitivity to reactive electrophiles and to a wide variety of other chemical irritants. This chapter summarizes the current knowledge on the molecular mechanisms underlying the activation of these channels by their most relevant agonists. Elucidation of these mechanisms is essential for understanding the bases of sensory phenomena and behaviors triggered by chemical stimulation and for designing effective therapeutic strategies targeting sensory disorders.

TRP proteins are crucial players in the detection and transduction events required for mechano-, osmo-, thermo-, and photo-sensation, as well as for smell, and taste (Hardie, 2014; Plant, 2014; Roper, 2014; Shimizu *et al.*, 2014; Voets, 2014; Vriens *et al.*, 2014; Zanini and Gopfert, 2014; Zufall, 2014). Multiple chemical species found in food activate TRP channels expressed at the endings of trigeminal nerve fibers. This leads to membrane depolarization and firing of action potentials in patterns that ultimately encode for sensory information different from that carried by anatomically distinct taste and smell nerves. Examples include the freshness of menthol, the pungency of chilis, wasabi and mustard, and the tingling sensation caused by carbonated drinks. In addition, there is evidence for the expression of some chemosensory TRP channels in oronasal keratinocytes (Kawashima *et al.*, 2012; Kido *et al.*, 2003; Kun *et al.*, 2012; Seki *et al.*, 2006; Stander *et al.*, 2004), which suggests that additional signaling events take place between these cells and sensory nerve endings. A number of excellent reviews have discussed the pathophysiological and therapeutic relevance of chemosensory TRP channels (Moran *et al.*, 2011; Nilius and Appendino, 2013; Nilius and Szallasi, 2014; Roper, 2014; Sexton *et al.*, 2014; Shimizu *et al.*, 2014). In this chapter, we provide the reader with a detailed update on the mechanisms underlying the role of TRP channels in chemesthesis. Notwithstanding that other TRP channels are also expressed throughout the oronasal cavities, we focus here on TRPM8, TRPV1, and TRPA1. These are the most relevant for chemesthesis triggered by eating and drinking experiences and are the best understood in terms of molecular mechanisms of chemically induced activation.

TRPM8, TRPV1, and TRPA1 channels are formed by homotetrameric proteins with six transmembrane segments (S1–S6) and intracellular *N*- and *C*-termini. The S5 and S6 segments and the S5–S6 linker (P-loop) form a pore with high selectivity for cations (Owsianik *et al.*, 2006). In contrast with TRPM8, both TRPV1 and TRPA1 contain several ankyrin domains in the *N*-terminus, which are required for channel function. The *C*-termini of these channels comprise key structural features, such as the TRP domain, and are responsible for tetramerization and channel functionality. These and other aspects of the structure, biophysical properties, expression patterns, and pathophysiological relevance of TRPM8, TRPV1, and TRPA1 can be found in recent reviews (Almaraz *et al.*, 2014; Bevan *et al.*, 2014; Zygmunt and Hogestatt, 2014). The role of these channels in chemesthesis is largely based on their expression in sensory nerve endings innervating the mucosa of the oronasal cavities. It should be emphasized, however, that the vast majority of anatomical studies have been performed in mice. The precise localization of functional TRPM8, TRPV1, and TRPA1 and their relative roles in the detection of non-specific, but very relevant agonists, such as menthol and allyl isothiocyanate (also known as mustard oil) in the human nose, mouth, and throat are still to be determined.

4.2 TRPM8

TRPM8 was originally found to be activated by the natural cooling compound menthol, which is found in mint plants such as *Mentha piperita* (Nilius and Appendino, 2013). Other natural agonists of TRPM8 relevant for human chemesthesis are menthone, geraniol, linalool, eucalyptol, hydroxycitronellal, and camphor (Behrendt *et al.*, 2004; McKemy *et al.*, 2002; Selescu *et al.*, 2013). In addition, TRPM8 can be activated by the synthetic compounds, such as the super-cooling agent icilin (see Almaraz *et al.*, 2014).

Here, it must be pointed out that menthol has been employed as a specific agonist of TRPM8 in multiple studies despite that it is known to also activate mouse TRPV3 (Macpherson *et al.*, 2006) and mouse TRPA1 (Karashima *et al.*, 2007), and inhibit chicken and rat voltage-gated Ca^{2+} channels (Swandulla *et al.*, 1987) and rat voltage-gated Na^+ channels (Gaudioso *et al.*, 2012) in sensory neurons. In addition, menthol is known to potentiate the currents through GABA and glycine receptors (Hall *et al.*, 2004; Zhang *et al.*, 2008). Reports on the functional expression of TRPM8 that are only based on menthol effects (without proper controls with *Trpm8* knock-out mice or highly selective TRPM8 inhibitors) must be taken with extreme caution.

4.2.1 Mathematical models of TRPM8 function: heated debate over a cool channel

Surprisingly, very little is known about the effects of menthol on TRPM8 at the single-channel level. It has been shown that, in the presence of menthol, TRPM8 channels have various levels of conductance (Hui *et al.*, 2005; McKemy *et al.*, 2002), as well as several open states and a shallow voltage dependence of the open probability (Hui *et al.*, 2005). However, no comparison was made with the properties of the channel under control conditions, hindering the possibility of determining the actual effects of menthol on channel gating. In a recent study by Gkika *et al.* (2015) it was shown that menthol increased the frequency of channel opening by mainly affecting transitions between closed states. However, a close inspection of the results of their analysis indicates that menthol duplicates the rate of the transition to the open state and reduces by ~20% the rate of the closing transition.

TRPM8 is thus far the best-studied chemosensory TRP channel in terms of mathematical description of gating properties. A number of models have been put forward to account for the effects of voltage, temperature, and menthol. These fall into two categories. On one hand there is a model based on the Monod–Wyman–Changeux (MWC) formalism (Changeux, 2012), which considers that menthol binding stabilizes the open states of the channel (Janssens and Voets, 2011; Voets *et al.*, 2007). Opening and closing of the channel is governed by voltage- and temperature-dependent transitions (Fig. 4.1a, and see later for further explanation). The voltage dependence of the open probability, $P_o(V)$, can be described by a classical Boltzmann-type function that tends to zero at very negative potentials and to 1 at very positive potentials:

$$P_O(V) = \frac{1}{\left(1 + \exp\left(-\frac{zF}{RT}(V - V_{1/2})\right)\right)} \tag{4.1}$$

where z is the apparent gating valence of the channel, F is the Faraday constant ($9.64 \cdot 10^4$ C mol^{-1}), R is the gas constant (8.31 J K^{-1}), T is the absolute temperature, V is the membrane potential, and $V_{1/2}$ is the voltage half-maximal activation.

On the other hand, other models are based on modular (or tiered) schemes where temperature, voltage, and menthol are thought to activate the channel independently from each other. Another salient property from these tiered models is that depolarization is only a partial channel activator (Brauchi *et al.*, 2004; Matta and Ahern, 2007; Raddatz *et al.*, 2014; Taberner *et al.*, 2014). The reader should note here that the modular models are often laxly referred to as "allosteric models"

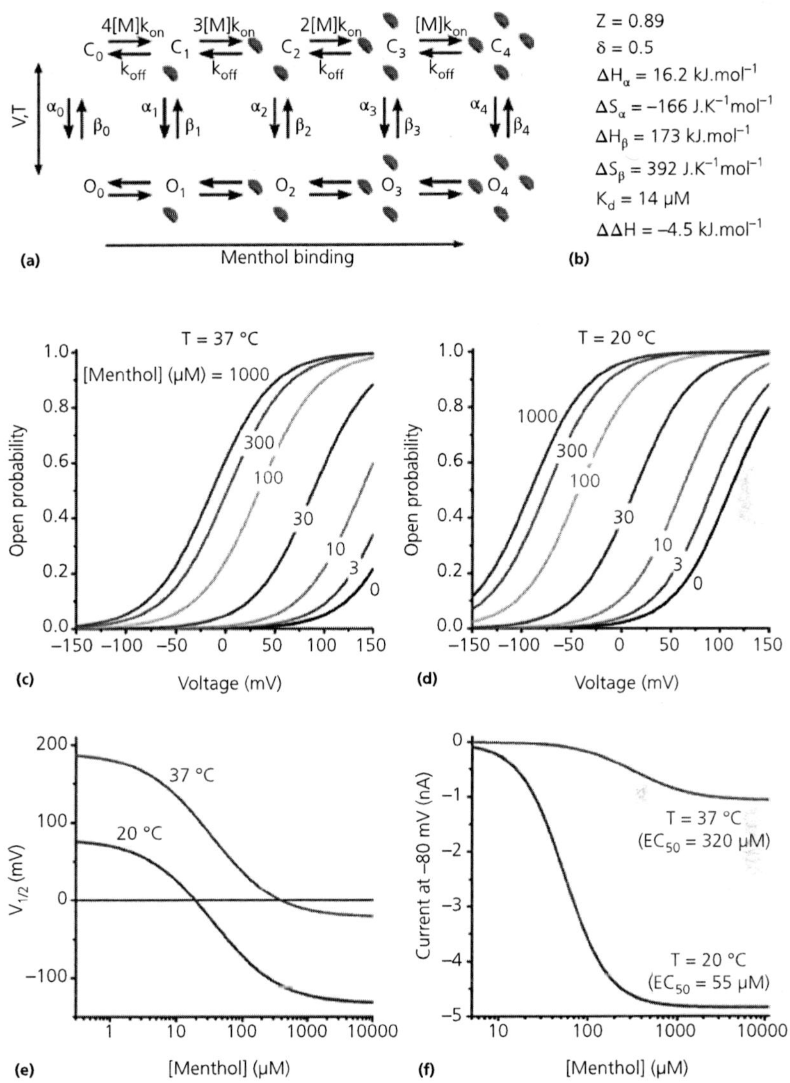

Fig. 4.1 Modeling the effects of menthol on TRPM8 gating. (a) Monod–Wyman–Changeux model describing the effects of voltage, temperature, and menthol on the gating of TRPM8. C_i and O_i represent the different closed and open states of the channel (i = 0, 1, 2, 3 or 4), respectively. The number of menthol molecules bound to the channel in the different states is represented by the menthol leaves. The rate constants k_{on} and k_{off} are associated with menthol binding and unbinding from single TRPM8 channel monomers, respectively. The rate constants α_i and β_i are associated with channel opening (transitions from C_i to O_i) and closing (transitions from O_i to C_i), respectively, and have an explicit dependence on voltage and temperature (see V,T on the left). (b) List of the values of the parameters defining the thermodynamic model of TRPM8 (Voets *et al.*, 2007). (c and d) Voltage dependence of TRPM8 activation at different menthol concentrations simulated with the MWC model (see eqn. 4.14). The simulations were run with a temperature of 37 °C (c) or 20 °C (d). Menthol concentration (μM), right to left: 0, 3, 10, 30, 100, 300, and 1000. (e) Dependence of the voltage for half-maximal activation of TRPM8 ($V_{1/2} = V_{act} - \Delta V_{Menthol}$, see eqn. 4.14) with the menthol concentration simulated at 37 °C (upper trace) or 20 °C (lower trace). (f) Comparison of the concentration-dependence of the effect of menthol on the inward TRPM8 current at −80 mV measured at 37 °C (upper trace) or 20 °C (lower trace). The currents were calculated as $I = N \cdot P_o \cdot g \cdot (-80 - V_{rev})$, where N is the number of channels (set at 1000), P_o is the open probability at −80 mV calculated using eqn. 4.14, g is the single-channel conductance (= 80 pS) and V_{rev} is the reversal potential (=0 mV).

(as opposed to the MWC). This can be confusing, because it neglects the fact that the MWC model implies, by definition, an allosteric effect of menthol on channel gating (Janssens and Voets, 2011; Voets *et al.*, 2004; Voets *et al.*, 2007).

The suitability of the MWC and modular models have been a matter of heated debate and, in our opinion, none of them are completely satisfactory. For instance, these models do not take into account the rates of menthol binding and unbinding and therefore cannot fully describe the effects of this compound on current kinetics. In addition, the MWC model assumes that, in the absence of menthol, the gating of human TRPM8 can be described by a two-state closed–open model (Janssens and Voets, 2011; Voets *et al.*, 2007). However, although this serves to describe accurately the kinetic and steady-state properties of macroscopic currents, single-channel data demonstrate that several open and closed states do exist (Fernandez *et al.*, 2011; Hui *et al.*, 2005).

Given their complexity, the modular models could in principle provide a better alternative for describing the details of single-channel kinetics. However, the most detailed modular model currently available does not include the effects of menthol (Raddatz *et al.*, 2014). Alternatively, Matta and Ahern (2007) proposed a modular model for the action of menthol on the gating of rat TRPM8 transfected in HEK 293F cells. They emphatically rejected the MWC model by stating that the whole-cell conductance, *G(V)*, should be described as:

$$G(V) = \frac{I(V)}{V} = \frac{G_{Max} - G_{Min}}{\left(1 + \exp\left(-\frac{zF}{RT}\left(V - V_{1/2}\right)\right)\right)} + G_{Min} \tag{4.2}$$

where $I(V)$ represents the steady-state currents measured at different membrane potentials and G_{Max} and G_{Min} are the whole-cell conductance determined at extreme positive or negative membrane potentials, respectively. They reported that both G_{Max} and G_{Min} were larger in the presence of 0.8 or 6 mM menthol than in the control and took this as proof for a voltage-independent effect of menthol and as evidence for depolarization to be a partial channel activator.

However, a number of issues may have obscured the interpretation of their experimental data (Zhu, 2007). In particular, the currents measured at very extreme voltages and high menthol concentrations (1 mM) could have been contaminated with leak and/or background components, yielding $G_{Min} > 0$. Of note, a recent report by Taberner *et al.* (2014) showed G_{Min} was equal to zero in the presence of 1 mM menthol. Furthermore, an eventual increase in G_{Max} could be explained, not only by a larger maximal open probability, but also by an increase in the number of functional channels expressed at the plasma membrane or in the single-channel conductance. The data presented by Matta and Ahern cannot distinguish between these possibilities and therefore do not unequivocally demonstrate that menthol activates TRPM8 in a voltage-independent manner. Importantly, the study of Fernandez *et al.* (2011), which is by far the best single-channel characterization of TRPM8, clearly shows that, at 20 °C, the voltage dependence of the open probability virtually saturates to 1 at very positive potentials. This means that the well-known strong potentiating effect of menthol cannot be explained by an increase in the maximal open probability, but by a negative shift of the voltage dependence of channel activation, as predicted by the MWC model (Janssens and Voets, 2011; Voets *et al.*, 2007). Furthermore, more detailed analyses of single-channel kinetics indicated that cold and depolarization have similar effects on the rate constants that determine TRPM8 gating (Fernandez *et al.*, 2011), which supports the idea that these stimuli affect the same gating steps as considered in the MWC model (Janssens and Voets, 2011; Voets *et al.*, 2007). Finally, it remains possible that differences in the effects of menthol on the human isoform, used by the group of Voets (Janssens and Voets, 2011; Voets *et al.*, 2007), and rat isoform, used by Matta and Ahern (2007), contribute to the divergences between the data presented in these studies.

Taking all this into account, we consider the MWC model as the simplest framework accounting for the effects of menthol on TRPM8 gating. In the following, we provide a textbook description of the mathematical derivations required to obtain equations that can be used to simulate and understand the functional properties of TRPM8 from a mechanistic point of view.

As shown in Fig. 4.1(a), the MWC model requires that all four monomers forming the TRPM8 structure undergo simultaneous transitions between the closed and open states and that menthol binds in a stepwise manner to each monomer.

The probability for the channels to be open P_O is given by:

$$P_O = P_{O0} + P_{O1} + P_{O2} + P_{O3} + P_{O4} \tag{4.3}$$

where P_{Oi} is the probability of the channel being in the i-th open state (Fig. 4.1a).

The equilibria between open states depend on the menthol concentration $[M]$ and on the rates of menthol binding (k_{on}) and unbinding (k_{off}). Assuming that the binding of menthol to any monomer does not affect the binding to others, the relationship between the probabilities of finding the channel in contiguous open states can be described by:

$$P_{Oi} = P_{Oi-1}\frac{(5-i)\cdot[M]k_{on}}{i\cdot k_{off}} = P_{Oi-1}\frac{(5-i)\cdot[M]}{i\cdot K_d} \tag{4.4}$$

where $i = 0$ to 4 and $K_d = k_{off}/k_{on}$ is the binding affinity of each channel monomer in the open conformation.

Combining eqns (4.3) and (4.4) yields:

$$P_O = P_{O0}\cdot\left(1+4\frac{[M]}{K_d}+6\left(\frac{[M]}{K_d}\right)^2+4\left(\frac{[M]}{K_d}\right)^3+\left(\frac{[M]}{K_d}\right)^4\right) \tag{4.5}$$

or:

$$P_O = P_{O0}\cdot\left(1+\frac{[M]}{K_d}\right)^4 \tag{4.6}$$

A similar analysis for the closed states yields:

$$P_C = P_{C0}\cdot\left(1+\frac{[M]}{K'_d}\right)^4 \tag{4.7}$$

where K'_d is the binding affinity of each channel monomer for menthol in the closed conformation.

The equilibrium between the states O_0 and C_0 can be described by:

$$P_{C0} = P_{O0} \cdot \exp\left(-\frac{zF}{RT}(V - V_{act})\right) \tag{4.8}$$

where V_{act} is the voltage at which the open probability in the absence of menthol is 0.5.

Substituting P_{C0} into eqn. (4.7) we obtain:

$$P_C = P_{O0} \cdot \left(1 + \frac{[M]}{K'_d}\right)^4 \cdot \exp\left(-\frac{zF}{RT}(V - V_{act})\right) \tag{4.9}$$

If we then isolate P_{O0} from this equation and introduce it into eqn. (4.6) and isolate P_C we find:

$$P_C = P_O \cdot \frac{\left(1 + \frac{[M]}{K'_d}\right)^4}{\left(1 + \frac{[M]}{K_d}\right)^4} \cdot \exp\left(-\frac{zF}{RT}(V - V_{act})\right) \tag{4.10}$$

Finally, substituting P_C into the normalization condition $P_O + P_C = 1$ and isolating P_O we find:

$$P_O(V) = \frac{1}{\left(1 + \frac{\left(1 + \frac{[M]}{K'_d}\right)^4}{\left(1 + \frac{[M]}{K_d}\right)^4} \cdot \exp\left(-\frac{zF}{RT}(V - V_{act})\right)\right)} \tag{4.11}$$

The factor before the exponential can be written as:

$$\frac{\left(1 + \frac{[M]}{K'_d}\right)^4}{\left(1 + \frac{[M]}{K_d}\right)^4} = \exp\left(-4 \cdot \ln\left(\frac{1 + \frac{[M]}{K_d}}{1 + \frac{[M]}{K'_d}}\right)\right) \tag{4.12}$$

Thus,

$$P_O(V) = \frac{1}{\left(1 + \exp\left(-\frac{zF}{RT}(V - V_{act}) - 4 \cdot \ln\left(\frac{1 + \frac{[M]}{K_d}}{1 + \frac{[M]}{K'_d}}\right)\right)\right)} \tag{4.13}$$

This equation can be written as:

$$P_o(V) = \frac{1}{\left(1 + \exp\left(-\frac{zF}{RT}\left(V - V_{act} + \Delta V_{Menthol}\right)\right)\right)} \tag{4.14}$$

where:

$$\Delta V_{Menthol} = 4\frac{RT}{zF}\ln\left(\frac{1 + \frac{[M]}{K_d}}{1 + \frac{[M]}{K'_d}}\right) \tag{4.15}$$

A relationship between K_d and K'_d can be found by taking into account the principle of microscopic reversibility, which states that:

$$\alpha_0 \cdot 4k_{on} \cdot \beta_1 \cdot k'_{off} = \beta_0 \cdot 4k'_{on} \cdot \alpha_1 \cdot k_{off} \tag{4.16}$$

Thus,

$$K'_d = K_d \frac{\alpha_1 \cdot \beta_0}{\alpha_0 \cdot \beta_1} \tag{4.17}$$

The rate constants α_i and β_i can be determined using Eyring's rate theory by:

$$\alpha_i = \frac{k_B T}{h}\exp\left(\frac{-\Delta H_{\alpha_i} + T\Delta S_{\alpha_i} + z_i F\delta V}{RT}\right) \tag{4.18}$$

and

$$\beta_i = \frac{k_B T}{h}\exp\left(\frac{-\Delta H_{\beta_i} + T\Delta S_{\beta_i} - z_i F(1-\delta)V}{RT}\right) \tag{4.19}$$

where k_B is the Boltzmann constant ($1.38 \cdot 10^{-23}$ J K^{-1}), h is Planck's constant ($6.63 \cdot 10^{-34}$ J s), δ accounts for the coupling between the local electric field sensed by the gating charge and the membrane potential V and $\Delta H_{\alpha i}$, $\Delta S_{\alpha i}$, $\Delta H_{\beta i}$, and $\Delta S_{\beta i}$ are the corresponding enthalpic and entropic changes associated with the opening and closing transitions (see Talavera *et al.*, 2007).

By introducing the expressions for the corresponding rate constants in eqn. (4.17) we obtain:

$$K'_d = K_d \exp\left(\frac{-\Delta H_1^{\ddagger} + T\Delta S_1^{\ddagger} + \Delta H_0^{\ddagger} - T\Delta S_0^{\ddagger} + FV(z_1 - z_0)}{RT}\right) \tag{4.20}$$

where $\Delta H^{\ddagger}_0 (= \Delta H_{\alpha 0} - \Delta H_{\beta 0})$, $\Delta S^{\ddagger}_0 (= \Delta S_{\alpha 0} - \Delta S_{\beta 0})$, $\Delta H^{\ddagger}_1 (= \Delta H_{\alpha 1} - \Delta H_{\beta 1})$, $\Delta S^{\ddagger}_1 (= \Delta S_{\alpha 1} - \Delta S_{\beta 1})$ are the differences in enthalpy and entropy between the open and closed states, and z_0 and z_1 are the corresponding valences of the apparent gating charges.

Given that menthol was shown to have no significant effect on the slope of the activation curves (Janssens and Voets, 2011; Voets *et al.*, 2007) we can consider that all apparent gating charges have the same valence ($z_i = z$). In addition, menthol does not alter the slope of the dependence between the voltage for half-maximal activation and temperature (Janssens and Voets, 2011; Voets *et al.*, 2007), indicating that $\Delta S^{\ddagger}_i$ is the same for all transitions (= $\Delta S^{\ddagger}$).

Thus,

$$K'_d = K_d \exp\left(\frac{-\Delta H_1^{\ddagger} + \Delta H_0^{\ddagger}}{RT}\right) = K_d \exp\left(\frac{-\Delta\Delta H}{RT}\right) \tag{4.21}$$

where $\Delta\Delta H$ is the change in enthalpy difference between the open and closed states that is induced by the binding of menthol to each channel monomer.

Then $\Delta V_{Menthol}$ can be written as:

$$\Delta V_{Menthol} = 4\frac{RT}{zF}\ln\left(\frac{1+\frac{[M]}{K_d}}{1+\frac{[M]}{K_d}\cdot\exp\left(\frac{\Delta\Delta H}{RT}\right)}\right) \tag{4.22}$$

From this we can determine that, at saturating menthol concentrations, $\Delta V_{Menthol}$ reaches a maximal value, $\Delta V_{Menthol,Max}$, which is given by:

$$\Delta V_{Menthol,Max} = -4\frac{\Delta\Delta H}{zF} \tag{4.23}$$

This expression is similar to that previously derived for the effect of temperature on the voltage for half-maximal activation (Nilius *et al.*, 2005):

$$\frac{dV_{act}}{dT} = -\frac{\Delta S^{\ddagger}}{zF} \tag{4.24}$$

This equation indicates that the large sensitivity of voltage-gated TRP channels to temperature is due to the combination of a large difference in entropy between the open and closed states and a low gating charge (Voets *et al.*, 2007). Indeed, in contrast to classical voltage-gated channels, which have a large gating charge (7–14 unitary charges; Hille, 2001) and are extremely sensitive to variations in membrane potential, all known voltage-gated thermoTRPs have very low gating charge (<1 unitary charge; Nilius *et al.*, 2005; Talavera *et al.*, 2007; Voets *et al.*, 2007).

Interestingly, eqn. (4.23) leads to two analogous conclusions. Firstly, the maximal shift that can be induced by menthol is proportional to the amount of energy ($\Delta\Delta H$) involved in the stabilization of the open state upon menthol binding. Secondly, $\Delta V_{Menthol,Max}$ is inversely proportional to z, suggesting that at least part of the high sensitivity of TRPM8 to menthol is determined by its low gating charge. It is tempting to speculate that the same principle may apply to other voltage- and ligand-gated TRP channels.

Equations (4.14), (4.23), and (4.24) allow further interpretation of the potential of the MWC model to reproduce experimental data. The modular models have been repeatedly portrayed as having the possibility to associate the actions of different stimuli (depolarization, temperature changes, and chemical agonists) to different channel domains (Almaraz *et al.*, 2014; Brauchi *et al.*, 2006; Matta and Ahern, 2007). This feature has been taken as a sign of superiority of these models over the MWC formalism to account for the fact that certain structural alterations modify the response to some stimuli while leaving the response to others unaffected (see later). However, it is important to notice that the MWC formalism can also easily accommodate that possibility. Indeed, eqns. (4.14), (4.23), and (4.24) clearly show that changes in z, $\Delta\Delta H$, K_d, and $\Delta S^{\ddagger}$, which could result from modification of totally different channel structures, may affect independently the sensitivity to voltage, menthol efficacy, menthol affinity, and the sensitivity to temperature, respectively. In other words, while the MWC model features a tight link between temperature and voltage, it can be modular in terms of structural determinants for the detection of different stimuli.

Figure 4.1 shows some of the simulations that can be obtained from the MWC model using the set of parameters displayed in panel (B). In the absence of menthol, at $T = 35$ °C, the channels activate at very positive potentials, with a voltage for half-maximal activation $V_{1/2} = V_{act} = +188$ mV (Fig. 4.1c). At 20 °C the channels activate at more negative potentials than at 37 °C. At this temperature and in the absence of menthol, $V_{1/2} = V_{act} = +78$ mV (Fig. 4.1d). Increasing menthol concentrations shift the activation curve to negative potentials (Fig. 4.1c–e), which results in significant channel activation in the physiological range of the membrane potential. The maximal shift induced by menthol, $\Delta V_{Menthol,Max}$, is −210 mV (see eqn. 4.23). These effects are reflected in the steady-state currents calculated for these temperatures as a function of the menthol concentration (Fig. 4.1f). In summary, the model accounts for the main features of TRMP8 modulation, such as the strong shift of the voltage dependence of channel activation to negative potentials induced by menthol and the potentiating effect of cooling. Thus, the MWC model could be quite useful to explore the role of TRPM8 in mathematical models describing the effects of cooling and menthol on the excitability of sensory neurons.

4.2.2 Structural determinants of activation of TRPM8 by menthol

Structure–function studies have identified several determinants for the chemical activation of TRPM8. Using a high-throughput random mutagenesis screening, Bandell *et al.* (2006) found that mutation of a residue located in transmembrane segment S2 (Y745) to histidine, resulted in a strong shift of the dependence of channel activation by menthol to higher concentrations. On the other hand, mutation of L1009 (in the *C*-terminal domain) to arginine decreased menthol efficacy (Fig. 4.2). Both mutations reduced the sensitizing effect of menthol on the response to cold, but did not affect the voltage dependence of channel activation, the response to cold, nor the modulation by PIP_2.

However, this study also showed that residues Y745 and L1009 are not the sole determinants of menthol effects on TRPM8. Indeed, the comparison of the amino

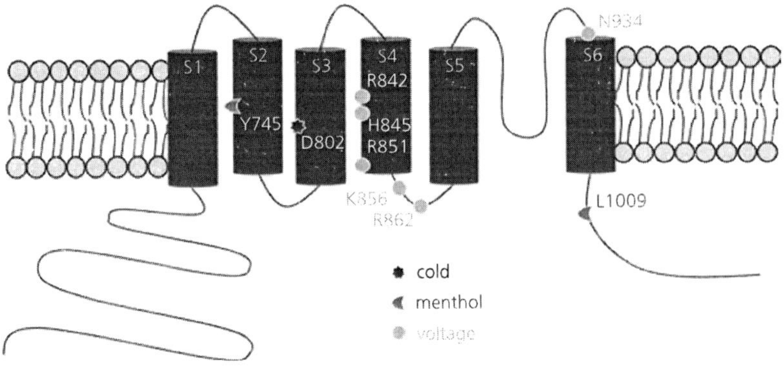

Fig. 4.2 Scheme of the structure of TRPM8 illustrating the position of residues important for the effects of cold (center), menthol (left), and changes in membrane potential (right).

acid sequence of TRPM8 with that of the close relative but menthol-insensitive TRPM2 revealed that Y745 is conserved and mutation of L1009 in TRPM8 to the corresponding proline found in TRPM2 did not reduce the sensitivity. Furthermore, several channel chimeras, in which short fragments of the *C*-terminal sequence of TRPM2 were substituted into corresponding locations around L1009 of TRPM8, were also sensitive to menthol. Finally, the transfer of the S2 segment and the TRP domain of TRPM8 to TRPM2 resulted in chimeric channels that were menthol insensitive.

A later study by Voets *et al.* (2007) was specifically designed to shed light on the relationship between the structural determinants of voltage-, cold-, and menthol-dependent activation of TRPM8. It was shown that mutation of positively charged residues in the S4 segment and the putative S4–S5 linker to neutral ones yielded channels with smaller gating charge than the wild type. These residues, R842 and K856, are therefore thought to form part of the voltage sensor (Fig. 4.2). Notably, the mutant R842A showed a decreased sensitivity to menthol, with a higher concentration for the half-maximal effect on the shift of the voltage dependence of channel activation ($EC_{50,\Delta V}$). On the other hand, the mutant K856A was more sensitive to menthol than wild-type TRPM8, displaying a lower $EC_{50,\Delta V}$ and a higher maximal effect ($\Delta V_{Menthol,Max}$). These changes were paralleled by modifications in the menthol binding affinity with respect to wild-type TRPM8, with a decrease for R842A and an increase for K856A. Interestingly, the mutations L1009R and Y745H were found not to alter and to decrease menthol binding, respectively.

Taken together, these studies strongly suggest that residues in the S2 (Y745) and S4 (R842) segments form the binding site for menthol and that other mutations, such as those in the *C*-terminal TRP domain, only affect menthol efficacy though events that occur downstream of menthol binding. This contention was supported by a thermodynamic analysis of the wild-type and mutant channels using the MWC model. Fitting the model parameters to the experimental data yielded that channels with mutations in the residue R842 have increased the affinity constant K_d, while the model for mutant K856A features lower K_d than the wild type. On the other hand, none of these mutations altered $\Delta\Delta H$ (i.e., menthol efficacy).

This analysis unveils a latent methodological issue that should be taken into account when addressing the chemical sensitivity of TRP channels. As argued previously, only the analysis of the effects of chemicals on mechanistically relevant gating parameters allows for the unequivocal identification of structural determinants of channel function (Meseguer *et al.*, 2011). For example, the fact that menthol induces a larger maximal current increase and larger $\Delta V_{Menthol,Max}$ in the mutant K856A could be interpreted as a higher efficacy of menthol than in wild-type TRPM8. However, the careful characterization of both channels using the MWC model indicates that the mutant K856A shows larger $\Delta V_{Menthol,Max}$ than the wild type, not because of an increased menthol efficacy ($\Delta\Delta H$), but because of its lower gating charge (see eqn. 4.21). Taking this into account, the only TRPM8 structural alteration that has been shown to yield channels with lower menthol efficacy than wild-type TRPM8 is the substitution of a fragment in the middle of the S4 segment by the corresponding sequence of the menthol insensitive relative TRPM5 (Voets *et al.*, 2007). Of note, two other fragment substitutions in S4 yielded inactive membrane proteins, indicating that specific sequences of TRPM8 in this region are essential for channel function. Similar findings were reported in another study in which the pore region (S5–S6) of TRPM2 was substituted into TRPM8, yielding a chimera with virtually no channel activity (Kühn *et al.*, 2010).

The question of how residues of the S2 and S4 segments are required for menthol binding was addressed by Pedretti *et al.*, using a structural model of TRPM8 built through a combination of homology comparisons with other proteins of known structure (Pedretti *et al.*, 2011; Pedretti *et al.*, 2009). Docking simulations suggested that the hydroxyl menthol group interacts with Y745 through an H-bond, whereas the rest of the structure is in contact with apolar residues through hydrophobic interactions. The role of R842 in menthol binding was less clear, as in the proposed structural model this residue is located relatively far away from Y745. These workers suggested that R842 might not be in direct contact with menthol, but could have a stabilizing effect on the binding site. According to a more recent study using molecular dynamics simulations, the same group proposed that binding of menthol to Y745 disrupts the interaction of this residue with D802, which, in turn, approaches R842 leading to drastic conformational changes in the S4 segment and channel activation.

As another possibility to explain the role of R842 in menthol binding, one could consider that in the real structure of TRPM8, R842 and Y745 are closer than predicted by the model. In that case, the hydroxyl menthol group could alternate between interacting directly with R842 or through an H-bond with Y745, so that both residues are required for stabilizing the menthol binding.

The predictions of the simulations mentioned earlier motivated Winking *et al.* (2012) to determine the role of residues of the S3 segments in the sensitivity to menthol. They found that mutation D802N did not reduce the sensitivity to menthol, but severely affected the response to cold. Other mutations in this region did affect menthol (and cold) sensitivity, but this was associated with a loss of the glycosylated form of the protein at the plasma membrane, indicating the compromised general functionality of the mutant channels. Of note, it has also been shown that, in comparison with wild-type TRPM8, the unglycosylated mutant N934Q displays a voltage dependence of channel activation that is shifted to more positive potentials and reduced responses to cold and menthol (Pertusa *et al.*, 2012).

The existence of interactions between residues of segments S3 and S4 proposed by Pedretti *et al.* (2009 and 2011) were further put to the test by Kühn *et al.* (2010) through the characterization of mutants in which the residues were changed to others of opposite electric charge. They found that the mutants R842D and R842E were not functional channels, probably because of deficient protein maturation. Interestingly, the functionality of these mutants was rescued by the additional substitution D802R. More surprisingly, the resulting double mutants were virtually indistinguishable from the wild-type channel in terms of voltage-dependent gating and retained almost full responsiveness to cold and menthol. Importantly, these results indicate that R842 does not form part of the binding site of menthol, and that the mutation found to prevent menthol binding, R842A (Voets *et al.*, 2007), probably destabilizes the interaction of this compound with residue Y745. In addition, these data could indicate that the existence of amino acid pairs of opposite charge in S3 and S4 is required for channel function. However, the single substitution D802R resulted in a channel with similar voltage dependence and sensitivity to menthol (although with decreased sensitivity to cold) when compared with the wild type. The simplest explanation of these results is that with an intact voltage sensor (S4), the charge of the amino acid at position 802 is not a determinant for channel function. However, if the charge of the voltage-sensing residue at position 842 is inverted, then a compensatory change has to be introduced at position 802 to yield a functional channel.

In two recent studies Pertusa *et al.* (2014) and Taberner *et al.* (2014) analyzed the role of the *N*-terminus and the proximal region of the *C*-terminus in TRPM8 gating, respectively. Led by an initial report on the importance of the *N*-terminal domain in localization and tetramerization of TRPM8 (Phelps and Gaudet, 2007), Pertusa *et al.* (2012) investigated the functional consequences of mutations in these regions. They found that channel constructs bearing specific structural modifications in these regions displayed either decreased or enhanced responses to menthol with respect to the control channel construct (mTRPM8-myc).

On the other hand, Taberner *et al.* (2014) found significant changes in basal gating properties and menthol effects in several chimeras and mutants in which parts of TRPV1 were substituted into the *C*-terminus of TRPM8. However, we do not agree with how these workers analyzed the effects of menthol. The reason for this is that, under their recording conditions (30 °C), the voltage dependence of channel activation is very much shifted to positive potentials and, clearly, no saturation of the open probability was reached in the absence of menthol. In our opinion, this leads to an underestimation of the maximal open probability (i.e., $G_{Max} < 1$), and to legitimize the modular model used to compare the gating properties of different channel mutants and chimeras. Nevertheless, the main conclusion of their study remains valid, as the data do show that the *C*-terminal region is involved in the mechanism leading to channel opening upon menthol binding.

Taken together, the available data indicate that menthol interacts with residues in the vicinity of the voltage-sensing core of TRPM8 and that other channel regions are important for gating events downstream of menthol binding. However, it is clear that direct structural insights and further exploration using molecular dynamics simulations are needed to understand the mechanism of menthol binding and its consequences on channel gating.

4.3 TRPV1

TRPV1 is expressed in abundance in somatic and visceral sensory neurons, though restricted to small- and medium-size neurons in the dorsal root, vagal, and trigeminal ganglia (Caterina *et al.*, 1997; Helliwell *et al.*, 1998). TRPV1 is a polymodal receptor, responsive to different exogenous activators including several vanilloids, heat, protons, and membrane depolarization. *In vivo*, TRPV1 is implicated in the detection and integration of nociceptive chemical and thermal stimuli (Caterina *et al.*, 2000; Caterina *et al.*, 1997; Davis *et al.*, 2000; Tominaga *et al.*, 1998). This property is particularly crucial in the oronasal cavity, where TRPV1 is involved in pungency, salty-taste sensation, and the metallic taste of some non-nutritive sweeteners (Riera *et al.*, 2007).

TRPV1 is by far the most studied of all TRP channels. Studies on the pharmacological properties and pathophysiological relevance of this channel are numerous and there are high-resolution structures of this channel in the presence or absence of capsaicin (Cao *et al.*, 2013; Hellmich and Gaudet, 2014). However, this does not imply that TRPV1 is the best characterized in terms of the activation mechanisms, as there is a lack of quantitative biophysical models describing the effects of chemical agonists on the gating properties of this channel.

Capsaicin, the main pungent ingredient of the hot chili pepper (*Capsicum* genus; Thresh, 1876) has been by far the most prominent activator of TRPV1 and, thus, the structural determinants for capsaicin-induced TRPV1 activation have been investigated extensively. Chimeric constructs with the vanilloid-insensitive TRPV2 channel and capsaicin-insensitive TRPV1 orthologous (e.g., avian and rabbit) served to identify S2 and S3 as being important for capsaicin binding (Jordt and Julius, 2002). Using rat/rabbit chimeras led to the identification of a single amino acid substitution, I550T, capable of conferring capsaicin sensitivity to the naturally insensitive rabbit orthologue. Additionally, the inverse substitution, T550I, was sufficient to reduce capsaicin responses in human and rat TRPV1 (Gavva *et al.*, 2004). Alternatively, mutational analyses have also led to support an arrangement where, presumably, the vanillyl moiety of capsaicin interacts with the cytosolic aromatic residue Y511 and other polar residues such as S512, R491, and M547 (Chou *et al.*, 2004; Gavva *et al.*, 2004; Jordt and Julius, 2002). Taken together, this suggests that T550 is crucial in accommodating the vanillyl residue, whereas Y511 participates in the interaction with the hydrophobic tail (Gavva *et al.*, 2004). Consistent with this, a modeling study indicated that π–π and hydrophobic interactions occur between the vanillyl group of capsaicin and Y511, while the adjacent S512 also participates via hydrogen bonds. Likewise, the carbonyl group interacts also via hydrogen bonds with Y511 and K571 (Lee *et al.*, 2011). More recently, Cao *et al.* (2013) reported a model of the structure of TRPV1 bound to capsaicin, based on cryo-electron microscopy images of the rat isoform embedded in amphipathic

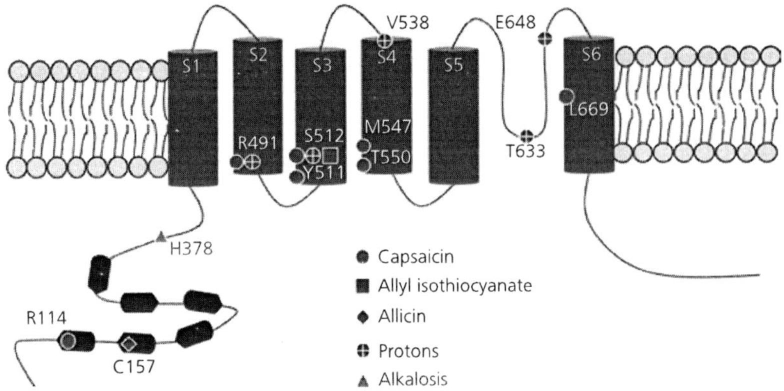

Fig. 4.3 Scheme of the structure of TRPV1 showing the position of residues important for the effects of capsaicin, allyl isothiocyanate, allicin, protons, and alkalosis.

polymers. This model confirmed the involvement of S3 and S4 as the pocket to host capsaicin (Fig. 4.3) and indicates that Y511 assumes two distinct rotamers depending on the presence of the ligand, suggesting that vanilloids may follow an "induced fit" mechanism. In addition, they found a contribution of S4–S5 linker and the L669 from the S6 belonging to an adjacent subunit. Thus, the putative capsaicin site simultaneously affects two regions directly involved in the structure of the pore domain, namely the S4–S5 linker and S6. Specifically, capsaicin-induced conformational changes widen the selectivity filter in the outer pore region by increasing the distance between the carbonyl oxygens of the residue G643. Additionally, the downward movement of M644 further contributes to the widening of the filter. Likewise, I679 in S6 rotates away from the central axis of the canal, expanding the lower gate. All these events are thought to allow pore opening.

Interestingly, current evidence excluded R114 within the *N*-terminus as the binding pocket for capsaicin, as initially proposed (Jung *et al.*, 2002). However, this residue may be required for the hydrophilic interaction of capsaicin, or it may be crucial in the regulatory domain near the ankyrin domain 1, a region in which multiple other regulatory ligands bind. R114 and Y199 are known to participate in the ATP-binding site of the TRPV1 channel (Lishko *et al.*, 2007).

Similar to capsaicin, several plant-derived molecules have a vanillyl moiety and activate the channel with different potencies. These are alkaloids present in black or white pepper, piperine and piperinoyl-piperidine (Izzo *et al.*, 2001; Mandadi and Roufogalis, 2008); eugenol, a phenol derived from clove and cinnamon leaf oil; gingerols and shogaols, commonly found in ginger (Holzer, 2008; Mandadi and Roufogalis, 2008); and various components present in some essential oils (thyme, rose, palmarosa), including citronellol and geraniol (Nilius and Appendino, 2013). Although the activation mechanisms for these compounds have not been elucidated, they most likely coincide in the binding pocket for capsaicin described earlier.

Besides vanilloids, several plant-derived compounds are TRPV1 agonists. Allicin, an organosulfur compound found in garlic (Salazar *et al.*, 2008), and allyl isothiocyanate (AITC), the pungent agent in mustard and wasabi, also activate TRPV1

(Everaerts *et al.*, 2011; Ohta *et al.*, 2007), although through different mechanisms. Allicin activates TRPV1 by covalent modification of a single cysteine residue in the *N*-terminus domain, namely C157 within the inner helix 2 in the second ankyrin domain (Salazar *et al.*, 2008). On the other hand, AITC-induced activation of TRPV1 was proposed to occur through S512 (Everaerts *et al.*, 2011; Gees *et al.*, 2013), a residue crucial for capsaicin sensitivity. As for capsaicin-induced activation, AITC induces a shift of the voltage dependence of channel activation towards negative voltages, a feature often found along TRP channel agonists (Nilius *et al.*, 2005; Talavera *et al.*, 2008; Voets *et al.*, 2005). How these residue-specific interactions lead to channel gating remains to be resolved.

All TRPV1 agonists mentioned here are highly lipophilic and, presumably, they interact with TRPV1 transmembrane and/or intracellular domains after having inserted in or crossed the membrane. Nonetheless, TRPV1 activators are not restricted to hydrophobic molecules. The most commonly used non-nutritive sweeteners (aspartame, saccharin, acesulfame-K, and sodium cyclamate) have been shown to activate TRPV1 in a dose-dependent manner (Riera *et al.*, 2007). Although a specific mechanism has yet to be elucidated, mutating Y511 is known to affect acesulfame-K and sodium cyclamate responses whereas aspartame and saccharin potency remain unaltered (Riera *et al.*, 2008). The identification of intracellular Y511 raises the interesting yet controversial question of whether these compounds can traverse the membrane in the absence of any specific transport pathway. At physiological pH, non-nutritive sweeteners are predominantly in their anionic forms, making them relatively membrane impermeable. This conundrum may have a tentative explanation in the known permeation of amphiphilic anions across bilayers (Thomae *et al.*, 2005). Nevertheless, the permeation rate of anionic species is much lower than the uncharged form of the acid and this is in apparent contradiction with the fast TRPV1 current induced by these compounds (Riera *et al.*, 2007). Sulfate salts of Cu^{2+}, Mg^{2+}, Fe^{2+}, and Zn^{2+}, which produce a metallic taste sensation, have also been reported to activate TRPV1 (Riera *et al.*, 2007). However, a putative interaction site or mechanism has not yet been put forward.

Acidic residues in the extracellular loop are susceptible to protonation following extracellular pH decrease (below 6.2), conferring TRPV1 with acid sensitivity (Bevan and Yeats, 1991; Caterina *et al.*, 2000; Jordt *et al.*, 2000; Tominaga *et al.*, 1998). TRPV1 mutagenesis have revealed that distinct residues are required for proton-driven activation and acid-sensitization (Jordt *et al.*, 2000). For instance, mutations of the aforementioned S512 and R491 also established the involvement of these residues in the responses to protons (Jordt and Julius, 2002). Likewise, E648 confers TRPV1 direct activation by protons, since its mutation to neutral amino acids selectively abrogates proton-induced responses, whereas capsaicin sensitivity is preserved (Jordt *et al.*, 2000). Proton-induced activation of TRPV1 occurs by a shift in the voltage dependence of activation to negative membrane potentials, a process fully dependent on F660 within the pore region (Aneiros *et al.*, 2011). Interestingly, the apolar residues V538 in the linker between S3 and S4 as well as T633 in the pore helix are also necessary for proton-induced activation of TRPV1 (Ryu *et al.*, 2007). The lack of a protonation site in these two residues places them in a regulatory function, most likely influencing the coupling of proton binding to channel gating (Bevan *et al.*, 2014). Remarkably, intracellular alkalosis can also activate TRPV1. Increases in cytosolic pH deprotonates H378, a histidine residue located between the ankyrin domain and the first transmembrane segment (Dhaka *et al.*, 2009).

At physiological temperatures, ethanol concentrations ranging from 0.1 to 3% (0.017–0.5 M) induce a pronounced concentration-dependent increase in intracellular calcium, in both primary sensory neurons and in recombinantly-expressed TRPV1 (Trevisani *et al.*, 2002). The agonistic attribute of ethanol is closely related to its sensitization effects on TRPV1 responses to heat stimuli. Ethanol induces a decrease in the apparent threshold to heat responses, from 42 to 34 °C, therefore shifting TRPV1 towards an activated state. Hence, TRPV1-expressing sensory afferents become activated by body temperature upon ethanol exposure. Electrophysiological recordings at lower temperatures (~23 °C) failed to find ethanol-mediated TRPV1 activation (Trevisani *et al.*, 2002), further supporting the "sensitization-mediated activation" hypothesis. Unfortunately, ethanol-induced TRPV1 sensitization has caught more of the attention of researchers, leaving mechanistic insights and molecular determinants for activation largely understudied.

4.3.1 Cross-sensitization between TRPV1 agonists

In nature, TRPV1 channel agonists are rarely presented as separated stimuli. Foods and drinks are typical examples of where a blend of chemical and thermal activators can converge on the same channel. Cross-sensitizations among the different TRPV1 agonists are well documented. For instance, in addition to the aforesaid lowering of heat threshold activation, ethanol sensitizes the channel to capsaicin and protons (Trevisani *et al.*, 2002). Likewise, acidosis sensitizes TRPV1 to capsaicin (Jordt *et al.*, 2000; McLatchie and Bevan, 2001), but also to AITC (Gees *et al.*, 2013). Cross-sensitization between acidosis and AITC occurs through the induction of additive shifts of the voltage dependence of the channel (Gees *et al.*, 2013), as described for capsaicin and heat sensitization (Voets *et al.*, 2004). Indeed, the sensitization effect of extracellular acidosis was fully abrogated by mutating E600, the residue that determines the shift of the activation curve by protons (Aneiros *et al.*, 2011). This glutamic acid in the outer pore loop of the channel interacts with main-chain Y653 and D654 through potential hydrogen bonding, restricting the outer pore in a non-conductive conformation. Protonation of E600 might eventually disrupt this triangular interaction, facilitating gating-associated movements of the pore helix and dilation of the selectivity filter (Cao *et al.*, 2013). On the other hand, AITC induces a shift of the activation curve to more negative potentials, sensitizing to lower pH, but also decreasing the apparent threshold for heat-induced activation (Alpizar *et al.*, 2014).

The addition of shifts of the voltage dependence of the channel, from a non-physiologically relevant voltage range to a functionally relevant voltage, has been proposed to rule this sensitization effect (Voets *et al.*, 2004). Here, TRPV1 is assumed to move between open and close states, complying with changes in free energy (Voets *et al.*, 2007). An alternative approach suggests an allosteric mechanism by adding two voltage-independent sensor entities, one reflecting temperature dependency (Brauchi *et al.*, 2004) and the other related to ligand-induced changes

(Matta and Ahern, 2007). Thus, voltage, temperature, and ligands are able to induce channel gating independently, but will also cooperate allosterically. However, as argued earlier for TRPM8, this model relies on data that do not demonstrate that capsaicin can activate TRPV1 in a voltage-independent manner.

4.4 TRPA1

TRPA1 was first identified under the name ANKTM1 (from the ankyrin repeat and transmembrane domain) as a cold-activated cation channel, highly enriched in sensory neurons (Story *et al.*, 2003). The sensitivity of mammalian TRPA1 to cold and its pathophysiological relevance have been matters of heated debate. In our opinion, most of the controversy comes from the unfortunate choice of TRP channel agonists to determine their functional expression in sensory neurons, for example, the use of menthol as a specific agonist of TRPM8. From a balanced review of the literature, it is now very clear that TRPA1 is sensitive to cold and that it contributes to certain (not all) behaviors and physiological responses induced by cold (Aubdool *et al.*, 2014; Karashima *et al.*, 2007; Voets, 2014).

As initially reported (Story *et al.*, 2003) and later confirmed (Kobayashi *et al.*, 2005), TRPA1 is expressed in a subset of mouse sensory neurons largely contained in the TRPV1-expressing population and is negligibly co-expressed with TRPM8. However, the expression patterns in human oronasal tissues have been little explored. Nevertheless, anatomical studies on a limited number of human samples indicate that TRPA1 is not expressed in sensory nerves of the tongue epithelium, but in those innervating the pharynx and the nasal epithelium (Peyrot des Gachons *et al.*, 2011). Notably, this supports the possibility that AITC irritates the human tongue, not by activating TRPA1, but TRPV1 (Everaerts *et al.*, 2011).

TRPA1 is arguably the most promiscuous chemosensory ion channel known so far. Since the first reports on mustard oils, cannabinoids, and cinnamaldehyde (Bandell *et al.*, 2004; Jordt *et al.*, 2004), a very large number of chemical species have been reported to activate this channel. Notably, these agonists vary greatly in terms of their chemical properties and therefore in the mechanisms underlying channel activation. The most well known in this respect is the group of electrophilic compounds, with AITC and cinnamaldehyde as prime examples relevant for human chemesthesis. These are known to activate the channel mainly via covalent modification of cysteine residues located in the *N*-terminus. A first study showed that a mutant of human TRPA1 carrying substitutions of three cysteines (C619, C639, and C663) displayed weaker responses to electrophilic compounds, while preserving the responsiveness to non-electrophilic agonists (Δ-9-tetrahydrocannabinol and 2-APB), with respect to the wild-type channel (Hinman *et al.*, 2006) (Fig. 4.4). In addition, mutation of a lysine residue (K708Q) was shown to abrogate the residual response of the triple cysteine human TRPA1 mutant to AITC. However, direct evidence for covalent modification of TRPA1 only came in a later paper by Macpherson *et al.* (2007) showing that the immunoprecipitated mouse protein can indeed be covalently labeled by alkyne-tagged agonists (iodoacetamide, AITC, and super cinnamaldehyde). Furthermore, it was shown that pretreatment with

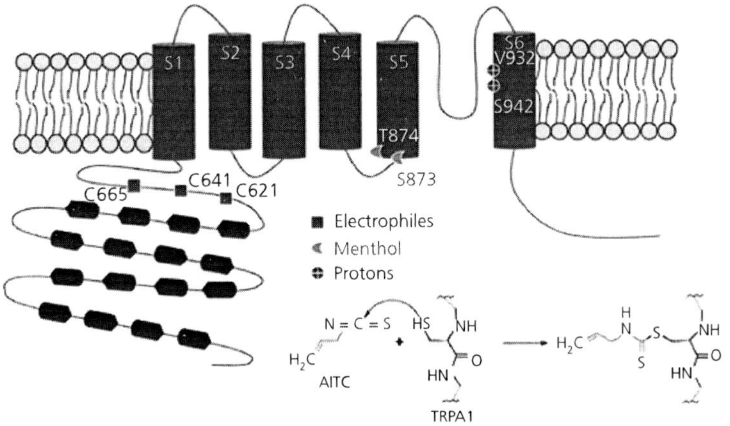

Fig. 4.4 Scheme of the structure of human TRPA1 illustrating the position of residues important for the effects of electrophilic compounds such as allyl isothiocyanate, menthol, and protons. The reaction of covalent modification of a cysteine residue by allyl isothiocyanate is represented in the bottom right.

untagged iodoacetamide, AITC, cinnamaldehyde, and super cinnamaldehyde, but not with the non-electrophilic agonist icilin, reduced the tagging with alkyne iodoacetamide. Using mass spectrometry, at least 14 cysteine residues were found to be modified by iodoacetamide, three of which were confirmed to be required for channel activation (C415, C422, and C622) with AITC, but not with icilin.

Of note, Macpherson *et al.* (2007) exposed a significant divergence from the data and interpretation given by Hinman *et al.* (2006). The former found that AITC induced somewhat irreversible channel activation over more than 10 min, whereas the latter showed the effects of this compound to be reversible within minutes of washout. The reason for this discrepancy remains unclear, but it could be related to the differences in recording conditions, expression systems (*Xenopus laevis* oocytes versus mammalian cells), and/or to the use of different TRPA1 isoforms (human versus mouse). Of note, according to the schematics of Hinman *et al.* (2006), all three cysteine residues required for activation of human TRPA1 are thought to be located between the last ankyrin repeat and the first transmembrane segment. In contrast, the key cysteine residues of mouse TRPA1 are located very close to or within the ankyrin repeats. We consider these differences not to be mere details, as they could have obvious implications for the use of the mouse as a surrogate for the study of human chemesthesis triggered by electrophilic irritants.

Finally, Macpherson *et al.* (2007) pointed out the reasonable possibility that mutations of these residues may have caused structural changes that could interfere with channel function. This issue has recently been raised by Zygmunt and Högestätt (2014), by recapping that *N*-terminal cysteines may form disulfide bridges that are important for the stabilization of the native channel structure. Indeed, using mass spectrometry, Wang *et al.* (2012) found four possible pairs in mouse TRPA1, C666–C622, C666–C463, C666–C193, and C622–C609. The fact that two of these residues can be shared in different bridges suggested that these represent different channel conformations. Furthermore, the finding that the reactivity of several cysteine residues to *N*-methylmaleimide was concentration dependent suggested that dynamic conformational changes alter the accessibility of these residues to electrophilic compounds. It is important to note, however, that the events leading to channel activation after cysteine modification are yet to be clarified. In addition, as summarized by Zygmunt and Högestätt (2014), there are other possibilities for the modulation of TRPA1 by electrophilic compounds. Firstly, some of these compounds activate the human wild type and the triple cysteine mutant. Secondly, other compounds may activate the channel by covalent modification and by disulfide formation. Thirdly, additional non-covalent interaction may lead to channel activation. In addition, it has been suggested that AITC induces vesicle exocytosis, increasing the membrane expression of TRPA1 (Schmidt *et al.*, 2009). Finally, we have reported that the canonical electrophilic agonists AITC and cinnamaldehyde have an inhibitory effect at high, but pharmacologically relevant, concentrations (Alpizar *et al.*, 2013; Everaerts *et al.*, 2011).

In addition to the electrophilic compounds, the very long list of known TRPA1 agonists contains many other chemical species. These are, for instance, intracellular Ca^{2+} (Doerner *et al.*, 2007; Zurborg *et al.*, 2007), irritant heavy metals such as Zn^{2+}, Cd^{2+}, Cu^{2+} (Gu and Lin, 2010; Hu *et al.*, 2009), reactive oxygen (H_2O_2, O_3), and gasotransmitters (H_2S, NO, O_2, CO_2) (Nilius *et al.*, 2012; Shimizu *et al.*, 2014; Zygmunt and Hogestatt, 2014).

From these, Ca^{2+} and CO_2 deserve separate attention for their relevance to human chemesthesis. Indeed, although extracellular Ca^{2+} *per se* is not expected to be a natural stimulus of TRPA1 and does not affect channel function, Ca^{2+} permeating the channel strongly potentiates channel activation (Doerner *et al.*, 2007; Zurborg *et al.*, 2007), but also induces a fairly irreversible desensitization (Wang *et al.*, 2008). Of note, it has been shown that during activation of TRPA1 by electrophilic compounds the channel undergoes an increase in pore diameter and an enhanced relative permeability to Ca^{2+} (Banke, 2011; Banke *et al.*, 2010; Chen *et al.*, 2008; Karashima *et al.*, 2010). All these Ca^{2+}-related mechanisms are expected to affect the chemosensory role of TRPA1 in ways that are yet to be clarified. At this point, one can hypothesize that the Ca^{2+}-induced activation and the increase in Ca^{2+} permeability serve as initial amplifiers of the TRPA1 current, whereas the Ca^{2+}-dependent desensitization could serve to shut down the response, allowing the recovery of the responsiveness of the sensory nerve endings and avoiding intracellular Ca^{2+} overload and the consequent cellular stress.

Regarding CO_2, it was shown that TRPA1 is activated by the intracellular acidification induced by high extracellular concentration of this gas (Wang *et al.*, 2010). Thus, it was proposed that activation of TRPA1 mediates the pain (or pungent) sensation caused by high CO_2 concentrations. The reader is referred to the recent review by Roper (2014) for a thorough discussion on the role of sensory TRPs in the sensations of carbonation, acidity, and pungency elicited by carbonated drinks.

Possibly the largest group of known TRPA1 agonists is formed by non-electrophilic compounds. These are, among others, cannabinoids (Jordt *et al.*, 2004), local anesthetics (Leffler *et al.*, 2011; Piao *et al.*, 2009), general anesthetics (Fischer *et al.*, 2010; Matta *et al.*,

2008), antifungal compounds (Meseguer *et al.*, 2008), dihydropyridines (Fajardo *et al.*, 2008), non-steroidal anti-inflammatory drugs (Hu *et al.*, 2010; Peyrot des Gachons *et al.*, 2011), primary alcohols (Komatsu *et al.*, 2012), polyunsaturated fatty acids (Materazzi *et al.*, 2008; Motter and Ahern, 2012), menthol derivatives (Karashima *et al.*, 2007), and plant alkaloids such as caffeine (Nagatomo and Kubo, 2008) and nicotine (Talavera *et al.*, 2009). Furthermore, TRPA1 was recently found to be activated by certain lipopolysaccharides (e.g., from *E. coli*) and to mediate acute pain and neurogenic inflammation caused by these bacterial endotoxins (Meseguer *et al.*, 2014). Very little is known about the mechanisms underlying the action of these TRPA1 agonists, but some studies have raised the possibility that these are based on the detection of mechanical alterations in the plasma membrane produced by the insertion of the compounds (Hill and Schaefer, 2007; Meseguer *et al.*, 2014). Further research is needed to validate this hypothesis. Next we will discuss the case of menthol, which is arguably the best studied non-electrophilic compound known to modulate TRPA1. As shown, this discussion illustrates key features of TRPA1 modulation, including the bimodal effect, the promiscuity inside and outside the TRP channel family, and, importantly, interspecies differences.

Menthol is one of the most emblematic TRP channel agonists because its stimulatory effect on TRPM8 established the concept that TRP channels are able to detect cold and compounds producing cool sensations, in addition to heat and burning compounds (McKemy *et al.*, 2002; Peier *et al.*, 2002). Menthol is also prominent because it was (and unfortunately by some still is) considered as a specific agonist of TRPM8. The discovery of menthol as a modulator of TRPA1 was based on the chemesthetic properties of this compound. Although menthol is indeed known to produce cooling and soothing effects that are probably mediated in part by TRPM8 activation (the TRPM8-dependent anesthetic effect of menthol), it also induces sensations qualified as warmth, burning, irritation, and pain. The latter are not likely to be mediated by TRPM8, because this channel shows little or no expression in nociceptors. The possible role of TRPA1 in the noxious effects of menthol was initially masked by reports of TRPA1 being insensitive to (Story *et al.*, 2003) or inhibited by (Macpherson *et al.*, 2006) this compound. However, a later study by Karashima *et al.* (2007) on the effects on basal mouse TRPA1 currents revealed a stimulatory action at concentrations below ~100 μM, whereas higher concentrations induced a reversible block. Notably, washout of menthol applied at blocking concentrations consistently induced a sudden increase in current amplitude above the basal level. This rebound effect gave further support to the notion that at low concentrations menthol behaves as a TRPA1 agonist.

Single-channel recordings revealed that, at concentrations between 10 and ~300 μM, menthol induced an increase in burst channel activity indicative of a stimulatory effect on channel activation. However, menthol also induced a fast flickering block, which led to a significant reduction of average currents and to the impossibility to resolve channel openings at 1 mM (Fig. 4.5). Based on these findings and on the analysis of the effects of menthol on the voltage dependence of channel activation, Karashima *et al.* (2007) hypothesized that this compound binds to two sites, one determining a shift of the activation curve to more negative potentials (as described for menthol in TRPM8 and for capsaicin in TRPV1), and the other

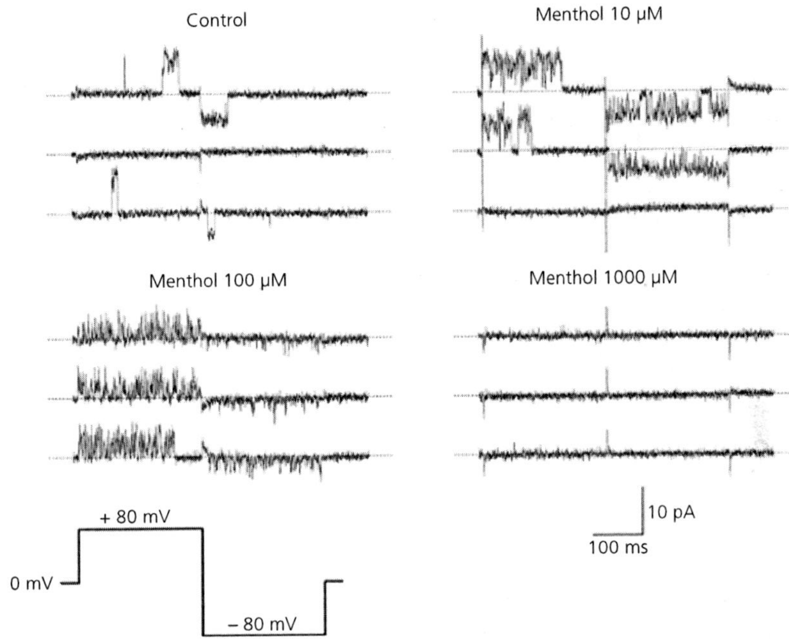

Fig. 4.5 Bimodal action of menthol on TRPA1. Single TRPA1 channel currents recorded in a CHO cell, in a control and in the presence of menthol (10, 100 or 1000 μM). The dotted lines indicate the zero-current levels. Three representative current traces are shown per condition. Menthol induces an increase in channel activity, but also a flickering block of the channel, which results in a strong block at high concentrations. The left-bottom panel shows the voltage protocol applied to induce channel activity at +80 (outward currents) and –80 mV (inward currents). For more details see (Karashima *et al.*, 2007).

causing a rapid block. The latter was proposed to be due to a block of the pore, but an inhibitory allosteric mechanism could not be discarded.

Importantly, Karashima *et al.* (2007) showed that, at concentrations at which menthol increases basal TRPA1 currents, this compound inhibits currents previously enhanced by AITC. This is consistent with the original observation that menthol inhibited TRPA1 currents previously activated by cinnamaldehyde (Macpherson *et al.*, 2006) and underscores the complexity of the chemical modulation of this channel. However, probably the most valuable conclusion of this study was that menthol is not a specific agonist of TRPM8. This had, for instance, strong implications for the understanding of the role of TRPA1 (versus TRPM8) as a cold sensor (see for details Karashima *et al.*, 2009) and must be considered for the interpretation of the effects of menthol in humans.

Interestingly, Xiao *et al.* (2008) reported that, in contrast to mouse TRPA1, human TRPA1 is only activated but not inhibited by menthol and that several TRPA1 isoforms from insects are insensitive to this compound. Taking advantage of these functional differences, these workers used a chimeric approach to identify the molecular determinants of the inhibitory action of menthol on mTRPA1. They found that nine specific residues in transmembrane segments S5 and S6, which are thought to form part of the channel pore, are required for the blocking effect of menthol on mTRPA1. A similar approach based on a comparison of the mammalian TRPA1 sequences to that of *Drosophila* TRPA1 helped in identifying the residues S876 and T877 of the S5 segment as critical determinants for mTRPA1 and hTRPA1 activation by menthol. Molecular docking simulations performed using a structural model of TRPM8 that was built by analogy with the structure of the bacterial K^+ channel KcsA suggested that the pore could provide a binding site for menthol. Direct structural information and further functional analyses are required to confirm this prediction and to understand the differential action of menthol on the different TRPA1 isoforms.

Notably, the bimodal action on TRPA1 is not exclusive for menthol. This has been reported for many other compounds, including, among others, the canonical electrophilic agonists AITC (Everaerts *et al.*, 2011) and cinnamaldehyde (Alpizar *et al.*, 2013), several menthol derivatives (Karashima *et al.*, 2007), citral (Stotz *et al.*, 2008), nicotine (Talavera *et al.*, 2009), CO_2 (Wang *et al.*, 2010), umbellulone and related compounds (Zhong *et al.*, 2011a), ligustilide (Zhong *et al.*, 2011b) and camphor (Alpizar *et al.*, 2013). As can be noted, this list includes compounds with very different chemical properties, which may indicate that the inhibitory effect could be due to an intrinsic property of the channel gating and not due to a specific effect of the inhibitors. Regardless of the mechanism, it is clear that this effect should be taken into account when addressing the chemosensory role of TRPA1 *in vivo*.

Of note, Paulsen *et al.* (2015) recently reported a structural model of human TRPA1 based on cryo-electron microscopy images of proteins embedded in an amphipathic polymer. This structural model covers about half of the protein and allowed confirmation of the supposed tetrameric channel structure and identification of particular structural features, such as interaction sites with stabilizing polyphosphates and the inhibitor A-967079. However, besides the confirmation of the intracellular location of the cysteines known to play a role in the activation by electrophilic compounds, no mechanistic insight for the action of these compounds could be extracted.

4.5 Concluding remarks

TRP channels are at the center of current research on the molecular mechanisms of chemosensation. The expression patterns and ability of these channels to respond to a wide variety of compounds and to integrate the actions of multiple stimuli makes them unique players in chemesthesis. Chemosensory TRP channels are not only important for the detection of compounds present in food, drinks, cosmetics, and so on, but also play a fundamental role in the mechanisms underlying pain and analgesia. The main objective of this chapter is to illustrate how the mechanisms of activation of these channels by natural agonists have been explored through different approaches. For instance, TRPM8 is the best studied in terms of mathematical modeling of its gating properties, TRPV1 is the only TRP channel for which partial high-resolution structures are available to deduce mechanisms of action of agonists, whereas TRPA1 is the best understood in terms of the nature of the interactions between some agonists and specific residues of the channel. Hopefully, future studies will combine all these approaches, and, with the much-needed anatomical characterization of the expression of these channels in human chemosensory tissues, they will enhance our understanding of how spices stimulate and modulate our chemical senses with their complex mixtures of TRP channel agonists. However, do we really want to spoil the magic of those sensations by revealing the underlying molecular mechanisms?

Acknowledgments

This work was supported by grants from the Research Council of the KU Leuven (OT/12/091, GOA/14/011, EF/95/010 and PF-TRPLe) and the Flemish Research Foundation (FWO, G.0702.12 and G.0C77.15N).

References

Almaraz, L., Manenschijn, J.A., de la Peña, E., and Viana, F. 2014. TRPM8. *Handb. Exp. Pharmacol.*, **222**, 547–579.

Alpizar, Y.A., Boonen, B., Gees, M., *et al.* 2014. Allyl isothiocyanate sensitizes TRPV1 to heat stimulation. *Pflügers Arch. EJP*, **466**, 507–515.

Alpizar, Y.A., Gees, M., Sanchez, A., *et al.* 2013. Bimodal effects of cinnamaldehyde and camphor on mouse TRPA1. *Pflügers Arch. EJP*, **465**, 853–864.

Aneiros, E., Cao, L., Papakosta, M., *et al.* 2011. The biophysical and molecular basis of TRPV1 proton gating. *EMBO J.*, **30**, 994–1002.

Aubdool, A.A., Graepel, R., Kodji, X., *et al.* 2014. TRPA1 is essential for the vascular response to environmental cold exposure. *Nat. Commun.*, **5**, 5732.

Bandell, M., Dubin, A.E., Petrus, *et al.* 2006. High-throughput random mutagenesis screen reveals TRPM8 residues specifically required for activation by menthol. *Nat. Neurosci.*, **9**, 493–500.

Bandell, M., Story, G.M., Hwang, S.W., *et al.* 2004. Noxious cold ion channel TRPA1 is activated by pungent compounds and bradykinin. *Neuron*, **41**, 849–857.

Banke, T.G. 2011. The dilated TRPA1 channel pore state is blocked by amiloride and analogues. *Brain Res.*, **1381**, 21–30.

Banke, T.G., Chaplan, S.R., and Wickenden, A.D. 2010. Dynamic changes in the TRPA1 selectivity filter lead to progressive but reversible pore dilation. *Am. J. Physiol. Cell. Physiol.*, **298**, C1457–1468.

Behrendt, H.J., Germann, T., Gillen, C., Hatt, H., and Jostock, R. 2004. Characterization of the mouse cold-menthol receptor TRPM8 and vanilloid receptor type-1 VR1 using a fluorometric imaging plate reader (FLIPR) assay. *Br. J. Pharmacol.*, **141**, 737–745.

Bevan, S., Quallo, T., and Andersson, D.A. 2014. TRPV1. *Handb. Exp. Pharmacol.*, **222**, 207–245.

Bevan, S., and Yeats, J. 1991. Protons activate a cation conductance in a sub-population of rat dorsal root ganglion neurones. *J. Physiol.*, **433**, 145–161.

Brauchi, S., Orio, P., and Latorre, R. 2004. Clues to understanding cold sensation: thermodynamics and electrophysiological analysis of the cold receptor TRPM8. *Proc. Natl. Acad. Sci. USA*, **101**, 15494–15499.

Brauchi, S., Orta, G., Salazar, M., Rosenmann, E., and Latorre, R. 2006. A hot-sensing cold receptor: C-terminal domain determines thermosensation in transient receptor potential channels. *J. Neurosci.*, **26**, 4835–4840.

Cao, E., Liao, M., Cheng, Y., and Julius, D. 2013. TRPV1 structures in distinct conformations reveal activation mechanisms. *Nature*, **504**, 113–118.

Caterina, M.J., Leffler, A., Malmberg, A.B., *et al.* 2000. Impaired nociception and pain sensation in mice lacking the capsaicin receptor. *Science*, **288**, 306–313.

Caterina, M.J., Schumacher, M.A., Tominaga, M., *et al.* 1997. The capsaicin receptor: a heat-activated ion channel in the pain pathway. *Nature*, **389**, 816–824.

Changeux, J.P. 2012. Allostery and the Monod-Wyman-Changeux model after 50 years. *Annu. Rev. Biophys.*, **41**, 103–133.

Chen, J., Zhang, X.F., Kort, M.E., *et al.* 2008. Molecular determinants of species-specific activation or blockade of TRPA1 channels. *J. Neurosci.*, **28**, 5063–5071.

Chou, M.Z., Mtui, T., Gao, Y.D., Kohler, M., and Middleton, R.E. 2004. Resiniferatoxin binds to the capsaicin receptor (TRPV1) near the extracellular side of the S4 transmembrane domain. *Biochemistry*, **43**, 2501–2511.

Davis, J.B., Gray, J., Gunthorpe, M.J., *et al.* 2000. Vanilloid receptor-1 is essential for inflammatory thermal hyperalgesia. *Nature*, **405**, 183–187.

Dhaka, A., Uzzell, V., Dubin, A.E., *et al.* 2009. TRPV1 is activated by both acidic and basic pH. *J. Neurosci.*, **29**, 153–158.

Doerner, J.F., Gisselmann, G., Hatt, H., and Wetzel, C.H. 2007. Transient receptor potential channel A1 is directly gated by calcium ions. *J. Biol. Chem.*, **282**, 13180–13189.

Everaerts, W., Gees, M., Alpizar, Y.A., *et al.* 2011. The capsaicin receptor TRPV1 is a crucial mediator of the noxious effects of mustard oil. *Curr. Biol.*, **21**, 316–321.

Fajardo, O., Meseguer, V., Belmonte, C., and Viana, F. 2008. TRPA1 channels: novel targets of 1,4-dihydropyridines. *Channels (Austin)*, **2**, 429–438.

Fernandez, J.A., Skryma, R., Bidaux, G., *et al.* 2011. Voltage- and cold-dependent gating of single TRPM8 ion channels. *J. Gen. Physiol.*, **137**, 173–195.

Fischer, M.J., Leffler, A., Niedermirtl, F., *et al.* 2010. The general anesthetic propofol excites nociceptors by activating TRPV1 and TRPA1 rather than GABAA receptors. *J. Biol. Chem.*, **285**, 34781–34792.

Gaudioso, C., Hao, J., Martin-Eauclaire, M.F., Gabriac, M., and Delmas, P. 2012. Menthol pain relief through cumulative inactivation of voltage-gated sodium channels. *Pain*, **153**, 473–484.

Gavva, N.R., Klionsky, L., Qu, Y., *et al.* 2004. Molecular determinants of vanilloid sensitivity in TRPV1. *J. Biol. Chem.*, **279**, 20283–20295.

Gees, M., Alpizar, Y.A., Boonen, B., *et al.* 2013. Mechanisms of transient receptor potential vanilloid 1 activation and sensitization by allyl isothiocyanate. *Mol. Pharmacol.*, **84**, 325–334.

Gkika, D., Lemonnier, L., Shapovalov, G., *et al.* 2015. TRP channel-associated factors are a novel protein family that regulates TRPM8 trafficking and activity. *J. Cell. Biol.*, **208**, 89–107.

Gu, Q., and Lin, R.L.2010. Heavy metals zinc, cadmium, and copper stimulate pulmonary sensory neurons via direct activation of TRPA1. *J. Appl. Physiol.* (1985), **108**, 891–897.

Hall, A.C., Turcotte, C.M., Betts, B.A., *et al.* 2004. Modulation of human GABAA and glycine receptor currents by menthol and related monoterpenoids. *Eur. J. Pharmacol.*, **506**, 9–16.

Hardie, R.C. 2014. Photosensitive TRPs. *Handb. Exp. Pharmacol.*, **223**, 795–826.

Helliwell, R.J., McLatchie, L.M., Clarke, M., *et al.* 1998. Capsaicin sensitivity is associated with the expression of the vanilloid (capsaicin) receptor (VR1) mRNA in adult rat sensory ganglia. *Neurosci. Lett.*, **250**, 177–180.

Hellmich, U.A., and Gaudet, R. 2014. High-resolution views of TRPV1 and their implications for the TRP channel superfamily. *Handb. Exp. Pharmacol.*, **223**, 991–1004.

Hill, K., and Schaefer, M. 2007. TRPA1 is differentially modulated by the amphipathic molecules trinitrophenol and chlorpromazine. *J. Biol. Chem.*, **282**, 7145–7153.

Hille, B. 2001. *Ion Channels of Excitable Membranes*, Sunderland, MA: Sinauer Associates.

Hinman, A., Chuang, H.H., Bautista, D.M., and Julius, D. 2006. TRP channel activation by reversible covalent modification. *Proc. Natl. Acad. Sci. USA*, **103**, 19564–19568.

Holzer, P. 2008. The pharmacological challenge to tame the transient receptor potential vanilloid-1 (TRPV1) nocisensor. *Br. J. Pharmacol.*, **155**, 1145–1162.

Hu, H., Bandell, M., Petrus, M.J., Zhu, M.X., and Patapoutian, A. 2009. Zinc activates damage-sensing TRPA1 ion channels. *Nat. Chem. Biol.*, **5**, 183–190.

Hu, H., Tian, J., Zhu, Y., *et al.* 2010. Activation of TRPA1 channels by fenamate nonsteroidal anti-inflammatory drugs. *Pflügers Arch. EJP*, **459**, 579–592.

Hui, K., Guo, Y., and Feng, Z.P. 2005. Biophysical properties of menthol-activated cold receptor TRPM8 channels. *Biochem. Biophys. Res. Commun.*, **333**, 374–382.

Izzo, A.A., Capasso, R., Pinto, L., *et al.* 2001. Effect of vanilloid drugs on gastrointestinal transit in mice. *Br. J. Pharmacol.*, **132**, 1411–1416.

Janssens, A., and Voets, T. 2011. Ligand stoichiometry of the cold- and menthol-activated channel TRPM8. *J. Physiol.*, **589**, 4827–4835.

Jordt, S.E., Bautista, D.M., Chuang, H.H., *et al.* 2004. Mustard oils and cannabinoids excite sensory nerve fibres through the TRP channel ANKTM1. *Nature*, **427**, 260–265.

Jordt, S.E., and Julius, D. 2002. Molecular basis for species-specific sensitivity to "hot" chili peppers. *Cell*, **108**, 421–430.

Jordt, S.E., Tominaga, M., and Julius, D. 2000. Acid potentiation of the capsaicin receptor determined by a key extracellular site. *Proc. Natl. Acad. Sci. USA*, **97**, 8134–8139.

Jung, J., Lee, S.Y., Hwang, S.W., *et al.* 2002. Agonist recognition sites in the cytosolic tails of vanilloid receptor 1. *J. Biol. Chem.*, **277**, 44448–44454.

Karashima, Y., Damann, N., Prenen, J., *et al.* 2007. Bimodal action of menthol on the transient receptor potential channel TRPA1. *J. Neurosci.*, **27**, 9874–9884.

Karashima, Y., Prenen, J., Talavera, K., *et al.* 2010. Agonist-induced changes in Ca^{2+} permeation through the nociceptor cation channel TRPA1. *Biophys. J.*, **98**, 773–783.

Karashima, Y., Talavera, K., Everaerts, W., *et al.* 2009. TRPA1 acts as a cold sensor in vitro and in vivo. *Proc. Natl. Acad. Sci. USA*, **106**, 1273–1278.

Kawashima, M., Imura, K., and Sato, I. 2012. Topographical organization of TRPV1-immunoreactive epithelium and CGRP-immunoreactive nerve terminals in rodent tongue. *Eur. J. Histochem.*, **56**, e21.

Kido, M.A., Muroya, H., Yamaza, T., Terada, Y., and Tanaka, T. 2003. Vanilloid receptor expression in the rat tongue and palate. *J. Dent. Res.*, **82**, 393–397.

Kobayashi, K., Fukuoka, T., Obata, K., *et al.* 2005. Distinct expression of TRPM8, TRPA1, and TRPV1 mRNAs in rat primary afferent neurons with adelta/c-fibers and colocalization with trk receptors. *J. Comp. Neurol.*, **493**, 596–606.

Komatsu, T., Uchida, K., Fujita, F., Zhou, Y., and Tominaga, M. 2012. Primary alcohols activate human TRPA1 channel in a carbon chain length-dependent manner. *Pflügers Arch. EJP*, **463**, 549–559.

Kühn, F.J., Witschas, K., Kuhn, C., and Luckhoff, A. 2010. Contribution of the S5-pore-S6 domain to the gating characteristics of the cation channels TRPM2 and TRPM8. *J. Biol. Chem.*, **285**, 26806–26814.

Kun, J., Helyes, Z., Perkecz, A., *et al.* 2012. Effect of surgical and chemical sensory denervation on non-neural expression of the transient receptor potential vanilloid 1 (TRPV1) receptors in the rat. *J. Mol. Neurosci.*, **48**, 795–803.

Lee, J.H., Lee, Y., Ryu, H., *et al.* S. 2011. Structural insights into transient receptor potential vanilloid type 1 (TRPV1) from homology modeling, flexible docking, and mutational studies. *J. Comput. Aided Mol. Des.*, **25**, 317–327.

Leffler, A., Lattrell, A., Kronewald, S., Niedermirtl, F., and Nau, C. 2011. Activation of TRPA1 by membrane permeable local anesthetics. *Mol. Pain*, **7**, 62.

Lishko, P.V., Procko, E., Jin, X., Phelps, C.B., and Gaudet, R. 2007. The ankyrin repeats of TRPV1 bind multiple ligands and modulate channel sensitivity. *Neuron*, **54**, 905–918.

Macpherson, L.J., Dubin, A.E., Evans, M.J., *et al.* 2007. Noxious compounds activate TRPA1 ion channels through covalent modification of cysteines. *Nature*, **445**, 541–545.

Macpherson, L.J., Hwang, S.W., Miyamoto, T., *et al.* 2006. More than cool: promiscuous relationships of menthol and other sensory compounds. *Mol. Cell. Neurosci.*, **32**, 335–343.

Mandadi, S., and Roufogalis, B.D. 2008. ThermoTRP channels in nociceptors: taking a lead from capsaicin receptor TRPV1. *Curr. Neuropharmacol.*, **6**, 21–38.

Materazzi, S., Nassini, R., Andre, E., *et al.* 2008. Cox-dependent fatty acid metabolites cause pain through activation of the irritant receptor TRPA1. *Proc. Natl. Acad. Sci. USA*, **105**, 12045–12050.

Matta, J.A., and Ahern, G.P. 2007. Voltage is a partial activator of rat thermosensitive TRP channels. *J. Physiol.*, **585**, 469–482.

Matta, J.A., Cornett, P.M., Miyares, R.L., *et al.* 2008. General anesthetics activate a nociceptive ion channel to enhance pain and inflammation. *Proc. Natl. Acad. Sci. USA*, **105**, 8784–8789.

McKemy, D.D., Neuhausser, W.M., and Julius, D. 2002. Identification of a cold receptor reveals a general role for TRP channels in thermosensation. *Nature*, **416**, 52–58.

McLatchie, L.M., and Bevan, S. 2001. The effects of pH on the interaction between capsaicin and the vanilloid receptor in rat dorsal root ganglia neurons. *Br. J. Pharmacol.*, **132**, 899–908.

Meseguer, V., Alpizar, Y.A., Luis, E., *et al.* 2014. TRPA1 channels mediate acute neurogenic inflammation and pain produced by bacterial endotoxins. *Nat. Commun.*, **5**, 3125.

Meseguer, V., Karashima, Y., Talavera, K., *et al.* 2008. Transient receptor potential channels in sensory neurons are targets of the antimycotic agent clotrimazole. *J. Neurosci.*, **28**, 576–586.

Meseguer, V.M., Denlinger, B.L., and Talavera, K. 2011. Methodological considerations to understand the sensory function of TRP channels. *Curr. Pharm. Biotechnol.*, **12**, 3–11.

Moran, M.M., McAlexander, M.A., Biro, T., and Szallasi, A. 2011. Transient receptor potential channels as therapeutic targets. *Nat. Rev. Drug Discov.*, **10**, 601–620.

Motter, A.L., and Ahern, G.P. 2012. TRPA1 is a polyunsaturated fatty acid sensor in mammals. *PLoS One*, **7**, e38439.

Nagatomo, K., and Kubo, Y. 2008. Caffeine activates mouse TRPA1 channels but suppresses human TRPA1 channels. *Proc. Natl. Acad. Sci. USA*, **105**, 17373–17378.

Nilius, B., and Appendino, G. 2013. Spices: the savory and beneficial science of pungency. *Rev. Physiol. Biochem. Pharmacol.*, **164**, 1–76.

Nilius, B., Appendino, G., and Owsianik, G. 2012. The transient receptor potential channel TRPA1: from gene to pathophysiology. *Pflügers Arch. EJP*, **464**, 425–458.

Nilius, B., and Szallasi, A. 2014. Transient receptor potential channels as drug targets: from the science of basic research to the art of medicine. *Pharmacol. Rev.*, **66**, 676–814.

Nilius, B., Talavera, K., Owsianik, G., *et al.* 2005. Gating of TRP channels: a voltage connection? *J. Physiol.*, **567**, 35–44.

Ohta, T., Imagawa, T., and Ito, S. 2007. Novel agonistic action of mustard oil on recombinant and endogenous porcine transient receptor potential V1 (pTRPV1) channels. *Biochem. Pharmacol.*, **73**, 1646–1656.

Owsianik, G., Talavera, K., Voets, T., and Nilius, B. 2006. Permeation and selectivity of TRP channels. *Annu. Rev. Physiol.*, **68**, 685–717.

Paulsen, C.E., Armache, J.P., Gao, Y., Cheng, Y., and Julius, D. 2015. Structure of the TRPA1 ion channel suggests regulatory mechanisms. *Nature*, **520**, 511–517.

Pedretti, A., Labozzetta, A., Lo Monte, M., *et al.* 2011. Exploring the activation mechanism of TRPM8 channel by targeted MD simulations. *Biochem. Biophys. Res. Commun.*, **414**, 14–19.

Pedretti, A., Marconi, C., Bettinelli, I., and Vistoli, G. 2009. Comparative modeling of the quaternary structure for the human TRPM8 channel and analysis of its binding features. Biochim. *Biophys. Acta*, **1788**, 973–982.

Peier, A.M., Moqrich, A., Hergarden, A.C., *et al.* 2002. A TRP channel that senses cold stimuli and menthol. *Cell*, **108**, 705–715.

Pertusa, M., Gonzalez, A., Hardy, P., Madrid, R., and Viana, F. 2014. Bidirectional modulation of thermal and chemical sensitivity of TRPM8 channels by the initial region of the N-terminal domain. *J. Biol. Chem.*, **289**, 21828–21843.

Pertusa, M., Madrid, R., Morenilla-Palao, C., Belmonte, C., and Viana, F. 2012. N-glycosylation of TRPM8 ion channels modulates temperature sensitivity of cold thermoreceptor neurons. *J. Biol. Chem.*, **287**, 18218–18229.

Peyrot des Gachons, C., Uchida, K., *et al.* 2011. Unusual pungency from extra-virgin olive oil is attributable to restricted spatial expression of the receptor of oleocanthal. *J. Neurosci.*, **31**, 999–1009.

Phelps, C.B., and Gaudet, R. 2007. The role of the N terminus and transmembrane domain of TRPM8 in channel localization and tetramerization. *J. Biol. Chem.*, **282**, 36474–36480.

Piao, L.H., Fujita, T., Jiang, C.Y., *et al.* 2009. TRPA1 activation by lidocaine in nerve terminals results in glutamate release increase. *Biochem. Biophys. Res. Commun.*, **379**, 980–984.

Plant, T.D. 2014. TRPs in mechanosensing and volume regulation. *Handb. Exp. Pharmacol.*, **223**, 743–766.

Raddatz, N., Castillo, J.P., Gonzalez, C., Alvarez, O., and Latorre, R. 2014. Temperature and voltage coupling to channel opening in transient receptor potential melastatin 8 (TRPM8). *J Biol. Chem.*, **289**, 35438–35454.

Riera, C.E., Vogel, H., Simon, S.A., Damak, S., and le Coutre, J. 2008. The capsaicin receptor participates in artificial sweetener aversion. *Biochem. Biophys. Res. Commun.*, **376**, 653–657.

Riera, C.E., Vogel, H., Simon, S.A., and le Coutre, J. 2007. Artificial sweeteners and salts producing a metallic taste sensation activate TRPV1 receptors. *Am. J. Physiol. Regul. Integr. Comp. Physiol.*, **293**, R626–634.

Roper, S.D. 2014. TRPs in taste and chemesthesis. *Handb. Exp. Pharmacol.*, **223**, 827–871.

Ryu, S., Liu, B., Yao, J., Fu, Q., and Qin, F. 2007. Uncoupling proton activation of vanilloid receptor TRPV1. *J. Neurosci.*, **27**, 12797–12807.

Salazar, H., Llorente, I., Jara-Oseguera, A., *et al.* 2008. A single N-terminal cysteine in TRPV1 determines activation by pungent compounds from onion and garlic. *Nat. Neurosci.*, **11**, 255–261.

Schmidt, M., Dubin, A.E., Petrus, M.J., Earley, T.J., and Patapoutian, A. 2009. Nociceptive signals induce trafficking of TRPA1 to the plasma membrane. *Neuron*, **64**, 498–509.

Seki, N., Shirasaki, H., Kikuchi, M., Sakamoto, T., Watanabe, N., and Himi, T. 2006. Expression and localization of TRPV1 in human nasal mucosa. *Rhinology*, **44**, 128–134.

Selescu, T., Ciobanu, A.C., Dobre, C., Reid, G., and Babes, A. 2013. Camphor activates and sensitizes transient receptor potential melastatin 8 (TRPM8) to cooling and icilin. *Chem. Senses*, **38**, 563–575.

Sexton, J.E., Vernon, J., and Wood, J.N. 2014. TRPs and pain. *Handb. Exp. Pharmacol.*, **223**, 873–897.

Shimizu, S., Takahashi, N., and Mori, Y. 2014. TRPs as chemosensors (ROS, RNS, RCS, gasotransmitters). *Handb. Exp. Pharmacol.*, **223**, 767–794.

Stander, S., Moormann, C., Schumacher, M., *et al.* 2004. Expression of vanilloid receptor subtype 1 in cutaneous sensory nerve fibers, mast cells, and epithelial cells of appendage structures. *Exp. Dermatol.*, **13**, 129–139.

Story, G.M., Peier, A.M., Reeve, A.J., *et al.* 2003. ANKTM1, a TRP-like channel expressed in nociceptive neurons, is activated by cold temperatures. *Cell*, **112**, 819–829.

Stotz, S.C., Vriens, J., Martyn, D., Clardy, J., and Clapham, D.E. 2008. Citral sensing by Transient [corrected] receptor potential channels in dorsal root ganglion neurons. *PLoS One*, **3**, e2082.

Swandulla, D., Carbone, E., Schafer, K., and Lux, H.D. 1987. Effect of menthol on two types of Ca currents in cultured sensory neurons of vertebrates. *Pflügers Arch. EJP*, **409**, 52–59.

Taberner, F.J., Lopez-Cordoba, A., Fernandez-Ballester, G., Korchev, Y., and Ferrer-Montiel, A. 2014. The region adjacent to the C-end of the inner gate in transient receptor potential melastatin 8 (TRPM8) channels plays a central role in allosteric channel activation. *J. Biol. Chem.*, **289**, 28579–28594.

Talavera, K., Gees, M., Karashima, Y., *et al.* 2009. Nicotine activates the chemosensory cation channel TRPA1. *Nat. Neurosci.*, **12**, 1293–1299.

Talavera, K., Nilius, B., and Voets, T. 2008. Neuronal TRP channels: thermometers, pathfinders and life-savers. *Trends Neurosci.*, **31**, 287–295.

Talavera, K., Voets, T., and Nilius, B. 2007. Mechanisms of thermosensation in TRP channels, in *Sensing with Ion Channels*, (ed. B. Martinac), Springer.

Thomae, A.V., Wunderli-Allenspach, H., and Kramer, S.D. 2005. Permeation of aromatic carboxylic acids across lipid bilayers: the pH-partition hypothesis revisited. *Biophys. J.*, **89**, 1802–1811.

Thresh, C.F. 1876. Capsaicin, the active principle in Capsicum fruits. *The Analyst*, **1**, 148–149.

Tominaga, M., Caterina, M.J., Malmberg, A.B., *et al.* 1998. The cloned capsaicin receptor integrates multiple pain-producing stimuli. *Neuron*, **21**, 531–543.

Trevisani, M., Smart, D., Gunthorpe, M.J., *et al.* 2002. Ethanol elicits and potentiates nociceptor responses via the vanilloid receptor-1. *Nat. Neurosci.*, **5**, 546–551.

Voets, T. 2014. TRP channels and thermosensation. *Handb. Exp. Pharmacol.*, **223**, 729–741.

Voets, T., Droogmans, G., Wissenbach, U., *et al.* 2004. The principle of temperature-dependent gating in cold- and heat-sensitive TRP channels. *Nature*, **430**, 748–754.

Voets, T., Owsianik, G., Janssens, A., Talavera, K., and Nilius, B. 2007. TRPM8 voltage sensor mutants reveal a mechanism for integrating thermal and chemical stimuli. *Nat. Chem. Biol.*, **3**, 174–182.

Voets, T., Talavera, K., Owsianik, G., and Nilius, B. 2005. Sensing with TRP channels. *Nat. Chem. Biol.*, **1**, 85–92.

Vriens, J., Nilius, B., and Voets, T. 2014. Peripheral thermosensation in mammals. *Nat. Rev. Neurosci.*, **15**, 573–589.

Wang, L., Cvetkov, T.L., Chance, M.R., and Moiseenkova-Bell, V.Y. 2012. Identification of in vivo disulfide conformation of TRPA1 ion channel. *J. Biol. Chem.*, **287**, 6169–6176.

Wang, Y.Y., Chang, R.B., and Liman, E.R. 2010. TRPA1 is a component of the nociceptive response to CO_2. *J. Neurosci.*, **30**, 12958–12963.

Wang, Y.Y., Chang, R.B., Waters, H.N., McKemy, D.D., and Liman, E.R. 2008. The nociceptor ion channel TRPA1 is potentiated and inactivated by permeating calcium ions. *J. Biol. Chem.*, **283**, 32691–32703.

Winking, M., Hoffmann, D.C., Kuhn, C., *et al.* 2012. Importance of a conserved sequence motif in transmembrane segment S3 for the gating of human TRPM8 and TRPM2. *PLoS One*, **7**, e49877.

Xiao, B., Dubin, A.E., Bursulaya, B., *et al.* 2008. Identification of transmembrane domain 5 as a critical molecular determinant of menthol sensitivity in mammalian TRPA1 channels. *J. Neurosci.*, **28**, 9640–9651.

Zanini, D., and Gopfert, M.C. 2014. TRPs in hearing. *Handb. Exp. Pharmacol.*, **223**, 899–916.

Zhang, X.B., Jiang, P., Gong, N., *et al.* 2008. A-type GABA receptor as a central target of TRPM8 agonist menthol. *PLoS One*, **3**, e3386.

Zhong, J., Minassi, A., Prenen, J., *et al.* 2011a. Umbellulone modulates TRP channels. *Pflügers Arch. EJP*, **462**, 861–870.

Zhong, J., Pollastro, F., Prenen, J., *et al.* 2011b. Ligustilide: a novel TRPA1 modulator. *Pflügers Arch. EJP*, **462**, 841–849.

Zhu, M.X. 2007. Understanding the role of voltage gating of polymodal TRP channels. *J. Physiol.*, **585**, 321–322.

Zufall, F. 2014. TRPs in olfaction. *Handb. Exp. Pharmacol.*, **223**, 917–933.

Zurborg, S., Yurgionas, B., Jira, J.A., Caspani, O., and Heppenstall, P.A. 2007. Direct activation of the ion channel TRPA1 by Ca^{2+}. *Nat. Neurosci.*, **10**, 277–279.

Zygmunt, P.M., and Hogestatt, E.D. 2014. TRPA1. *Handb. Exp. Pharmacol.*, **222**, 583–630.

CHAPTER 5

Anatomy and physiology of chemesthesis

Cecil J. Saunders[1] and Wayne L. Silver[2]

[1] *Department of Otorhinolaryngology, Perelman School of Medicine, University of Pennsylvania, Philadelphia, PA, USA*

[2] *Department of Biology, Wake Forest University, Winston-Salem, NC, USA*

5.1 Introduction

The idea that olfaction and gustation were not responsible for the detection of chemical irritants began in 1912 with the observation that fish responded to acid placed on their flank (Parker, 1912). Parker (1912) chose the term common chemical sense to describe the system responsible for detecting irritating chemicals. Subsequent anatomical studies demonstrated that the common chemical sense was mediated through free nerve endings found in epithelia and mucous membranes of vertebrates (Parker, 1922). Although these nerve endings resembled pain receptors, Parker (1922) concluded that "the common chemical sense is a true sense with an independent set of receptors and a sensation quality entirely its own". Today we know that the nerve fibers that respond to irritating chemicals are not part of a separate, independent sensory system but rather are part of the general somatosensory system. These fibers are a subset of pain and temperature sensitive fibers, which terminate in free nerve endings throughout the skin and mucosal membranes of the nose, mouth, respiratory tract, eye, and anal and genital orifices. In 1990, the term chemesthesis replaced common chemical sense to describe the sensations elicited by the chemical stimulation of somatosensory nerves (see Chapter 1). More recently, solitary chemoreceptor cells have been discovered that respond to chemicals and synapse onto these somatosensory axons.

In this chapter we present an overview of the anatomy and physiology of the nose and the mouth as they relate to chemesthesis.

5.2 Anatomy

Somatosensory nerves mediating chemesthesis innervate the entire body surface as well as the epithelia lining body orifices. Small, slow-conducting, thinly myelinated A-δ (1–5 μm diameter; 5–30 m s^{-1}) and small, slow-conducting, unmyelinated

C (0.2–1.5 μm diameter; 0.5–2 m s^{-1}) nerve fibers typically mediate chemesthesis. The cell bodies for these nerve fibers are found in the dorsal root and Gasserian (trigeminal, Vth cranial nerve) ganglia and to a lesser extent the petrous (glossopharyngeal, IXth cranial nerve) and nodose (vagus, Xth cranial nerve) ganglia. The dorsal root cells project to the body, while the trigeminal, petrous and nodose cells project to the head and face. Although chemical irritants from food may stimulate somatosensory nerves in the cornea (e.g., when you cut onions) or the anus (where it has been noted that a spicy meal may burn twice), we focus here on the oral and nasal cavities, as they are the most common sites of chemesthetic stimulation.

5.2.1 Oral cavity

The oral cavity houses three types of stratifying squamous epithelia: keratinized epithelium in the gums and hard palate, non-keratinized epithelium in the buccal regions and floor of the mouth, and a specialized epithelium (mixture of keratinized and non-keratinized epithelia) on the tongue (Finger and Simon, 2000). The nerve fibers subserving chemesthesis do not project past these epithelial layers but rather terminate in free nerve endings just below tight junctions that bind the cells of the epithelial layer together. While solitary chemoreceptor cells (discussed later) have been observed in a variety of mucus membranes, they have never been observed in the oral epithelium of adult mammals (Sbarbati and Osculati, 2003). Therefore, in the oral cavity most chemesthetic stimuli must be lipophilic to defuse through the epithelial cell's membranes and reach the free nerve endings below.

The oral cavity contains branches of the trigeminal nerve: buccal and lingual nerves (from the mandibular division) and palatine nerves (from the maxillary branch). The buccal nerve innervates the cheeks and some teeth. The lingual nerve projects to the anterior two-thirds of the tongue and surrounding tissue while the palatine nerves innervate the roof of the mouth and soft palate. In the tongue, free nerve endings are found in high abundance surrounding the taste buds, especially in the fungiform papillae (Farbman and Hellekant, 1978; Suemune *et al.*, 1992). Branches of the glossopharyngeal and vagus nerves, which do not innervate taste receptor cells, are also found in the oral cavity. These nerve fibers may respond to chemicals and are located in the posterior one-third of the tongue, the pharynx (IX), the larynx, the base of the tongue, and the epiglottis (X).

It should be obvious that, in humans, the oral and nasal cavities are connected by the nasopharynx. Therefore, anything volatile placed in the oral cavity will have access to the nasal cavity. The act of chewing food and moving it around with the tongue helps release volatile compounds. This is a major contributor to the flavor of food. While capsaicin – the burning chemical found in chili peppers – is among the most commonly encountered pungent compounds in food, it is not volatile and remains in the oral cavity. However, many other chemesthetic stimuli are delivered to the nasal cavity retronasally, including ally isothiocyanate (wasabi), menthol (mint), and eugenol (cloves). When we breathe out, air from our lungs is forced through the nasopharynx where it picks up these compounds and conveys them into the nasal cavity.

5.2.2 Nasal cavity

The nose is the primary avenue that gases are admitted into and then expelled out of the respiratory tract. As such, specific adaptations in the anatomy of the nasal cavity allow for the sampling, filtering, and conditioning of inhaled air before it reaches the lower airways (Harkema *et al.*, 2006). During inhalation air passes through the nares into the nasal vestibule, past the posterior nasal valves into the posterior nasal cavity (Djupesland, 2013). The nasal valves are the narrowest section of the upper airway (Messerklinger, 1978; Youngs *et al.*, 2006). Throughout the day erectile mucosa near one nasal valve will swell, causing the opposite side of the nasal cavity to be favored during inhalation (Cole, 2003). This cycling between nostrils has been implicated in odor perception (Zhou and Chen, 2009; Zhou *et al.*, 2010) but also may limit the exposure of noxious chemicals to one side of the nasal cavity.

Several large bony projections, interchangeably referred to as either turbinates or concha, extend from the lateral wall of the nasal cavity and curve inferiorly towards the nares (Messerklinger, 1978; Youngs *et al.*, 2006). The turbinates are covered with highly vascularized mucosal epithelium, which provides a reservoir of humidity that prevents air from desiccating and irritating the fragile tissues of the olfactory epithelium found at the roof of the nasal cavity (Welch *et al.*, 2002; Harkema *et al.*, 2006). In addition to moistening inhaled air, the nasal labyrinth is also capable of scrubbing potentially damaging reactive chemicals from the environment (Morris *et al.*, 2005). As many chemesthetic compounds are perceived as noxious because of their reactivity (Sadofsky *et al.*, 2011), this property further emphasizes the role of the nasal cavity in the protection of the airway from noxious chemicals.

The labyrinth of the nasal cavity is the first location where inhaled air is checked for noxious chemicals before being passed on to the lower airway. As such, the nasal cavity contains sensory branches of the trigeminal nerve: the ethmoid nerve (from the ophthalmic division) and the nasopalatine nerve (from the maxillary division) (Bryant and Silver, 2000). The ethmoid nerve innervates the anterior nasal mucosa and external nasal surface, while the nasopalatine nerve innervates the posterior portion of the nasal cavity (Richards *et al.*, 2010). The nerve fibers from the trigeminal ganglion reach the nasal cavity and splay out across the nasal epithelium (Finger *et al.*, 1990).

The nasal cavity is lined with three types of epithelia. The first of these three types is a thin, squamous epithelium found in the nasal vestibule, which due to its moderate keratinization is somewhat protected from chemesthetic stimuli (Harkema *et al.*, 2006). Conversely, the olfactory epithelium, found in small patches at the roof of the nasal cavity in humans, is a pseudostratified mucosal tissue required for the detection of odors (Schild and Restrepo, 1998; Brann and Firestein, 2014). In humans, these two epithelial types constitute a minority of the nasal cavities' total epithelium (Harkema *et al.*, 2006).

The majority of the upper airway is covered by respiratory epithelium, a pseudostratified non-keratinized epithelial barrier that consists of several major cell types (Basbaum and Jany, 1990). Traditionally the role of these epithelial cells in chemesthesis was thought to be limited to their role as a physical barrier but current research has challenged this paradigm (Green, 2012). Specifically, the discovery of solitary chemosensory cells, a neuro-epithelial cell that forms synapses with trigeminal c-fibers, has forced a reevaluation of the previous nerve-centric view of chemesthesis (Tizzano and Finger, 2013). However, even the better known components of the respiratory epithelium, such as ciliated cells, have also recently been ascribed roles in the detection and response to chemical irritants (Lee and Cohen, 2013).

5.2.3 Solitary chemosensory cells

Solitary chemosensory cells (SCCs) provide a mechanism by which chemicals in the lumen of the airway can stimulate the underlying somatosensory nerve fibers (Finger, 1997; Merigo *et al.*, 2005). The characterization of SCCs in the early 2000s challenged the dogma that chemesthetic substances were required to diffuse through the epithelial layer to activate the underlying nerve fibers (Finger *et al.*, 2003). While the anatomy of SCCs suggests a function as sensory end-organs of trigeminal nerves, only recently has their role in detecting xenobiotics and triggering inflammation begun to be fully appreciated (Tizzano *et al.*, 2010; Lee *et al.*, 2014b; Saunders *et al.*, 2014).

The morphology of SCCs allows them to monitor the nasal cavity and send signals via trigeminal nerve fibers (Silver and Finger, 2009). SCCs resemble the bottle shape of taste receptors cells in that they project at least one of the apical processes from a wider rounded cell body into the lumen above the epithelial layer (Fig. 5.1) (Gulbransen *et al.*, 2008a; Gulbransen *et al.*, 2008b). These apical processes, of which there may be several, terminate in a microvillus tuft (Finger *et al.*, 2003). Some SCCs also have a basal process that appears to be reaching out to make contact with trigeminal nerve fibers (Saunders *et al.*, 2013b, 2014). This unique morphology allows for SCCs to relay information about the chemical contents of the nasal cavity to the underlying trigeminal nerve fibers.

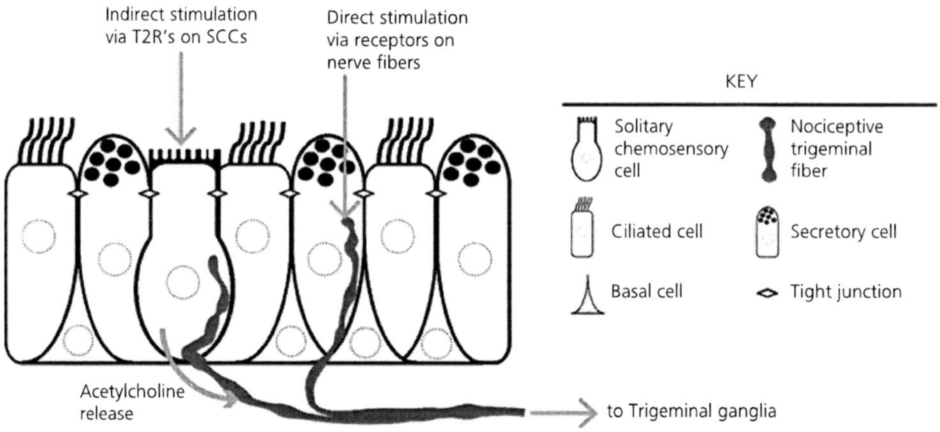

Fig. 5.1 The nasal epithelium is composed of several different cells types and is innervated by nociceptive trigeminal nerve fibers. Chemesthetic sensations can be produced either by directly stimulating receptors (such as TRPV1 or TRPA1) on nerve fibers or by activating bitter taste receptors (T2Rs) on solitary chemosensory cells (SCCs), which then release acetylcholine to activate nerve fibers. Both SCCs and ciliated cell can detect "bitter" substances via T2Rs but only SCCs form synapses with nerve fibers. Stimulation of T2Rs on either SCCs or ciliated cells triggers a local defensive response, such as increased ciliary beat frequency, production of nitrogen oxide, and the secretion of antimicrobial peptides.

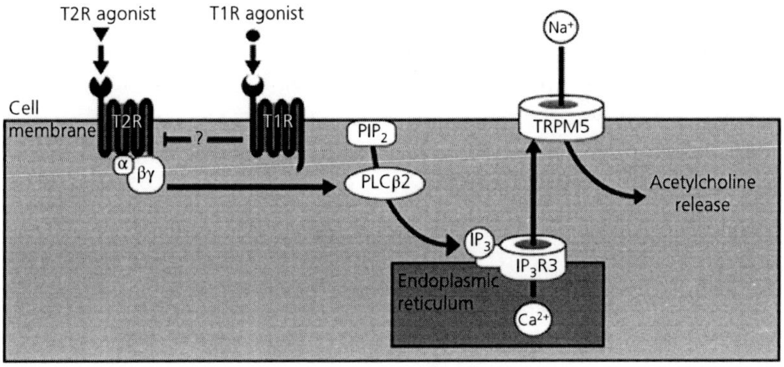

Fig. 5.2 Solitary chemosensory cells express both "bitter" T2R and "sweet" T1R receptors. T2R activation results in the PLCβ2 mediated production of IP_3, which then triggers the type 3 IP_3 receptors (IP_3R3). Activation of IP_3R3 causes the release of calcium from the endoplasmic reticulum, which in turn causes opening the monovalent cation channel TRPM5. The opening of TRPM5 allows sodium to enter and depolarize the cell. Activation of this signaling cascade results in the release of acetylcholine onto nociceptive trigeminal fibers creating sensations of irritation. Activation of T1R receptors inhibits the T2R pathway though a yet to be determined mechanism.

To relay information about potentially dangerous chemicals, SCCs utilize receptors and downstream signaling components originally associated with taste receptor cells (Finger *et al.*, 2003; Tizzano *et al.*, 2010). Although they share many similarities, one of the most profound differences between SCCs and taste cells is that SCCs co-express both bitter sensitive taste receptor proteins (T2Rs) and sweet sensitive taste receptor proteins (T1Rs), but taste receptors cells typically express only one or the other (Fig. 5.2) (Lee *et al.*, 2014b). SCCs in mice express the bitter sensitive taste receptor proteins mT2R8 and mT2R19 (Finger *et al.*, 2003), while those in humans express hT2R47 (Lee *et al.*, 2014b). To report activation of T2Rs, SCCs utilize the same collection of secondary messaging proteins as taste cells: G-α-gustducin, PLCβ2, IP_3R3, and TRPM5 (Finger *et al.*, 2003; Gulbransen *et al.*, 2008a; Tizzano *et al.*, 2010; Barham *et al.*, 2013). Interestingly, recent results suggest that co-activation of SCC sweet receptors inhibit the SCC's response to bitter substances, although the exact signaling pathway responsible for this inhibition is currently unknown (Lee *et al.*, 2014b). While the intracellular signaling components of SCCs share much in common with taste receptor cells and we refer to compounds that stimulate SCCs by their gustatory descriptor (e.g., bitter or sweet) for lack of a more precise descriptor, it is important to always keep in mind that SCCs only form synapses with trigeminal nerve fibers and thus when stimulated by "bitter" substances are only capable of producing sensations associated with chemesthesis (Tizzano and Finger, 2013) (see also Chapter 6 for a discussion of the "doctrine of specific nerve energies").

SCCs form many repeating, intimate contacts with trigeminal c-fibers (Finger *et al.*, 2003). On the light microscopy level, peptidergic trigeminal fibers appear to intertwine and enwrap both SCCs cell bodies and processes (Gulbransen, 2008a; Saunders *et al.*, 2014). On the electron microscopy level, the cell membrane of SCCs is invaginated to partially surround the nerve fibers at most sites of contact (Finger *et al.*, 2003). Additionally, the points of contact between SCCs and nerve fibers contain many features associated with synapses, including small vesicles positioned for neurotransmitter release from the SCCs onto the nerve fiber (Finger *et al.*, 2003). These vesicles presumably contain acetylcholine, as SCCs produce two of the enzymes required for the vesicular release of that neurotransmitter, choline acetyltransferase and the vesicular acetylcholine transporter (Lin *et al.*, 2008; Krasteva *et al.*, 2011; Saunders *et al.*, 2014). The presence of cholinergic synapses between SCCs and trigeminal nerve fibers allows the SCCs to monitor the lumen of the nasal cavity and activate trigeminal nerve fibers in response to chemicals that are not capable of penetrating the epithelium.

5.2.4 Other chemosensory epithelial cells

While SCCs are intimately associated with trigeminal nerve fibers, they also interact with other cell types that comprise the respiratory epithelium. SCCs usually appear as singlets, or less commonly doublet and triplets and constitute approximately 1% of the total epithelial cells (Saunders *et al.*, 2013b). This is in contrast to the more abundant ciliated, secretory and basal cells, each of which makes up approximately 30% of the epithelium (Cole *et al.*, 2010; Reynolds and Malkinson, 2010). A recent study conducted in human tissue culture has demonstrated that SCCs can signal through gap junctions to these neighboring epithelial cells (Lee *et al.*, 2014b). Thus, SCC stimulation by chemesthetic agents not only triggers a neural response but also produces a local protective response in the ciliated and secretory cells.

Ciliated cells are not just effectors of epithelial defense but also appear to be chemodetectors themselves (Kinnamon and Reynolds, 2009; Kinnamon, 2012). Ciliated cells also utilize many elements of the canonical taste transduction pathway to detect potentially dangerous chemicals (Shah *et al.*, 2009). Not only are the bitter taste receptors hT2R4, hT2R38, hT2R43, and hT2R46 and the transduction proteins α-gustducin PLCβ2 and TRPM5 present in ciliated cells, but many of these proteins appear to be enriched in or near the cilia (Shah *et al.*, 2009; Lee *et al.*, 2014a). Stimulation of this pathway results in increases in the frequency of ciliary beating and production of nitric oxide, presumably to dislodge and kill pathogenic bacteria. Currently, ciliated cells' and SCC-trigeminal system's response to "bitter" compounds are viewed as distinct pathways with little if any cross talk (Lee *et al.*, 2012; Lee *et al.*, 2014a; Lee *et al.*, 2014b). However, the ability of the ciliated cells to produce nitric oxide, which in addition to its bactericidal properties is also a diffusible signaling molecule (Dupont *et al.*, 2014), raises the possibility of interaction between the local inflammatory pathway and neurally mediated irritation.

In the current model of nasal inflammatory signaling, T2R ("bitter") receptors found on SCCs and ciliated cells detect bacteria metabolites (Lee *et al.*, 2014a; Saunders *et al.*, 2014). Certain bacteria, including the opportunistic respiratory pathogen *Pseudomonas*, secrete signaling molecules that they utilize in quorum sensing (Fuqua *et al.*, 1994; Folkesson *et al.*, 2012). These bacterial signals, particularly a class of molecules termed acyl homoserine lactones (AHLs) (Fuqua and Greenberg, 2002), activate bitter receptors in the airway and recruit an inflammatory and immune response to fight off the bacterial infection (Tizzano *et al.*, 2010; Krasteva *et al.*, 2012; Lee *et al.*, 2014a; Lee *et al.*, 2014b). In addition to bacterial metabolites, these T2R receptors also respond to bitter tastants placed on the epithelium (Finger *et al.*, 2003; Gulbransen *et al.*, 2008a; Lin *et al.*, 2008; Lee *et al.*, 2014b). Thus, a volatile bitter chemical from food in the mouth should be capable of stimulating SCCs after reaching the nasal cavity via retronasal flow.

5.3 Physiology

5.3.1 Reflexes

The activation of chemesthetic trigeminal nerve fibers elicits a wide variety of physiological reflexes (James and Daly, 1969). These are among the most powerful reflexes in the body, which serve to reduce the effects of chemical irritants and protect the body from further exposure. The reflexes are mediated primarily by the chemical activation of TRP channels (see Chapters 4 and 14) on trigeminal nerve fibers (Gerhold and Bautista, 2009; Julius, 2013).

The reflexes that may be elicited by chemical stimulation of chemesthetic nerves in the nasal cavity include a decrease in respiratory rate and nasal patency (Warren *et al.*, 1994), cough (Geppetti *et al.*, 2010), a decrease in heart rate (Nalivaiko *et al.*, 2003), bronchoconstriction (Eilers *et al.*, 2010), an increase in nasal secretion and sneezing (Geppetti *et al.*, 1988), tearing (Philip *et al.*, 1994), apnea (Vijayaraghavan *et al.*, 1993; Saunders *et al.*, 2013a), and an increase in cerebral blood flow (Kunkler *et al.*, 2011). Stimulation of the oral cavity with irritants may elicit similar reflexes, as well as increased salivary flow (Lawless, 1984; Takahashi *et al.*, 1995), flushing of the face (vasodilation) (Kashima and Hayashi, 2013; Ma *et al.*, 2014) and sweating (Lee, 1954).

Many of these reflexes come into play when you eat a spicy meal containing capsaicin, which activates TRPV1 receptors found on peptidergic trigeminal c-fibers (Tominaga *et al.*, 1998). When stimulated, these nociceptive fibers release many inflammatory peptides at the site of stimulation, a phenomenon known as axon reflex (Langley, 1900; Finger *et al.*, 1990; Julius, 2013). Two of the peptides, substance P and calcitonin gene related peptide (CGRP), act on blood vessels causing dilation, the extravasation of blood plasma and the flushing associated with capsaicin consumption (Rather, 1971; Rodger *et al.*, 1995). These inflammatory neuropeptides also act on the c-fibers themselves, sensitizing them, which results in the sensation of heat and triggers sweating (O'Connor *et al.*, 2004). Although it does not involve irritating chemicals, ice cream headache or "brain freeze" is thought to be produced by similar mechanisms (Selekler and Budak, 2004), that is, stimulation of trigeminal nerve fibers leading to a change in cerebral blood flow.

5.3.2 Neurophysiology of chemesthesis

Neural responses to chemesthetic stimuli are examined via three main approaches: (1) recording from single trigeminal ganglion cells, dorsal root ganglion cells or heterologous cells expressing chemesthetic receptors; (2) extracellular recording from trigeminal nerve fibers; and (3) extracellular recording and brain imaging from central trigeminal areas. Here, we focus mainly, but not exclusively, on stimuli relevant to food.

5.3.2.1 Single-cell recordings

Most of the recent work on chemesthetic responses to chemicals comes from patch clamping or calcium imaging dissociated cells. From these reports we know that TRP channels in trigeminal neurons (see Chapters 4 and 14) respond to a wide variety of pungent compounds found in food. These include: eugenol (cloves: TRPV1, TRPV3, and TRPA1) (Yang *et al.*, 2003; Bandell *et al.*, 2004; Xu *et al.*, 2006; Chung *et al.*, 2014), carvacrol (oregano: TRPA1 and TRPV3) (Xu *et al.*, 2006), gingerol (ginger: TRPV1 and TRPA1) (Bandell *et al.*, 2004), allyl isothiocyanate (mustard and horseradish: TRPA1) (Jordt *et al.*, 2004; Bandell *et al.*, 2004), cinnamaldehyde (cinnamon: TRPA1) (Bandell *et al.*, 2004), menthol (mint: TRPM8, TRPA1: inhibited at high concentrations) (McKemy *et al.*, 2002; Karashima *et al.*, 2007), capsaicin (chili pepper: TRPV1) (Caterina *et al.*, 1997), and piperine (black pepper: TRPV1 and TRPA1) (Okumura *et al.*, 2010). TRPV3 found on kertinocytes in the oral cavity responds to eugenol and carvacrol (Xu *et al.*, 2006). TRPV3 contributes to the warmth attributed to its agonists, although the mechanism is not well understood. Recent experiments suggest that ATP released from keratinocytes may activate nerve fibers commonly stimulated by chemesthetic compounds.

Interestingly, trigeminal ganglion cells expressing TRPV1 are also activated by the artificial sweeteners, saccharin, aspartame, sodium cyclamate, and acesulfame potassium, which may account, in part, for the unpleasant aftertaste of these compounds (Riera *et al.*, 2007; Riera *et al.*, 2008). Another channel family (the pH- and anesthetic-sensitive two-pore potassium channel) is responsible for the unusual tingling and buzzing sensations of the pungent compound hydroxyl-α-sanshool, found in Szechuan peppercorns (Bautista *et al.*, 2008) (see also Chapter 8).

Carbon dioxide, which gives carbonated beverages their pungent, stinging quality, also activates trigeminal nerve fibers (Komai and Bryant, 1993; Erickson and Silver, 1994). The CO_2 activates trigeminal nerve endings by diffusing across cell membranes. Once inside the cell it reacts with water in the presence of carbonic anhydrase to dissociate into bicarbonate and a proton, acidifying the intracellular environment and stimulating TRPA1 channels leading to a response (Wang *et al.*, 2010). Similarly, weak acids, such as acetic acid found in vinegar and propionic acid found in butter and cheese, pass though cell membranes acidifying the inside of the cell, activating TRPA1 channels in trigeminal nerve fibers (Wang *et al.*, 2011).

5.3.2.2 Nerve recording

The earliest chemesthetic trigeminal nerve recordings were multiunit, integrated responses, and primarily examined thresholds to a variety of stimuli (e.g., Ito, 1962; Tucker, 1971; Silver and Moulton, 1982). Rat ethmoid nerve thresholds ranged from 3 ppm for octanol to 3020 ppm for methanol. Lipid solubility plays an important role in chemesthesis with more lipid soluble compounds eliciting lower thresholds (Silver *et al.*, 1986). The same relationship appears to be true for oral chemesthesis as well (Bryant and Silver, 2000). Some compounds that are rated by humans as irritating (e.g., methanol, toluene, and butyl acetate) exhibited high thresholds, while others (e.g., propionic acid and menthol) had relatively low thresholds (Silver, 1992).

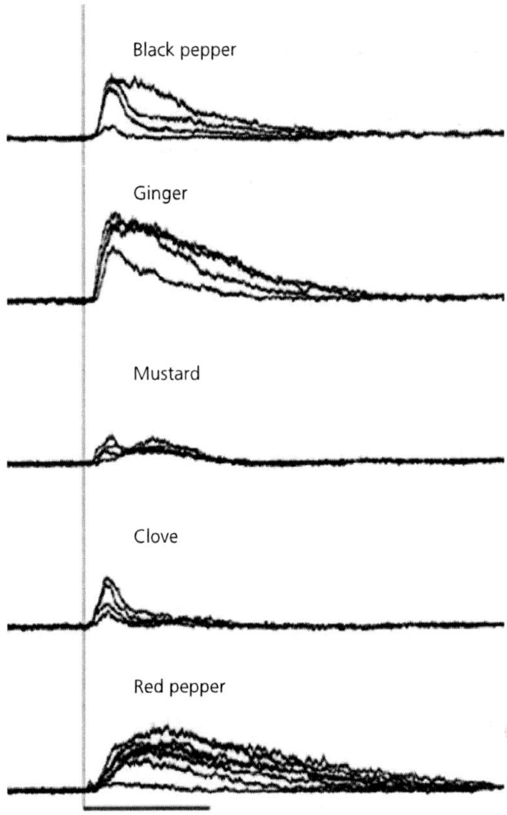

Fig. 5.3 Whole nerve integrated responses from the rat ethmoid nerve to spice solutions in rat Ringer's (pH 7.2) delivered to the nasal cavity. Responses to the five spice solutions at varying dilutions were recorded from one rat in the order shown (black to red pepper). For ease of presentation, response traces to the different dilutions (10, 40, 70, and 100% stock, plus 1, 4, and 7% stock for red pepper) are overlaid. The vertical grey line represents the time of bolus injection of stimulus. The horizontal bar at the bottom of the figure represents 30 seconds.

Oral trigeminal nerve fibers are activated by pungent spices (Okuni, 1978) although little work in this area is available. In other studies (H. Alimohammadi and W.L. Silver, unpublished results), nasal trigeminal nerve fibers were seen to respond to a variety of spices (Fig. 5.3). Each spice tested displayed a unique integrated response pattern, with marked variations in both the onset and offset of nerve response. Chilis were found to be the most stimulatory of the five spices tested, followed by ginger, black pepper, clove, and mustard.

5.3.2.3 Chemesthesis in the central nervous system

The sensations produced by spices that enhance flavor are commonly misattributed as "taste" or "smell". While it might be easy to accredit this misconception to ignorance of chemesthesis, this misstatement actually reveals the extent to which chemesthesis is integrated with the other chemical senses (see Chapter 9). As would be expected for somatosensory nerves, trigeminal neurons project into brain regions associated with touch, temperature, and pain; however, some of these neurons also convey chemesthetic signals to brain areas associated with olfaction and gustation.

Most of the sensory branches of the trigeminal nerve project to the trigeminal sensory nuclei, which are distributed from the mesencephalon to the medulla. Different somatosensory modalities are represented throughout each of these nuclei with many of the fibers responsible for detecting chemesthetic sensations converging to the subnucleus caudalis of the spinal trigeminal nucleus.

The neurons of the trigeminal subnucleus caudalis (Vc) receive converging input from fibers containing TRPV1, TRPA1, TRPM8, and TRPV3 receptors, which respond to pungent chemicals. Carstens' laboratory has conducted extensive extracellular recordings to determine the responses of cells in the Vc to chemesthetic stimuli delivered to the oral cavity (Carstens *et al.*, 1998; Carstens *et al.*, 2002; Simons *et al.*, 2004; Zanotto *et al.*, 2007; Zanotto *et al.*, 2008; Klein *et al.*, 2014). They demonstrated a high percentage of the neurons in the superficial lamina of the dorsomedial Vc are broadly tuned, responding to menthol, cinnamaldehyde, allyl isothiocyanate, capsaicin, carvacrol, eugenol, noxious heat, and cooling. In addition, the warmth evoking chemicals carvacrol and eugenol have a synergistic effect on heat-evoked electrical responses of Vc cells.

In contrast to extracellular recordings, fMRI and EEGs have been used to determine entire brain regions that are activated by chemesthetic stimulation. Many of these studies compared olfactory and trigeminal stimulation (usually CO_2) but a few have used pungent chemicals found in food as stimuli (Kobal and Hummel, 1988; Boyle *et al.*, 2007; Bensafi *et al.*, 2008; Hummel *et al.*, 2009; Huart *et al.*, 2011; Billot *et al.*, 2011). Trigeminal stimulation typically activates the brain stem, thalamus, caudate nucleus, anterior and dorsolateral orbitofrontal cortex, medial frontal gyrus, frontal operculum, superior temporal gyrus, cingulate, and the postcentral gyrus. Both olfactory and trigeminal stimulation result in common activation of several brain regions, including the medial orbitofrontal cortex, amygdala, parahippocampal gyrus, and cerebellum. This convergence in the neural pathways between the olfactory and trigeminal systems is suggestive of the degree to which the two systems are integrated. In addition to activating these brain regions, trigeminal ganglion neurons may affect cells in the olfactory bulbs. Trigeminal ganglia neurons send axons directly into the olfactory bulb where stimulation by chemesthetic compounds could modulate olfactory processing at its earliest central relays (Schaefer *et al.*, 2002).

Similar to the relationship with the olfactory system, the anatomy of the trigeminal system also allows for modulation of the gustatory system. Many trigeminal free nerve endings from the oral cavity project into the nucleus of the solitary tract (*nucleus tractus solitarii*, NTS), the initial relay for gustatory information in the brain (Braud *et al.*, 2012; Corson *et al.*, 2012; Felizardo *et al.*, 2009). Transection of mandibular trigeminal fibers results in reduced tastant evoked activity by NTS neurons; conversely, electrical stimulation of these trigeminal fibers can induce activity in these same neurons (Boucher *et al.*, 2003). Additionally, thermal trigeminal stimulation modulates the activity of NTS neurons in response to a variety of tastants (Wilson and Lemon, 2013).

While these results suggest trigeminal stimulation may have potent effects on gustatory processing, few studies have examined the role of chemesthetic stimuli directly. One such study determined that nicotine, a potent trigeminal irritant with a bitter taste, modulates NTS neurons by acting through both gustatory and trigeminal nerves (Simons *et al.*, 2006). Another similar study examined the ability of ethanol to evoke responses in NTS neurons and observed that it produced activity in cells tuned for sweet stimuli as well as an additional neuron type that was not tuned to any taste modality (Lemon *et al.*, 2011). Ethanol is a commonly ingested by humans and produces sensations of bitterness, sweetness, and burning pain. That many commonly ingested chemicals are capable of activating the gustatory and olfactory systems in addition to chemesthetic nerve fibers makes it difficult to separate chemesthesis from these other two senses.

5.4 Summary

Chemesthetic nerve fibers in the nasal and oral cavities respond to a variety of pungent compounds found in food. Stimulation of these nerve fibers contributes significantly to flavor. During eating, nasal trigeminal nerve fibers are activated by volatile chemicals in the oral cavity that move into the nasal cavity following exhalation or swallowing. Trigeminal nerve endings can either be stimulated directly via free nerve endings or indirectly through solitary chemoreceptor cells (Fig. 5.1). Activating trigeminal nerve fibers triggers the reflexes often associated with eating spicy food, including flushing, tearing, sneezing, sweating, and coughing. The molecular mechanisms involved in chemesthesis involve, primarily, TRP channels, which, in addition to detecting changes in temperature, also interact with many of the compounds found in piquant food.

References

Bandell, M., Story, G.M., Hwang, S.W., *et al.* 2004. Noxious cold ion channel TRPA1 is activated by pungent compounds and bradykinin. *Neuron*, **41**, 849–857.

Basbaum, C. and Jany, B. 1990. Plasticity in the airway epithelium. *Am. J. Physiol.*, **259** (2 Pt 1), L38–L46.

Barham, H.P., Cooper, S.E., Anderson, C.B., *et al.* 2013. Solitary chemosensory cells and bitter taste receptor signaling in human sinonasal mucosa. *Int. Forum Allergy Rhinol.*, **3** (6), 450–457.

Bautista, D.M., Sigal, Y.M., Milstein, A.D., *et al.* 2008. Pungent agents from Szechuan peppers excite sensory neurons by inhibiting two-pore potassium channels. *Nat. Neurosci.*, **11**, 772–779.

Bensafi, M., Iannilli, E., Gerber, J., and Hummel, T. 2008. Neural coding of stimulus concentration in the human olfactory and intranasal trigeminal systems. *Neuroscience*, **154**, 832–838.

Billot, P.E., Comte, A., Galliot, E. *et al.* 2011. Time course of odorant- and trigeminal-induced activation in the human brain: An event-related functional magnetic resonance imaging study. *Neuroscience*, **189**, 370–376.

Boyle, J. A., Heinke, M., Gerber, J., Frasnelli, J., and Hummel, T. 2007. Cerebral activation to intranasal chemosensory trigeminal stimulation. *Chem. Senses*, **32**, 343–353.

Boucher, Y., Simons, C.T., Faurion, A., Azerad, J., and Carstens, E. 2003. Trigeminal modulation of gustatory neurons in the nucleus of the solitary tract. *Brain Res.*, **973**, 265–274.

Brann, J. H. and Firestein, S.J. 2014. A lifetime of neurogenesis in the olfactory system. *Front. Neurosci.*, **8**, 182.

Braud, A., Vandenbeuch, A., Zerari-Mailly, F., and Boucher, Y. 2012. dental afferents project onto gustatory neurons in the nucleus of the solitary tract. *J. Dent. Res.*, **91**, 215–220.

Bryant, B. and Silver, W.L. 2000. Chemesthesis: The common chemical sense, in *Neurobiology of Taste and Smell*, 2nd edn (eds. T.E. Finger, W.L. Silver, and D. Restrepo), New York: Wiley-Liss, Inc., pp. 73–100.
Carstens, E., Kuenzler, N., and Handwerker, H.O. 1998. Activation of neurons in rat rigeminal subnucleus caudalis by different irritant chemicals applied to oral or ocular mucosa. *J. Neurophysiol.*, **80**, 465–492.
Carstens, E., Carstens, M.L., Dessirier, J.-M., *et al.* 2002. It hurts so good: oral irritation by spices and carbonated drinks and the underlying neural mechanisms. *Food Qual. Prefer.*, **13**, 431–443.
Caterina, M.J., Schumacher, M.A., Tominaga, M. *et al.* 1997. The capsaicin receptor: A heat-activated ion channel in the pain pathway. *Nature*, **389**, 816–824.
Chung, G., Im, S.T., Kim, Y.H. *et al.* 2014. Activation of transient receptor potential ankyrin 1 by eugenol. *Neuroscience*, **261**, 153–160.
Cole, B.B., Smith, R.W., Jenkins, K.M. *et al.* 2010. Tracheal Basal cells: a facultative progenitor cell pool. *Am. J. Pathol.*, **177**, 362–376.
Cole, P. 2003. The four components of the nasal valve. *Am. J. Rhinol.*, **17**, 107–110.
Corson, J., Aldridge, A., Wilmoth, K., and Erisir, A. 2012. A survey of oral cavity afferents to the rat nucleus tractus solitarii. *J. Comp. Neurol.*, **520**, 495–527.
Djupesland, P.G. 2013. Nasal drug delivery devices: characteristics and performance in a clinical perspective - a review. *Drug. Deliv. Transl. Res.*, **3**, 42–62.
Dupont, L.L., Glynos, C., Bracke, K.R., Brouckaert, P., and Brusselle, G.G. 2014. Role of the nitric oxide-soluble guanylyl cyclase pathway in obstructive airway diseases. *Pulm. Pharmacol. Ther.*, **29**, 1–6.
Eilers, H., Cattaruzza, F., Nassini R. *et al.* 2010. Pungent general anesthetics activate transient receptor potential-A1 to produce hyperalgesia and neurogenic bronchoconstriction. *Anesthesiology*, **112**, 1452–1463.
Erickson, J.L. and Silver, W.L. 1994. Acetazolamide inhibits ethmoid nerve responses to carbon dioxide but not other irritants. *Chem. Senses*, **19**, 466–467.
Farbman, A.I. and Hellekant, G. 1978. Quantitative analysis of the fiber population in rat chorda tympani nerves and fungiform papillae. *Am. J. Anat.*, **153**, 509–522.
Felizardo, R., Boucher, Y., Braud, A., *et al.* 2009. Trigeminal projections on gustatory neurons of the nucleus of the solitary tract: A double-label strategy using electrical stimulation of the chorda tympani and tracer injection in the lingual nerve. *Brain Res.*, **1288**, 60–68.
Finger, T.E., St Jeor, V.L., Kinnamon, J.C., and Silver, W.L. 1990. Ultrastructure of substance P- and CGRP-immunoreactive nerve fibers in the nasal epithelium of rodents. *J. Comp. Neurol.*, **294**, 293–305.
Finger, T.E. 1997. Evolution of taste and solitary chemoreceptor cell systems. *Brain Behav. Evol.*, **50**, 234–243.
Finger, T.E., Bottger, B., Hansen, A. *et al.* 2003. Solitary chemoreceptor cells in the nasal cavity serve as sentinels of respiration. *Proc. Natl. Acad. Sci. USA.*, **100**, 8981–8986.
Finger, T.E. and Simon, S.A. 2000. *Cell biology of taste epithelium, in Neurobiology of Taste and Smell*, 2nd edn (eds. T.E. Finger, W.L. Silver, and D. Restrepo), New York: Wiley-Liss, Inc., pp. 287–314.
Folkesson, A., Jelsbak, L., Yang, L. *et al.* 2012. Adaptation of *Pseudomonas aeruginosa* to the cystic fibrosis airway: an evolutionary perspective. *Nat. Rev. Microbiol.*, **10**, 841–851.
Fuqua, C. and Greenberg, E.P. 2002. Listening in on bacteria: Acyl-homoserine lactone signalling. *Nat. Rev. Mol. Cell Biol.*, **3**, 685–695.
Fuqua, W.C., Winans, S.C., and Greenberg, E.P. 1994. Quorum sensing in bacteria–the Luxr-Luxi family of cell density-responsive transcriptional regulators. *J. Bacteriol.*, **176**, 269–275.
Geppetti, P., Fusco, B.M., Marabini, S. *et al.* 1988. Secretion, pain and sneezing induced by the application of capsaicin to the nasal mucosa in man. *Br. J. Pharmacol.*, **93**, 509–514.
Geppetti, P., Patacchini, R., Nassini, R., and Materazzi, S. 2010. Cough: The emerging role of the TRPA1 channel. *Lung*, **188**, S63–S68.
Gerhold, K.A. and Bautista, D.M. 2009. Molecular and cellular mechanisms of trigeminal chemosensation. *Ann. NY Acad. Sci.*, **1170**, 184–189.
Green, B.G. 2012. Chemesthesis and the chemical senses as components of a "chemofensor complex". *Chem. Senses*, **37**, 201–206.
Gulbransen, B.D., Clapp, T.R., Finger, T.E., and Kinnamon, S.C. 2008a. Nasal solitary chemoreceptor cell responses to bitter and trigeminal stimulants in vitro. *J. Neurophysiol.*, **99**, 2929–2937.

Gulbransen, B.D., Silver, W.L., and Finger, T.E. 2008b. Solitary chemoreceptor cell survival is independent of intact trigeminal innervation. *J. Comp. Neurol.*, **508**, 62–71.

Harkema, J.R., Carey, S.A., and Wagner, J.G. 2006. The nose revisited: a brief review of the comparative structure, function, and toxicologic pathology of the nasal epithelium. *Toxicol. Pathol.*, **34**, 252–269.

Huart, C., Meusel,T., Gerber, J. *et al.* 2011. The depth of the olfactory sulcus is an indicator of congenital anosmia. *Am. J. Neuroradiol.*, **32**, 1911–1914.

Hummel, T., Iannilli, E., Frasnelli, J., Boyle, J., and Gerber, J. 2009. Central processing of trigeminal activation in humans. *Ann. N. Y. Acad. Sci.*, **1170**, 190–195.

Ito, K. 1962. Electrophysiological studies on the peripheral olfactory nervous system of mammalia. *Kitahanto J. Med. Sci.*, **18**, 405–417.

James, J.E.A. and DaleyM. de B. 1969. Nasal reflexes. *Proc. Roy. Soc. Med.*, **62**, 1287–1293.

Jordt, S.E., Bautista, D.M., Chuang, H.H. *et al.* 2004. Mustard oils and cannabinoids excite sensory nerve fibres through the TRP channel ANKTM1. *Nature*, **427**, 260–265.

Julius, D. 2013. TRP channels and pain. *Ann. Rev. Cell. Dev. Biol.*, **29**, 355–384.

Karashima, Y., Damann, N., Prenen, J. *et al.* 2007. Bimodal action of menthol on the transient receptor potential channel TRPA1. *J. Neurosci.*, **27**, 9874–9884.

Kashima, H. and Hayashi, N. 2013. Facial skin blood flow responses to irritant stimuli in the oral cavity. *Auton. Neurosci.*, **174**, 61–65.

Kinnamon, S.C. 2012. Taste receptor signalling-from tongues to lungs. *Acta. Physiol.*, **204**, 158–168.

Kinnamon, S.C. and Reynolds, S.D. 2009. Using taste to clear the air(ways). *Science*, **325**, 1081–1082.

Klein, A.H., Joe, C.L., Davoodi, A. *et al.* 2014. Eugenol and carvacrol excite first- and second-order trigeminal neurons and enhance their heat-evoked responses. *Neuroscience*, **271**, 45–55.

Kobal, G. and Hummel, C. 1988. Cerebral chemosensory evoked potentials elicited by chemical stimulation of the human olfactory and respiratory nasal mucosa. *Electroencephalogr. Clin. Neurophysiol.*, **71**, 241–250.

Komai, M. and Bryant, B.P. 1993. Acetazolamide specifically inhibits lingual trigeminal nerve responses to carbon dioxide. *Brain Res.*, **612**, 122–129.

Krasteva, G., Canning, B.J., Hartmann, P. *et al.* 2011. Cholinergic chemosensory cells in the trachea regulate breathing. *Proc. Natl. Acad. Sci. USA*, **108**, 9478–9483.

Krasteva, G., Canning, B.J., Papadakis, T., and Kummer, W. 2012. Cholinergic brush cells in the trachea mediate respiratory responses to quorum sensing molecules. *Life Sci.*, **91**, 992–996.

Kunkler, P. E., Ballard, C. J., Oxford, G. S., and Hurley, J. H. 2011. TRPA1 receptors mediate environmental irritant-induced meningeal vasodilatation *Pain*, **152**, 38–44.

Langley, J.N. 1900. On axon-reflexes in the pre-ganglionic fibres of the sympathetic system. *J. Physiol.*, **25**, 364–398.

Lawless, H. 1984. Oral chemical irritation psychophysical properties. *Chem. Senses*, **9**, 143–155.

Lee, R.J., Chen, B., Redding, K.M., Margolskee, R.F., and Cohen, N.A. 2013. Mouse nasal epithelial innate immune responses to Pseudomonas aeruginosa quorum-sensing molecules require taste signaling components. *Innate Immun.*, **20**, 606–617.

Lee, R.J. and Cohen, N.A. 2014a. The emerging role of the bitter taste receptor T2R38 in upper respiratory infection and chronic rhinosinusitis. *Am. J. Rhinol. Allergy*, **27**, 283–286.

Lee, R.J., Kofonow, J.M., Rosen, P.L. *et al.* 2014b. Bitter and sweet taste receptors regulate human upper respiratory innate immunity. *J. Clin. Invest.*, **124**, 1393–1405.

Lee, R.J., Xiong, G., Kofonow, J.M. *et al.* 2012. T2R38 taste receptor polymorphisms underlie susceptibility to upper respiratory infection. *J. Clin. Invest.*, **122**, 4145–4159.

Lee, T.S. 1954. Physiological gustatory sweating in a warm climate. *J. Physiol.*, **124**, 528–542.

Lemon, C.H., Wilson, D.M., and Brasser, S.M. 2011. Differential neural representation of oral ethanol by central taste-sensitive neurons in ethanol-preferring and genetically heterogeneous rats. *J. Neurophysiol.*, **106**, 3145–3156.

Lin, W., Ogura, T., Margolskee, R.F., Finger, T.E., and Restrepo, D. 2008. TRPM5-expressing solitary chemosensory cells respond to odorous irritants. *J. Neurophysiol.*, **9**, 1451–1460.

Ma, L., Lee, B.H., Mao, R. *et al.* 2014. Nicotinic acid activates the capsaicin receptor TRPV1: potential mechanism for cutaneous flushing. *Arterioscler. Thromb. Vasc. Biol.*, **34**, 1272–1280.

McKemy, D.D., Neuhausser, W.M., and Julius, D. 2002. Identification of a cold receptor reveals a general role for TRP channels in thermosensation. *Nature*, **416**, 52–58.

Merigo, F., Benati, D., Tizzano, M., Osculati, F. and Sbarbati, A. 2005. alpha-Gustducin immunoreactivity in the airways. *Cell Tissue Res.*, **319**, 211–219.

Messerklinger, W. 1978. *Endoscopy of the Nose*, Baltimore, MD: Urban and Schwarzenberg.

Morris, J.B., Wilki, W.S., and Shusterman, D.J. 2005. Acute respiratory responses of the mouse to chlorine. *Toxicol. Sci.*, **83**, 380–387.

Nalivaiko, E., De Pasquale, C.G., and Blessing, W.W. 2003. Electrocardiographic changes associated with the nasopharyngeal reflex in conscious rabbits: vago-sympathetic co-activation. *Auton. Neurosci.*, **105**, 101–104.

O'Connor, T.M., O'Connell, J., O'Brien, D.I. *et al.* 2004. The role of substance P in inflammatory disease. *J. Cell Physiol.*, **201**, 167–180.

Okumura, Y., Narukawa, M., Iwasaki, Y. *et al.* 2010. Activation of TRPV1 and TRPA1 by black pepper components. *Biosci. Biotech. Biochem.*, **74**, 1068–1072.

Okuni, Y. 1978. Response of lingual nerve fibers of the rat to pungent spices and irritants pungent spices. *Shika Gakuho.*, **78**, 1323–1349.

Parker, G.H. 1912. The relations of smell, taste and the common chemical sense in vertebrates, *J. Acad. Sci. Phila.*, **15**, 221

Parker, G.H. 1922. *Smell, Taste, and Allied Senses in Vertebrates*, Philadelphia: Lippincott.

Philip, G., Baroody, F.M., Proud, D., Naclerio, R.M., and Togias, A.G. 1994. The human nasal response to capsaicin. *J. Allergy Clin. Immunol.*, **94**, 1035–1045.

Rather, L.J. 1971. Disturbance of function (functio laesa): the legendary fifth cardinal sign of inflammation, added by Galen to the four cardinal signs of Celsus. *Bull. NY Acad. Med.*, **47**, 303–322.

Reynolds, S.D. and Malkinson, A.M. 2010. Clara cell: Progenitor for the bronchiolar epithelium. *Int. J. Biochem. Cell Biol.*, **42**, 1–4.

Richards, PM., Saunders, C.J., and Silver, W.L. 2010. Functional neuroanatomy of the upper airway in experimental animals, in *Toxicology of the Nose and Upper Airways* (eds. J.B. Morris and D.J. Shusterman), Informa Healthcare Books, **712** pp. 45–64.

Riera, C.E., Vogel, H., Simon, S.A., Damak, S., and le Coutre, J. 2008. The capsaicin receptor participates in artificial sweetener aversion. *Biochem. Biophys. Res. Commun.*, **376**, 653–657.

Riera, C.E., Vogel, H., Simon, S.A., and Coutre, J. 2007. Artificial sweeteners and salts producing a metallic taste sensation activate TRPV1 receptors. *Am. J. Physiol. Regul. Integr. Comp. Physiol.*, **293**, R626–R634.

Rodger, I.W., Tousignant, C., Young, D., Savoie, C., and Chan, C.C. 1995. Neurokinin receptors subserving plasma extravasation in guinea pig airways. *Can. J. Physiol. Pharmacol.*, **73**, 927–931.

Sadofsky, L.R., Boa, A.N., Maher, S.A. *et al.* 2011. TRPA1 is activated by direct addition of cysteine residues to the N-hydroxysuccinyl esters of acrylic and cinnamic acids. *Pharmacol. Res.*, **63**, 30–36.

Saunders, C.J., Li, W.Y., Patel, T.D., Muday, J.A., and Silver, W.L. 2013a. Dissecting the role of TRPV1 in detecting multiple trigeminal irritants in three behavioral assays for sensory irritation. [v1; ref status: indexed, http://f1000r.es/p8] F1000Research 2013, **2**:74, DOI: 10.12688/f1000research.2-74.v1 (accessed 22 September 2015).

Saunders, C.J., Reynolds, S.D., and Finger, T.E. 2013b. Chemosensory brush cells of the trachea: A stable population in a dynamic epithelium. *Am. J. Respir. Cell. Mol. Biol.*, **49**, 190–196.

Saunders, C.J., Christensen, M., Finger, T.E., and Tizzano, M. 2014. Cholinergic neurotransmission links solitary chemosensory cells to nasal inflammation. *Proc. Natl. Acad. Sci. USA*, **111**, 6075–6080.

Sbarbati, A. and Osculati, F. 2003. Solitary chemosensory cells in mammals? *Cells Tiss. Organs*, **175**, 51–55.

Schaefer, M.L., Bottger, B., Silver, W.L., and Finger, T.E. 2002. Trigeminal collaterals in the nasal epithelium and olfactory bulb: A potential route for direct modulation of olfactory information by trigeminal stimuli. *J. Comp. Neurol.*, **444**, 221–226.

Schild, D. and Restrepo, D. 1998. Transduction mechanisms in vertebrate olfactory receptor cells. *Physiol. Rev.*, **78**, 429–466.

Seleker, H.M. and Budak, F. 2004. Idiopathic stabbing headache and experimental ice cream headache (short-lived headaches) *Europ. Neurol.*, **51**, 6–9.

Shah, A.S., Ben-Shahar, Y., Moninger, T.O., Kline, J.N., and Welsh, M.J. 2009. Motile cilia of human airway epithelia are chemosensory. *Science*, **325**, 1131–1134.

Silver, W.L. 1992. Neural and pharmacological basis for nasal irritation, in *Sources of Indoor Air Contaminants: Characterizing Emissions and Health Impacts* (eds. W.G. Tucker, B.P. Leaderer, L., Mølhave, and W.S. Cain), New York: Ann. N.Y. Acad. Science, pp. 152–163.

Silver, W.L. and Moulton, D.G. 1982. Chemosensitivity of rat nasal trigeminal receptors. *Physiol. Behav.*, **28**, 927–931.

Silver, W.L., Mason, J.R., Adams, M.A., and Smeraski, C. 1986. Nasal trigeminal chemoreception: Responses to *n*-aliphatic alcohols. *Brain Res.*, **376**, 221–229.

Silver, W.L. and Finger, T.E. 2009. The anatomical and electrophysiological basis of peripheral nasal trigeminal chemoreception. *Ann. NY. Acad. Sci.*, **1170**, 202–205.

Simons, C.T., Sudo, S., Sudo, M., and Carstens, E. 2004. Mustard oil has differential effects on the response of trigeminal caudalis neurons to heat and acidity. *Pain*, **110**, 64–71.

Simons, C.T., Boucher, Y., Carstens, M.I., and Carstens, E. 2006. Nicotine suppression of gustatory responses of neurons in the nucleus of the solitary tract. *J. Neurophysiol.*, **96**, 1877–1886.

Suemune, S., Nishimori, T., Hosoi, M. *et al.* 1992. Trigeminal nerve endings of lingual mucosa and musculature of the rat. *Brain Res.*, **586**, 161–165.

Takahashi, H., Izumi, H., and Karita, K. 1995. Parasympathetic reflex salivary secretion in the cat parotid gland. *Jpn. J. Physiol.*, **45**, 475–490.

Tizzano, M., Gulbransen, B.D., Vandenbeuch, A. *et al.* 2010. Nasal chemosensory cells use bitter taste signaling to detect irritants and bacterial signals. *Proc. Natl. Acad. Sci. USA*, **107**, 3210–3215.

Tizzano, M. and Finger, T.E. 2013. Chemosensors in the nose: guardians of the airways. *Physiology (Bethesda)*, **28**, 51–60.

Tominaga, M., Caterina, M.J., Malmberg, A.B. *et al.* 1998. The cloned capsaicin receptor integrates multiple pain-producing stimuli. *Neuron*, **21**, 531–543.

Tucker, D. 1971. Nonolfactory responses from the nasal cavity: Jacobson's organ and the trigeminal system, in *Handbook of Sensory Physiology*, vol. IV, part I, (ed. L.M. Beidler), Berlin: Springer-Verlag, pp. 151–181.

Vijayaraghavan, R., Schaper, M., Thompson, R., Stock, M.F., and Alarie, Y. 1993. Characteristic modifications of the breathing pattern of mice to evaluate the effects of airborne chemicals on the respiratory tract. *Arch. Toxicol.*, **67**, 478–490.

Wang, Y.Y., Chang, R.B., and Liman, E.R. 2010. TRPA1 is a component of the nociceptive response to CO_2. *J. Neurosci.*, **30**, 12958–12963.

Wang, Y.Y., Chang, R.B., Allgood, S.D., Silver, W.L., and Liman, E.R. 2011. A TRPA1-dependent mechanism for the pungent sensation of weak acids. *J. Gen. Physiol.*, **137**, 493–505.

Warren, D.W., Walker, J.C., Drake, A.F., and Lutz, R.W. 1994. Effects of odorants and irritants on respiratory behavior. *Laryngoscope*, **104**, 623–626.

Welch, K.C., Foster, G.D., Ritter, C.T. *et al.* 2002. A novel volumetric magnetic resonance imaging paradigm to study upper airway anatomy. *Sleep*, **25** (5), 532–542.

Wilson, D.M. and Lemon, C.H. 2013. Modulation of central gustatory coding by temperature. *J. Neurophysiol.*, **110**, 1117–1129.

Xu, H., Delling, M., Jun, J.C., and Clapham, D. 2006. Oregano, thyme and clove-derived flavors and skin sensitizers activate specific TRP channels. *Nat. Neurosci.*, **9**, 628–635.

Yang, B.H., Piao, Z.G., Kim, Y.B. *et al.* 2003. Activation of vanilloid receptor 1 (VR1) by eugenol. *J. Dent. Res.*, **82**, 781–785.

Youngs, R., Evans, K., and Watson, M. 2006. *The Paranasal Sinuses: A Handbook of Applied Surgical Anatomy*, London and New York: Taylor and Francis Group.

Zanotto, K.L., Iodi Carstens, M., and Carstens, E. 2008. Cross-desensitization of responses of rat trigeminal subnucleus caudalis neurons to cinnamaldehyde and menthol. *Neurosci. Lett.*, **430**, 29–33.

Zanotto, K.L., Merrill, A.W., Carstens, M.I., and Carstens, E. 2007. Neurons in superficial trigeminal subnucleus caudalis responsive to oral cooling, menthol and other irritant stimuli. *J. Neurophysiol.*, **97**, 966–978.

Zhou, W. and Chen, D. 2009. Binaral rivalry between the nostrils and in the cortex. *Curr. Biol.*, **19**, 1561–1565.

Zhou, W., Jiang, Y., He, S., and Chen, D. 2010. Olfaction modulates visual perception in binocular rivalry. *Curr. Biol.*, **20**, 1356–1358.

CHAPTER 6

Types of chemesthesis I. Pungency and burn: historical perspectives, word usage, and temporal characteristics

John E. Hayes
Sensory Evaluation Center and Department of Food Science, College of Agricultural Sciences, The Pennsylvania State University, University Park, PA, USA

6.1 Introduction

6.1.1 Müller, Myers, and the doctrine of specific nerve energies

In 1840, Johannes Müller published his *Handbuch der Physiologie des Menchschen für Vorlesungen,* writing: "We have no sensation of external things that we do not perceive through the state of our nerves. Inner causes of sensation cause different sensations in different nerves according to the nature of the sense, namely the sensations those nerves mediate. The same outer stimulus excites different sensations in the different senses according to the nature of the sense, namely the sensations that these senses mediate. The unique sensations of each sensory nerve can be stimulated by a variety of inner and outer stimuli. [Thus] sensation is not the transduction of a quality or an object to the consciousness, but rather the communication of a quality or an object by a sensory nerve to the consciousness is caused by an external event, and the qualities, or 'sensory energies' are different in different sensory nerves".

This concept is sometimes referred to as the doctrine of specific nerve energies.

Later, the English physician and psychologist Charles Samuel Myers made a distinction between an "adequate stimulus" – that is, the specific stimulus a sense organ is adapted to respond to – and an "inadequate stimulus", which is any other stimulus that activates the sense organ (Myers, 1909). This can be easily demonstrated by pressing (gently!) on the side of the eye, which increases intraocular pressure and causes the optic nerve to fire. The resulting sensation is perceived as a flash of light, despite the fact that the photoreceptor was activated by a mechanical stimulus, not a photon. That is, returning to Müller, the specific quality of a sensation depends on the nature of the nerve which is stimulated, not the manner of stimulation.

These concepts, while antiquated, are still a useful place to begin when discussing oral pungency and "hot" sensations from chemesthetic stimuli. As outlined elsewhere in this book, the TRPV1 receptor (*née* VR1) is a polymodal nociceptor that responds to heat pain from temperatures above ~42.5 °C (Caterina *et al.*, 2000; Caterina *et al.*, 1997; McIntyre *et al.*, 2001). However, when exposed to capsaicin and other chemicals, such as piperine, the temperature at which the TRPV1 receptor allows calcium to enter the cell is substantially reduced (Caterina *et al.*, 1997; McNamara *et al.*, 2005), resulting in depolarization and eventually in sensations that we describe as "burning". Or in terms Myers might use, we can think of capsaicin and piperine as inadequate stimuli that nonetheless evoke sensations of heat pain, even in the absence of noxious thermal stimulation. From an evolutionary perspective, hijacking a heat pain sensor with secondary metabolites appears remarkably clever, although we should be careful not to give plants the agency for doing this intentionally. Still, these chemicals clearly confer evolutionary advantage, as they allow plants to prevent predation (Glendinning, 2007), or even select for predators in a species specific manner to encourage seed dispersal (Jordt and Julius, 2002; Tewksbury and Nabhan, 2001). As plants make these chemicals to prevent consumption, at least by mammals, this raises the obvious question of why humans would choose to intentionally seek out and consume plants that cause painful sensations. Reasons for this are discussed in depth in Chapter 2, so I will confine my remarks here to briefly discussing the ways we talk about these sensations and the plants that evoke them.

6.1.2 Columbian Exchange and the quest for spices

The Columbian Exchange refers to the process by which the Old World obtained cocoa, potatoes, tomatoes, and the chili from the New World, while the New World acquired coffee, smallpox, measles, and syphilis. From a culinary perspective, it is almost impossible to understate the importance of this event, for it is hard to conceive of the Belgians without the potato or chocolate, the Italians without the tomato, or the Thais without the chili. Plus, it should be recalled that Columbus was not in search of the New World for discovery's own sake, or solely for precious metals such as gold or silver; rather, the search for pungent spices was a major motivation in the hunt for an alternate route to India. Sailing in the opposite direction, Vasco da Gama's men reportedly yelled *Christos e espiciarias!* ("For Christ and spices!") upon landing at Kappakadavu on the west coast of India in May of 1498.

Originally native to Central and South America, the chili (*Capsicum* spp.) has been used by humans for at least 7 millennia. In Peru, dating of the preceramic levels of the Guitarrero Cave suggests habaneros (*Capsicum chinense*) may have been domesticated since 10 000 BP (Smith Jr, 1980), although this date has been disputed. Less controversially, there is evidence of multiple domesticated species (C. *chinense and C. baccatum*) at Huaca Prieta, Peru since ~4500 BP via analyses of fruits (Pickersgill, 1969) and seeds (Chiou and Hastorf, 2012). Recent work suggests the intentional cultivation of specific types of chilis (*C. baccatum*) in coastal South America began as far back as 7600 BP (Chiou *et al.*, 2014). These estimates are largely consistent with paleobiolinguistic evidence: analysis of proto-languages from Central and South America indicates there were specific words for chilis at least 6500 BP for south central Mexico and 4400 BP for South America (Brown *et al.*, 2013). Dates from paleobiolinguistics represent the most recent date a language was spoken, so speakers may have had a term for chilis for hundreds or thousands of years prior to this (Kraft *et al.*, 2014).

As noted previously by Szallasi and Blumberg (1999), the other common name given to the chili – "red pepper" – mistakenly associates the chili (*Capsicum* spp.) with a botanically distinct plant that also evokes hot burning sensations: black pepper (*Piper nigrum*), which contains the TRPV1 agonist piperine. Likewise, hua jiao (*Zanthoxylum simulans*) and sanshō (*Zanthoxylum piperitum*) are commonly referred to as Szechuan peppers and Japanese peppers, respectively; this is unsurprising as they contain sanshools, which elicit burn in addition to buzzing and tingling (Albin and Simons, 2010; Sugai *et al.*, 2005).

Presumably, this linguistic convergence is no accident: when exposed to novel stimuli that evoke sensations for which we lack rich descriptive language, it is likely we grope about for familiar examples within our prior experience, similar to the way we name odors with concrete references (e.g., "smells like cut grass"). Accordingly, it is not surprising that botanically diverse Old World and New World plants have come to share the common descriptor "pepper", at least in English.

6.2 Language usage

The literature is replete with studies where participants are asked to rate perceived burn or burning and stinging from capsaicin. Notably, these terms are typically supplied by the experimenter, with or without definitions, and there are few published studies that ask participants to characterize or describe the sensations from chemesthetic stimuli. When we explicitly asked English-speaking participants to identify the predominant sensation from whole mouth capsaicin from a list of seven different descriptors (burning, stinging/pricking, itching, tingling, warm/hot, numbness, and tickle), burning was endorsed most often (56% of all trials), followed by warm/hot (19%), and stinging/pricking (13%) (Bennett and Hayes, 2012). These data are largely consistent with earlier work: Lawless and Stevens reported that burning and stinging were the most endorsed thermal and tactile sensations, respectively, for capsaicin on the tongue tip (Lawless and Stevens, 1988), while Green reported that capsaicin was primarily stinging and burning, with slightly more endorsements for stinging (Green, 1991a). Likewise, descriptive panel data using consensus descriptors generated by the participants suggests capsaicin, piperine, and ginger oleoresin are primarily burning, with minimal tingling or numbing; conversely, ethanol was equally burning and tingling, with substantial numbing as well (Cliff and Heymann, 1992). For zingerone, Prescott and Stevenson reported that burning and warmth were endorsed significantly more often than numbness; endorsements of stinging and tingling were lower than those of burning and warmth, and greater than numbing, but these differences were not significant (Prescott and Stevenson, 1996b). For carvacrol, numbing and warming predominate, while numbing, warming, and tingle are the most endorsed qualities for eugenol (Klein *et al.*, 2013). In general, there is a surprising lack of work on the qualities evoked by chemesthetic stimuli. It may be that personal experience provides many researchers sufficient face validity for burning/stinging as an exhaustive descriptor, at least for chili and capsaicin, such that significant additional introspection and consideration do not seem warranted.

Research on whether other languages have additional richness to describe chemesthetic sensations is also lacking and has only been minimally explored. After having several cohorts of naïve participants sort nine chemesthetic stimuli and two tastants (e.g., capsaicin, zingerone, menthol, citric acid, eugenol, etc.) into groups, we asked them to generate descriptors for each group. The descriptors were then regressed as vectors into the multidimensional space generated from the sorting data. Among native English language speakers, words such as hot, burning, pricking, stinging, and sharp were used for stimuli like allyl isothiocyanate, zingerone, and capsaicin (Byrnes *et al.*, 2015), as would be expected. When 30 bilingual speakers completed the same task in Spanish, words such as *ardiente* (burning temperature)

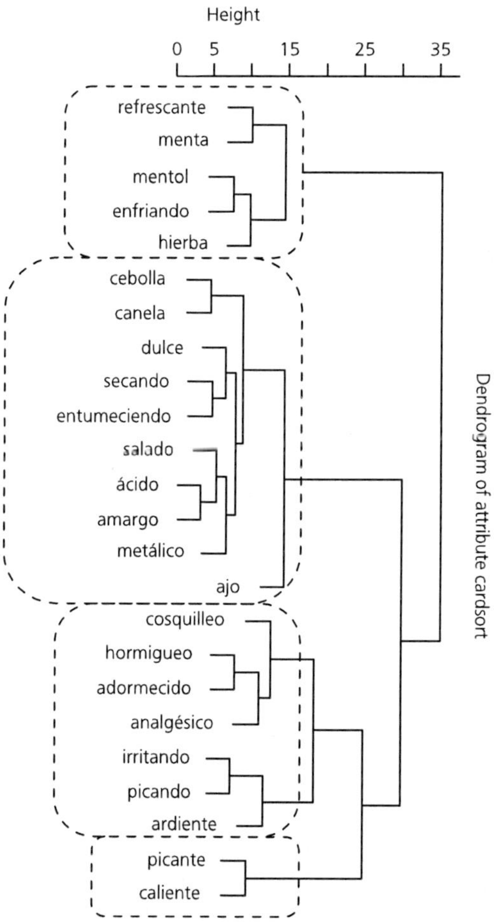

Fig. 6.1 Dendrogram of cluster analysis of 24 Spanish language descriptors sorted by 12 native Spanish speakers in an online card-sorting task. See text for English translations.

and *irritando* (irritating) were significant vectors in the multidimensional space; words like *picando* (stinging/biting), *picante* (spicy hot), and *caliente* (hot temperature) were also used, but surprisingly did not reach statistical significance (Boone, 2014).

To gain more insight into the nuanced meanings of all 24 descriptors generated by the Spanish speakers, approximately half of the same participants ($n = 12$) completed a follow-up study, grouping the verbal descriptors in a card-sorting task online. As shown in Fig. 6.1, four clusters were formed: a refreshing/cooling group, a group of specific foods and prototypical tastes, a group of touch sensations, and a hot/spicy group. The refreshing/cooling group included: *refrescante* (refreshing), *menta* (mint), *mentol* (menthol), *enfriando* (cooling), and *hierba* (herbal). The second group is a combination of many foods, flavors, and some touch sensations, including foods such as *cebolla* (onion), *canela* (cinnamon), and *ajo* (garlic), prototypical tastes such as *dulce* (sweet), *salado* (salty), *ácido* (sour), *amargo* (bitter), and *metálico* (metallic), and two touch sensations, *secando* (drying) and *entumeciendo* (numbing). The third group of touch sensations includes: *cosquilleo* (tickle/tingle), *hormigueo* (tickle/tingle), *adormecido* (numbing, literally asleep), *analgésico* (analgesic), *irritando* (irritating), *picando* (biting/stinging), and *ardiente* (burning). The final group is composed of only two attributes: *picante* (spicy, as a flavor) and *caliente* (hot, as an indication of temperature) (Boone, 2014). Although preliminary, these data suggest additional work is warranted to determine whether Spanish (or other languages) may contain more linguistic richness to describe the chemesthetic quality described in English as "hot" or "burning".

6.3 Differentiation from classical tastes

As reviewed in Chapter 1, it was long recognized that oral pungency/chemically induced heat was distinct from the "basic" tastes[1] almost a century before the canonical capsaicin receptor TRPV1 was cloned in 1997 (Caterina and Julius, 2001; Caterina *et al.*, 1997). Indeed, in 1909, Myers wrote "...early classifications included many tastes (e.g., dry, astringent, pungency...), which we now recognize to result from the excitation of the end organs not of taste but of touch, temperature, etc. which are contained in the tongue" (Myers, 1909). Accordingly, it is interesting to note that capsaicin falls well outside Henning's classical taste tetrahedron that has sweet, sour, salty, and bitter at the vertices (Lim and Green, 2007), although it must be noted that umami, now well accepted as a prototypical taste, also falls outside this space (Yamaguchi, 1987).

Oral pungency differs from prototypical taste percepts such as sweetness or bitterness in several critical ways, including the time course of the sensation (in terms of onset and dissipation), as well as sensitization, and desensitization. Generally, tastes are perceived relatively quickly, with suprathreshold reaction times on the order of 400–800 milliseconds (Halpern, 1986) and peak intensities that occur within 10 seconds (Lawless and Skinner, 1979), before decaying and disappearing in about a minute (e.g., Lawless and Skinner, 1979; Leach and

[1] See Delwiche (1996) for a thoughtful discussion of whether the concept of basic tastes is valid or useful.

Noble, 1986) (not withstanding the lingering of sucralose or some bitterants). Conversely, burning stimuli take substantially more time to reach their maximum (e.g., Cliff and Heymann, 1993) and may not dissipate for anywhere between 3 and 10 minutes, depending on concentration (e.g., Bennett and Hayes, 2012; Green and Hayes, 2004; Green and Schullery, 2003; Lawless, 1984).

Tastes typically adapt quite rapidly, where adaptation is defined as a reduced response to a specific stimulus with repeated exposure to that stimulus. For example, when given fresh quinine, sodium chloride or sucrose samples every minute for 5 minutes, the perceived intensity of each stimulus drops to less than half of the initial intensity within 2 minutes, and when the same sample is held in the mouth continuously, the magnitude of the reduction approaches 90% by the end of exposure (Meiselman, 1968). Critically, these responses are quality specific, and not general: adapting to sucrose does not reduce the intensity of quinine. Likewise, smells also show adaptation that is large, rapid, and specific (see Dalton, 2000). Conversely, pungent stimuli such as capsaicin and piperine show a pattern of increased intensity with repeated exposure, where the perceived burn continues to increase upon repeated, periodic exposure – this phenomenon is commonly referred to a sensitization[2] in the psychophysical literature.

6.4 Sensitization

Psychophysical sensitization in the oral cavity was first described in 1985 by Lawless and Gillette. When a 2 ppm vanillyl nonanamide solution was flowed over a small region of the tongue continuously using a flow chamber, the sensation magnitude increased, more than doubling by the end of 7 minutes (Lawless and Gillette, 1985). Subsequently, Stevens and Lawless reported sequential "enhancement" for whole mouth stimuli: when participants held two samples of capsaicin in the mouth for two 1 minute intervals, the second sample was approximately 50% stronger, and the same pattern was seen for piperine (Stevens and Lawless, 1987). When using 7 mm filter-paper disks to target specific oral regions such as the tongue, lip, palate, and cheek, responses to capsaicin and piperine both exhibit sensitization, with the most dramatic increases occurring on the tongue tip, with the magnitude doubling or even tripling by the end of 5 minutes (Lawless and Stevens, 1988). Notably, this process appears to depend, at least somewhat, on the sensory modality rather than the stimulus. For example, salts like magnesium chloride (Green and Hayes, 2003), potassium chloride (Green and Gelhard, 1989), or sodium chloride (Green and Gelhard, 1989) act as irritants even at relatively moderate concentrations. When sodium chloride is sipped repeatedly at 1 minute intervals without rinsing, perceived irritation continues to build, eventually doubling in intensity over 15 minutes; critically, however, ratings

[2] Confusingly, the term sensitization is also used to describe the physiological processes by which nociceptors in peripheral neurons or the central nervous system can show increased responses to painful stimuli. For example, protein kinase C and protein kinase A sensitize TRPV1 by phosphorylation of specific serine residues. Critically, the observation of psychophysical sensitization does not imply a specific physiological mechanism.

for saltiness do not show the same increase (Green and Gelhard, 1989). However, this is not to say that sensitization is normative for all chemesthetic stimuli that elicit burning sensations, as many burning stimuli appear not to sensitize.

When ethanol or sodium chloride impregnated filter-paper disks are placed on the anterior tongue at 30 second intervals, ratings of irritation for sodium chloride increase, eventually tripling over 10 minutes, while ethanol irritation does not increase with repeated exposure (Green, 1990). Whether the failure of ethanol burn to sensitize might reflect some sort of competing anesthetic process is unknown, but in light of the numbness described by Cliff and Heymann (1992), additional work in this area is warranted. Likewise, irritation from zingerone (Prescott and Stevenson, 1996a), nicotine (Dessirier *et al.*, 1997), cinnamaldehyde (Prescott and Swain-Campbell, 2000) and mustard oil (Simons *et al.*, 2003) does not appear to sensitize, at least for most participants. Extant data conflict on the exact pattern of responses with repeated exposure to ethanol (cf. Green, 1990 and Prescott and Swain-Campbell, 2000), perhaps due to differences in concentration or delivery (sip and spit versus filter-paper disks), but in neither case is there any evidence of sensitization.

For capsaicin, sensitization also depends strongly on the time course of exposure. When the interstimulus interval (ISI) is short (e.g., ~2 minutes or less), sensitization is robustly observed for capsaicin with both regional and whole mouth stimulation (e.g., Dessirier *et al.*, 1997; Green, 1989, 1991b; Green and Hayes, 2003; Prescott, 1999; Simons *et al.*, 2003). With a longer ISI, however, the effect is less pronounced or absent. Nasrawi and Pangborn looked at individual responses from participants to 5 ml of capsaicin with a 2.5 minutes ISI, observing that approximately half of the participants showed responses consistent with sensitization, while perceived burn declined for the remaining participants (Nasrawi and Pangborn, 1990). Subsequently, Green systematically explored the temporal aspects of capsaicin exposure, finding that, when averaged across all participants, sensitization occurred with ISIs of 1 or 2.5 minutes; longer ISIs of 5 or 10 minutes resulted instead in desensitization (Green, 1989).

Finally, in spite of the general robustness with which sensitization is observed for capsaicin, a caveat is necessary, as model systems may not generalize to real foods. In perhaps the only study assessing potential sensitization with actual foods (i.e., soup and chili) under conditions mimicking normal eating patterns (as opposed to regional or whole mouth stimulation with capsaicin solutions at fixed intervals), the evidence for sensitization is minimal (Prescott, 1999). Accordingly, more work is needed to determine: (a) if the failure to observe capsaicin or piperine sensitization generalizes to other foods and (b) if the absence of sensitization for some burning stimuli (e.g., cinnamaldehyde, mustard oil, ethanol, etc.) is merely the result of concentration or a non-optimal ISI.

6.5 Acute psychophysical desensitization

Evidence that exposure to capsaicin results in desensitization was first uncovered by Nicholas Jancsó and Aurelia Jancsó-Gábor in the 1940s and eventually published in 1955 in German (see Szolcsanyi, 2014). Around the same time, T.S. Lee incidentally reported what could be considered the first English language report of psychophysical desensitization in a study of the "gustatory sweating" reflex where he quantified facial sweat as a response to oral exposure to chili and other stimuli (Lee, 1954). Specifically, Lee painted chili paste on the tongue seven times sequentially at 15–20 minute intervals, leaving the paste on for 3.5 minutes in each trial. After increasing on the second trial, the number of active sweat glands decreased on later trials, and Lee states that anecdotal reports of subjective pain also decreased. Notably, Lee also observed the chili paste produced pain and sweating when applied to the lip, thereby supporting the independence of chemesthesis from taste, as the lip does not contain taste nerves. Subsequently, Jancsó's protégé and collaborator Szolcsanyi demonstrated that exposing the tongue to a 10 000 ppm capsaicin solution for 60 seconds ten times over a 40 minute period resulted in complete desensitization that lasted several hours. Critically, this phenomenon is selective: response to capsaicin, mustard oil, and zingerone are all reduced, with no effect on touch, cooling (either thermal or chemical cooling), or taste thresholds for glucose, quinine, salt, and ascorbic acid (Szolcsanyi, 1977). Nor are extreme doses needed to observe this effect. Frequent consumers of chili report lower burn ratings when given capsaicin (Lawless *et al.*, 1985) or zingerone (Prescott and Stevenson, 1996b) in the laboratory, which has been interpreted as evidence of desensitization from dietarily relevant doses. Other work indicates psychophysical desensitization can be induced in the laboratory with low dose stimuli that are not painful (Green, 1989).

Today, the dominant paradigm for intentionally inducing acute psychophysical desensitization in the laboratory involves repeatedly exposing the participant to low doses at short intervals (30–60 seconds), followed by a longer hiatus. When this is done, sensitization is frequently (but not always) observed during the initial conditioning block, followed by a substantial reduction in reported burn from a probe stimulus presented after the hiatus. For example, when applying capsaicin to the anterior tongue every 30 seconds for 5 minutes via cotton swabs, the mean burn rating for the tenth stimulus is ~52% greater than the first stimulus, peaking between "strong" and "very strong" on a generalized Labeled Magnitude Scale (gLMS); however, after a 20 minute break, the burn for the final stimulus is ~82% lower than the first stimulus (below "weak" on a gLMS) (Green and Hayes, 2003). On the posterior tongue, similar effects are seen: burn increases ~106% (i.e., more than doubles) between the first and tenth stimulus, while the final stimulus is ~60% lower than the first stimulus.

Numerous other studies show similar reductions in capsaicin burn when a 10–15 minute break is provided between the final conditioning stimulus and the probe trial: for whole mouth stimuli, reductions of between 45 and 60% are observed (Prescott and Swain-Campbell, 2000) for groups stratified by propylthiouracil taster status (Snyder *et al.*, 2008), while filter-paper disks on the tongue tip show reductions of ~88 or ~112%, depending on concentration (Green, 1996). Similar patterns of sensitization and desensitization are also seen for capsaicin on non-oral tissues, albeit over a different time course (Green, 1998).

Evidence of acute desensitization for other chemesthetic stimuli is mixed, and the time course of exposure appears to be critical. For capsaicin self-desensitization, a hiatus of ~3.5 or more minutes is required; otherwise sensitization occurs instead

(Green, 1989, 1991b). With breaks of 10–15 minutes, piperine self-desensitizes, and piperine and zingerone cross-desensitize with capsaicin on the anterior tongue (Green, 1993, 1996), and cinnamaldehyde self-desensitizes with whole mouth stimulation (Prescott and Swain-Campbell, 2000). Conversely, whole mouth ethanol does not show any appreciable self-desensitization following a 10 minute hiatus, although burn ratings drop steadily when ethanol is presented at 1 minute intervals (Prescott and Swain-Campbell, 2000). This may be why whisky or tequila do not seem to burn quite so much after the fourth or fifth sip.

Under conditions where capsaicin sensitizes, mustard oil containing the TRPA1 agonist allyl isothiocyanate appears to self-desensitize when presented at 1 minute intervals; following a 10 minute hiatus, responses to mustard oil are still greatly depressed (~70%), an indication of self-desensitization (Simons *et al.*, 2003). There is also evidence of symmetrical cross-desensitization between mustard oil and capsaicin, although the magnitude of the depression is substantially smaller. Regarding zingerone, a TRPV1 agonist, early work suggested it failed to self-desensitize (Green, 1989), but subsequent work revealed that zingerone self-desensitizes at 30 second intervals, and this desensitization persists after a 5 minute hiatus (Prescott and Stevenson, 1996a). When the TRPV3 agonists eugenol and carvacrol are presented on the anterior tongue via filter-paper disks at 1 minute intervals, each self-desensitizes within the first few applications, and their respective self-desensitization persists after a 10 minute break (Klein *et al.*, 2013). Repeated exposure to eugenol and carvacrol also reduces capsaicin burn after a 10 minute hiatus (Klein *et al.*, 2013), suggesting desensitization may be occurring at the level of the neuron and not the receptor.

Notably, psychophysical desensitization to capsaicin reduces the burning and stinging of ethanol and cinnamaldehyde without altering the numbing of both, or the warming of cinnamaldehyde (Green, 1991a). Likewise, capsaicin desensitization reduces burning/stinging ratings from magnesium chloride and urea on the anterior tongue without altering the sourness of urea or saltiness of magnesium chloride (Green and Hayes, 2003). Whether these quality specific effects generalize to whole mouth stimuli is unknown, and more work in this area is warranted.

Finally, no discussion of acute psychophysical desensitization is complete without mentioning what Green described as "stimulus-induced recovery". Namely, in stark contrast to the adaptation seen with tastants and odorants, after psychophysical desensitization is observed, additional stimulation with capsaicin at intervals on the order of seconds (Green, 1996; Green and Rentmeister-Bryant, 1998) or minutes (Green and Rentmeister-Bryant, 1998) can cause burning sensations to return, with intensities eventually recovering to magnitudes close to those elicited by the initial stimulus. However, looking only at means across the group potentially obscures individual differences in this phenomenon: in one report, 95% of participants showed evidence of stimulus-induced recovery (Green, 1996), while in another, 96% of participants showed desensitization on the anterior tongue, but only 33% showed complete recovery, with most participants (62%) showing only partial recovery with recurrent stimulation (Green, 1998).

Mechanistically, stimulus-induced recovery provides important insight, as it argues against the disruption of cellular function or the depletion of neurotransmitters such as Substance P as an explanation for acute desensitization. Nor are these effects limited to capsaicin, as stimulus-induced recovery is also observed with piperine (Green, 1996). Intriguingly, what appears to be acute sensitization across the first block of repeated stimulation in the laboratory may in fact be stimulus-induced recovery if the participant enters the laboratory in a chronically desensitized state from prior dietary exposure. There is some evidence to support this (Green, 1996), but additional confirmation is needed with dietarily relevant doses under more natural eating conditions.

6.6 Chronic psychophysical desensitization

Some of the initial evidence that psychophysical desensitization to relatively low doses of capsaicin might last across days comes from observations that frequent chili consumers give lower burn ratings to capsaicin (Cowart, 1987; Lawless *et al.*, 1985; Stevenson and Yeomans, 1993) and zingerone (Prescott and Stevenson, 1995) in the laboratory, although not all studies fail to find a relationship between intake frequency and perceived burn (Byrnes and Hayes 2015; Green, 1996; McBurney *et al.*, 1997). Controlled studies also suggest exposure to capsaicin in the laboratory leads to desensitization that can last across multiple days, irrespective of the delivery system. When 100 ppm capsaicin are repeatedly applied to the anterior tongue via filter-paper disks, after 24 hours the burn is 57% lower on the first trial (Green, 1996). The period of desensitization can even last multiple days, depending on concentration: in regional testing with cotton swabs, desensitization induced by 10 ppm capsaicin recovers within a day or two, while recovery from 100 ppm capsaicin may take up to 4 days (Karrer and Bartoshuk, 1991).

Multiple studies have used specially formulated candy as a capsaicin delivery system. McBurney and colleagues had participants rate the burn from 100 ppm capsaicin on filter-paper disks on days 1 and 5, asking them to suck on a hard candy containing 75 ppm capsaicin for 10–12 minutes a day on days 2–4. When comparing the peak burn from the paper disk on day 1 to day 5, peak burn was reduced by 48%; there were no specifics on the time delay between the final candy exposure on day 4 and the paper disk on day 5, but presumably it was ~24 hours (McBurney *et al.*, 1997). Conversely, Green and Rentmeister-Bryant failed to observe any evidence of desensitization across days when participants were given capsaicin-containing taffy (Green and Rentmeister-Bryant, 1998). Desensitization via capsaicin-containing taffy also reportedly reduces the pain from oral mucositis in cancer patients (Berger *et al.*, 1995).

While all of these results have generally been interpreted as evidence of desensitization, it remains possible that lower ratings could instead reflect a contextual shift from exposure. That is, if frequent use outside the laboratory or repeated exposure within the laboratory alters the internal reference participants use when rating "hot" sensations, their burn ratings may drop not because of desensitization, but rather because the participant now has a much larger range of more intense prior experiences with oral burn, causing them to rate the test stimuli less intensely (Stevenson and Prescott, 1994). As a masters student in Harry Lawless' laboratory, I tested this hypothesis with a protocol intended to mimic dietary exposure in 24 participants,

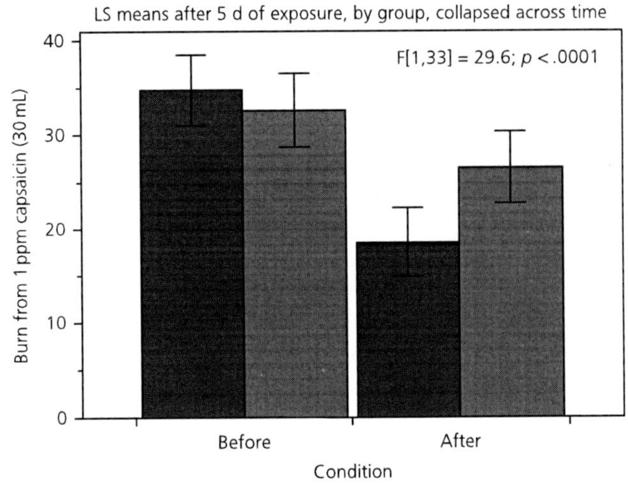

Fig. 6.2 Burn from 1 ppm capsaicin solutions was collected over 1 minute from 35 individuals assigned to a 2 ppm capsaicin ($n = 18$) or citric acid ($n = 17$) group. Individuals were randomized to a capsaicin exposure group or a citric acid control group, and exposed to treatment with the respective mouthwash for 5 days in the laboratory. Collapsing the ratings over 1 minute, the least square means for the capsaicin group (dark bars) showed a 47% drop in perceived burn following treatment, compared with a 19% drop in the citric acid control group (light bars). See text for additional details on timing. Data from (Hayes, 2000).

reasoning that changes in evoked salivary flow would reflect desensitization but not contextual effects. After randomization to a capsaicin or citric acid group, participants rinsed with 30 ml of mouthwash (2 ppm capsaicin or 238 mM citric acid) at home for 10 days. On the 2 days before and 2 days after exposure, capsaicin-evoked burn and salivary flow were measured in the laboratory. Perceived burn ($F[1,17] = 5.0$; $p < 0.04$) and salivary flow ($F[1,15] = 10.0$; $p < 0.007$) were reduced in the capsaicin group but not the citric acid group following the treatment period, suggesting lower burn ratings following low dose daily exposure is not merely due to altered scale usage (Hayes, 2000). In the follow-up study shown in Fig. 6.2, 35 participants rated capsaicin burn every 15 seconds for 1 minute on days 1 and 2 and 12 and 13; on days 5–9, they came to the laboratory, rinsing with either 2 ppm capsaicin ($n = 18$) or citric acid ($n = 17$) but made no ratings. There was no exposure on weekends (days 3/4 and 10/11). As shown in Fig. 6.2, even after several days off, the burn of 1 ppm capsaicin showed a significant group by period (before/after) interaction in mixed model ANOVA. These data suggest that a low dose 2 ppm capsaicin mouthwash induces desensitization that lasts at least ~72 hours.

6.7 Summary

Plants have evolved a wide range of secondary metabolites that are sensory active in humans and other animal species, presumably to prevent or selectively direct predation. Yet humankind seems to find many of these materials attractive and highly desirable, rather than aversive. Indeed, humankind has been captivated by these materials for millennia, driving very early crop domestication as well as the age of discovery. In many cases, they have unique temporal properties that clearly distinguish them from the better studied chemical senses of smell and taste, and the words we use to describe them often, but not always, draw parallels with thermal sensations. Integrated, transdisciplinary approaches ranging from linguistics to molecular biology to psychophysics are required to fully characterize these materials, the mechanisms by which they act, and their place in our food supply.

References

Albin, K.C. and Simons, C.T. 2010. Psychophysical evaluation of a sanshool derivative (alkylamide) and the elucidation of mechanisms subserving tingle. *Plos One*, **5** (3), DOI: ARTN e9520, DOI: 10.1371/journal.pone.0009520.

Bennett, S.M. and Hayes, J.E. 2012. Differences in the chemesthetic subqualities of capsaicin, ibuprofen, and olive oil. *Chem. Senses*, **37** (5), 471–478, DOI: bjr129 [pii], 10.1093/chemse/bjr129.

Berger, A., Henderson, M., Nadoolman, W. *et al.* 1995. Oral capsaicin provides temporary relief for oral mucositis pain secondary to chemotherapy/radiation therapy. *J. Pain Symptom. Manage.*, **10** (3), 243–248, DOI: 10.1016/0885-3924(94)00130-D.

Boone, L.M. 2014. Perceptual mapping of chemesthetic stimuli in Spanish and English speakers. BS Honors Thesis. Pennsylvania State University, University Park, PA. Retrieved from https://honors.libraries.psu.edu/paper/21396/ (accessed 20 January 2015).

Brown, C.H., Clement, C.R., Epps, P., Luedeling, E., and Wichmann, S.R. 2013. The paleobiolinguistics of domesticated chili pepper (*Capsicum* spp.). *Ethnobiol. Lett.*, **4**, 1–11.

Byrnes, N.K. and Hayes, J.E. (2015) Gender differences in the influence of personality traits on spicy food liking and intake. *Food Qual. Pref.*, **42**, 12–19.

Byrnes, N.K., Nestrud, M.A., and Hayes, J.E. 2015. Perceptual mapping of chemesthetic stimuli in naïve assessors. *Chemosens. Percept.*, **8** (1), 19–32.

Caterina, M.J., and Julius, D. 2001. The vanilloid receptor: a molecular gateway to the pain pathway. *Annu. Rev. Neurosci.*, **24**, 487–517.

Caterina, M.J., Leffler, A., Malmberg, A.B., *et al.* 2000. Impaired nociception and pain sensation in mice lacking the capsaicin receptor. *Science*, **288** (5464), 306 313.

Caterina, M.J., Schumacher, M.A., Tominaga, M., *et al.* 1997. The capsaicin receptor: a heat-activated ion channel in the pain pathway. *Nature*, **389** (6653), 816–824.

Chiou, K.L., and Hastorf, C.A. 2012. *Capsicum spp. at the Preceramic sites of Huaca Prieta and Paredones, Chicama Valley, Peru.*Berkeley, California, retrieved from http://archaeobotany.berkeley.edu/Research/LabReport/lab74/lab74.pdf (accessed 18 January 2015).

Chiou, K.L., Hastorf, C.A., Bonavia, D., and Dillehay, T.D. 2014. Documenting cultural selection pressure changes on chile pepper (*Capsicum baccatum* L.) seed size through time in coastal Peru (7,600 B.P.–present). *Econ. Bot.*, **68** (2), 190–202, DOI: 10.1007/S12231-014-9270-Y.

Cliff, M. and Heymann, H. 1992. Descriptive analysis of oral pungency. *J. Sens. Stud.*, **7** (4), 279–290.

Cliff, M., and Heymann, H. 1993. Time-intensity evaluation of oral burn. *J. Sens. Stud.*, **8**, 201–211.

Cowart, B.J. 1987. Oral chemical irritation: Does it reduce perceived taste intensity? *Chem. Senses*, **12** (3), 467–479.

Dalton, P. 2000. Psychophysical and behavioral characteristics of olfactory adaptation. *Chem. Senses*, **25** (4), 487–492.

Delwiche, J. 1996. Are there 'basic' tastes? *Trends Food Sci.Technol.*, **7** (12), 411–415.

Dessirier, J.M., O'Mahony, M., and Carstens, E. 1997. Oral irritant effects of nicotine: psychophysical evidence for decreased sensation following repeated application and lack of cross- desensitization to capsaicin. *Chem. Senses*, **22** (5), 483–492.

Glendinning, J.I. 2007. How do predators cope with chemically defended foods? *Biol. Bull.*, **213** (3), 252–266.

Green, B.G. 1989. Capsaicin sensitization and desensitization on the tongue produced by brief exposures to a low concentration. *Neurosci. Lett.*, **107** (1–3), 173–178.
Green, B.G. 1990. Effects of thermal, mechanical, and chemical stimulation on the perception of oral irritation, in *Irritation* (eds. B.G. Green, J.R. Mason, and M.R. Kare), New York: Dekker.
Green, B.G. 1991a. Capsaicin cross-desensitization on the tongue-psychophysical evidence that oral chemical irritation is mediated by more than one sensory pathway. *Chem. Senses*, **16** (6), 675–689, DOI: 10.1093/Chemse/16.6.675.
Green, B.G. 1991b. Temporal characteristics of capsaicin sensitization and desensitization on the tongue. *Physiol. Behav.*, **49** (3), 501–505.
Green, B.G. 1993. Evidence that removal of capsaicin accelerates desensitization on the tongue. *Neurosci. Lett.*, **150** (1), 44–48.
Green, B.G. 1996. Rapid recovery from capsaicin desensitization during recurrent stimulation. *Pain*, **68** (2–3), 245–253.
Green, B.G. 1998. Capsaicin desensitization and stimulus-induced recovery on facial compared to lingual skin. *Physiol. Behav.*, **65** (3), 517–523.
Green, B.G., and Gelhard, B. 1989. Salt as an oral irritant. *Chem. Senses*, **14** (2), 259–271.
Green, B.G., and Hayes, J.E. 2003. Capsaicin as a probe of the relationship between bitter taste and chemesthesis. *Physiol. Behav.*, **79** (4–5), 811–821.
Green, B.G., and Hayes, J.E. 2004. Individual differences in perception of bitterness from capsaicin, piperine and zingerone. *Chem. Senses*, **29** (1), 53–60.
Green, B.G., and Rentmeister-Bryant, H. 1998. Temporal characteristics of capsaicin desensitization and stimulus-induced recovery in the oral cavity. *Physiol. Behav.*, **65** (1), 141–149.
Green, B.G., and Schullery, M.T. 2003. Stimulation of bitterness by capsaicin and menthol: differences between lingual areas innervated by the glossopharyngeal and chorda tympani nerves. *Chem. Senses*, **28** (1), 45–55.
Halpern, B.P. 1986. Constraints imposed on taste physiology by human taste reaction-time data. *Neurosci. Biobehav. Rev.*, **10** (2), 135–151.
Hayes, J.E. 2000. Psychophysical and physiological response of humans to the oral irritant capsaicin. MS thesis. Cornell University.
Jordt, S.E., and Julius, D. 2002. Molecular basis for species-specific sensitivity to "hot" chili peppers. *Cell*, **108** (3), 421–430.
Karrer, T., and Bartoshuk, L. 1991. Capsaicin desensitization and recovery on the human tongue. *Physiol Behav*, **49** (4), 757–764.
Klein, A.H., Carstens, M.I., and Carstens, E. 2013. Eugenol and carvacrol induce temporally desensitizing patterns of oral irritation and enhance innocuous warmth and noxious heat sensation on the tongue. *Pain*, **154** (10), 2078–2087, DOI: 10.1016/j.pain.2013.06.025.
Kraft, K.H., Brown, C.H., Nabhan, G.P., *et al.* 2014. Multiple lines of evidence for the origin of domesticated chili pepper, Capsicum annuum, in Mexico. *Proc. Nat. Acad. Sci. USA*, **111** (17), 6165–6170, DOI: 10.1073/Pnas.1308933111.
Lawless, H.T. 1984. Oral chemical irritation: Psychophysical properties. *Chem. Senses*,**9** (2), 143–155.
Lawless, H.T., and Gillette, M. 1985. Sensory responses to oral chemical heat, in *Characterization and Measurement of Flavor Compounds* (eds. D.D. Bills and C.J. Mussinan), Washington, D.C.: American Chemical Society.
Lawless, H.T., Rozin, P., and Shenker, J. 1985. Effects of oral capsaicin on gustatory, olfactory and irritant sensations and flavor identification in humans who regularly or rarely consume chili pepper. *Chem. Senses*, **10** (4), 579–589.
Lawless, H.T., and Skinner, E.Z. 1979. The duration and perceived intensity of sucrose taste. *Percept. Psychophys.*, **25** (3), 180–184.
Lawless, H.T., and Stevens, D.A. 1988. Responses by humans to oral chemical irritants as a function of locus of stimulation. *Percept. Psychophys.*, **43** (1), 72–78.
Leach, E J., and Noble, A.C. 1986. Comparison of bitterness of caffeine and quinine by a time intensity procedure. *Chem. Senses*,**11** (3), 339–345, DOI: 10.1093/Chemse/11.3.339.
Lee, T.S. 1954. Physiological gustatory sweating in a warm climate. *J. Physiol.*, **124**, 528–542.
Lim, J.Y., and Green, B.G. 2007. The psychophysical relationship between bitter taste and burning sensation: Evidence of qualitative similarity. *Chem. Senses*, **32** (1), 31–39. DOI: 10.1093/Chemse/Bjl033.

McBurney, D.H., Balaban, C.D., Christopher, D.E., and Harvey, C. 1997. Adaptation to capsaicin within and across days. *Physiol. Behav.*, **61** (2), 181–190.

McIntyre, P., McLatchie, L.M., Chambers, A., *et al.* 2001. Pharmacological differences between the human and rat vanilloid receptor 1 (VR1). *Br. J. Pharmacol.*, **132** (5), 1084–1094.

McNamara, F.N., Randall, A., and Gunthorpe, M.J. 2005. Effects of piperine, the pungent component of black pepper, at the human vanilloid receptor (TRPV1). *Br. J. Pharmacol.*, **144** (6), 781–790.

Meiselman, H.L. 1968. Magnitude estimations of course of gustatory adaptation. *Percept. Psychophys.*, **4** (4), 193–196, DOI: 10.3758/Bf03206299.

Müller, J. 1840. *Handbuch der Physiologie des Menchschen für Vorlesungen.*Coblenz: Holscher.

Myers, C.S. 1909. *A text-book of Experimental Psychology*, Harlow: Longmans, Green.

Nasrawi, C.W., and Pangborn, R.M. 1990. Temporal gustatory and salivary responses to capsaicin upon repeated stimulation. *Physiol. Behav.*, **47** (4), 611–615.

Pickersgill, B. 1969. Archaeological record of chili-peppers (Capsicum spp) and sequence of plant domestication in Peru. *Am. Antiq.*, **34** (1), 54–61, DOI: 10.2307/278313.

Prescott, J. 1999. The generalizability of capsaicin sensitization and desensitization. *Physiol. Behav.*, **66** (5), 741–749.

Prescott, J., and Stevenson, R.J. 1995. Effects of oral chemical irritation on tastes and flavors in frequent and infrequent users of chili. *Physiol. Behav.*, **58** (6), 1117–1127.

Prescott, J., and Stevenson, R.J. 1996a. Desensitization to oral zingerone irritation: effects of stimulus parameters. *Physiol. Behav.*, **60** (6), 1473–1480.

Prescott, J., and Stevenson, R.J. 1996b. Psychophysical responses to single and multiple presentations of the oral irritant zingerone: relationship to frequency of chili consumption. *Physiol. Behav.*, **60** (2), 617–624.

Prescott, J., and Swain-Campbell, N. 2000. Responses to repeated oral irritation by capsaicin, cinnamaldehyde and ethanol in PROP tasters and non-tasters. *Chem. Senses*, **25**, 239–246.

Simons, C.T., Carstens, M.I., and Carstens, E. 2003. Oral irritation by mustard oil: self-desensitization and cross-desensitization with capsaicin. *Chem. Senses*, **28** (6), 459–465.

Smith Jr, C.E. 1980. Plant remains from Guitarrero cave, in *Guitarrero Cave: Early Man in the Andes*, New York: Academic Press, pp. 87–120.

Snyder, D.J., Duffy, V.B., Hayes, J.E., and Bartoshuk, L.M. 2008. Propylthiouracil (PROP) taste, in *The Senses: A Comprehensive Reference* (eds-in chief A.I. Basbaum, A. Kaneko, G.M. Shepherd, *et al.*), New York: Academic Press, pp. 391–399.

Stevens, D.A., and Lawless, H.T. 1987. Enhancement of responses to sequential presentation of oral chemical irritants. *Physiol. Behav.*, **39** (1), 63–65.

Stevenson, R.J., and Prescott, J. 1994. The effects of prior experience with capsaicin on ratings of its burn. *Chem. Senses*, **19** (6), 651–656.

Stevenson, R.J., and Yeomans, M.R. 1993. Differences in ratings of intensity and pleasantness for the capsaicin burn between chili likers and non-likers: Implications for liking development. *Chem. Senses*, **18** (5), 471–482.

Sugai, E., Morimitsu, Y., Iwasaki, Y., *et al.* 2005. Pungent qualities of sanshool-related compounds evaluated by a sensory test and activation of rat TRPV1. *Biosci. Biotechnol. Biochem.*, **69** (10), 1951–1957, DOI: 10.1271/Bbb.69.1951.

Szallasi, A., and Blumberg, P.M. 1999. Vanilloid (Capsaicin) receptors and mechanisms. *Pharmacol. Rev.*, **51** (2), 159–212.

Szolcsanyi, J. 1977. A pharmacological approach to elucidation of the role of different nerve fibres and receptor endings in mediation of pain. *J. Physiol. (Paris)*, **73** (3), 251–259.

Szolcsanyi, J. 2014. Capsaicin and sensory neurones: a historical perspective, in *Capsaicin as a Therapeutic Molecule* (ed. O. Abdel-Salam), Basel: Springer.

Tewksbury, J.J., and Nabhan, G.P. 2001. Seed dispersal. Directed deterrence by capsaicin in chilies. *Nature*, **412** (6845), 403–404.

Yamaguchi, S. 1987. Fundamental properties of umami in human taste sensation, in *Umami: A Basic Taste*, New York,: Dekker, pp. 41–73.

CHAPTER 7
Types of chemesthesis II: Cooling

Steven Pringle
Aroma Chemical Services International, Gmbh, Höxter-Stahle, Germany

7.1 Consumers and oral perception: where chemesthesis contributes to flavor

Despite its common usage, the term "flavor" is often misunderstood and difficult to define. Even within the scientific community, within different functions and various areas of expertize, the term flavor can often be used to describe different effects. The International Organization for Standardization defines flavor as: "Complex combination of olfactory, gustatory and trigeminal sensations perceived during tasting. The flavo(u)r may be influenced by tactile, thermal, painful and/or kinaesthetic effects".

7.1.1 Taste perception

As children we are often taught that we have five senses and that one of those, our sense of taste, comes from the interaction of food with different areas on our tongue. We are told that taste is made up of sweet, salty, sour, bitter, and savory, and that our taste buds pick up these tastes. As we continue through our education we learn that this isn't exactly accurate and that there are a number of factors that contribute to our perception of taste.

One of the major influences in our perception of flavor is odor or olfaction. Odor-active molecules are volatile chemical compounds that interact with the olfactory epithelium located in the roof of the two nasal cavities, just below and between the eyes. These odorants must possess certain molecular characteristics in order to provide the necessary sensory properties to be recognized. They must have some water solubility, a relatively high vapor pressure, low polarity, some lipophilicity, and surface activity. It has been reported (Demole and Wuest, 1967) that no known odorant has a molecular weight above 294, and the olfactory sense is able to distinguish between a practically infinite number of chemical compounds at low concentrations (Ohloff, 1994).

The availability of thousands of aroma chemicals all with distinctive odor profiles, to the trained observer, allows us to experience a plethora of different flavors. However, odor alone does not make flavor. For example, if you were to pick up a piece of licorice and hold it to your nose, you would most certainly be able to smell the distinctive anise-like odor we all associate with licorice. However, you would not, until you placed the licorice into your mouth, experience the sweetness of the glycyrrhizin contained within the licorice. Similarly, if you were to hold your nose and pop a couple of red jelly beans into your mouth you would be able to taste sweetness and notice the gelatinous texture, but you would not be able to distinguish whether the jelly bean was strawberry, raspberry or some other flavor until volatiles reach the olfactory epithelium. In the narrow technical meaning, taste consists of only five percepts – sweet, sour, salty, bitter, and umami (savory) – that are relayed by the gustatory nerves. However, even with input from olfactory and gustatory nerves, these two channels are not sufficient to fully explain the sensation of flavor.

It therefore must also be recognized that within the nasal cavity another sensory system is contained in the form of the "trigeminal nerve" receptors. The trigeminal nerve, which is the largest cranial nerve and the sensory nerve of the face, mouth, teeth, the majority of the scalp, and the motor nerve of the muscles of mastication, provides a second set of nerve endings that are responsible for pain, temperature, tactile, and pressure sensations in the areas of the eyes, mouth, and nasal cavity. It is in this area where certain chemical stimulants such as L-menthol (mint), camphor, allyl isothiocyanate (mustard, mustard oil), capsaicin (chili peppers), and diallyl sulfide (garlic) can stimulate irritating effects at high concentrations in the nasal cavity, which can be interpreted in different ways. It has also been claimed (Ohloff, 1994), that about 70% of all odors can stimulate the trigeminal nerve, although they may be less sensitive than olfactory receptors.

It seems, therefore, that how we perceive flavor is influenced by multiple neural pathways. However, to believe that taste is driven by only what is experienced by our tongue and nose is an oversimplification. As shown in Fig. 7.1, there are a number of factors that influence the perception of flavor, with the interaction with the olfactory epithelium and the trigeminal nerve being only part of the picture.

It is this stimulation beyond olfaction that is often referred to as chemesthesis (Bryant and Silver, 2000). Some odorants can elicit activity in the trigeminal nerve (which is associated with the perception of irritation) at levels that do not create sensations associated with irritation and can contribute to greater odor intensity. Irritants can alter the perception of odor intensity and taste, and odorants can alter the perception of the level of irritation. Chemesthesis (from "chem" pertaining to chemical and "esthesis" for the ability to perceive or feel) has become adopted to replace what had previously been known as the "common chemical sense" (Parker, 1912). The chemosensitive fibers are a subset of pain and temperature (nociceptive and thermal, respectively) sensitive fibers that occur throughout the body, not just in the oral cavity. In order to reach the relevant nerve receptors, compounds that exhibit these properties must pass through the lipid phase of the cell membrane of the aqueous phase. For this reason chemesthetic sensations have a slower onset before perception than those experienced through olfaction. Compounds that exhibit these effects generally require at least some binding to the relevant receptor and an ability to partition the fat/water boundary.

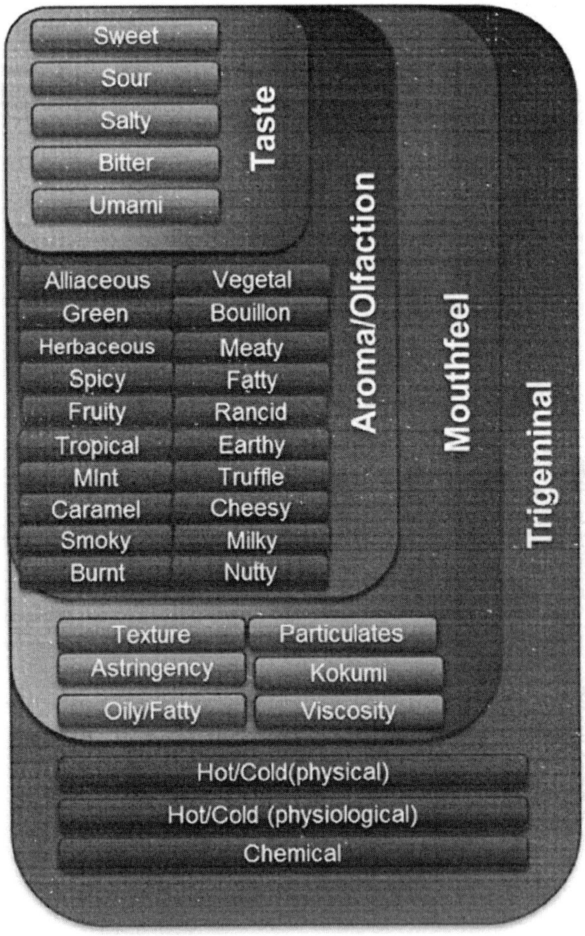

Fig. 7.1 Factors affecting taste perception.

Whether these receptor interactions are perceived as hot or cold, the underlying process is reliant on ion transfer across cellular membranes. With the exception of receptor protein ion channels, these receptor membranes are impermeable to ions such as K^+ or Ca^{2+}. In the last two decades, our knowledge of these receptors has increased significantly, and in the case of thermoreceptors (Table 7.1) we know that these are activated when thermal or chemical stimuli excite the sensory neurons of the dorsal or trigeminal ganglia (Jordt *et al.*, 2003). Other chapters of this book will describe in more detail the mechanisms by which these receptors function, however we can state here that these belong to the class of transient receptor potential channels (TRP) of which seven subsets exist. These are TRPC, TRPV, TRPM, TRPA, TRPP, TRPML, and TRPN. TRPV1 was the first of these receptors to be characterized and is often referred to as the vanilloid receptor due to it being activated by capsaicin as well as by heat. The cold receptor TRPM8, which is where this chapter will focus, was characterized by McKemy *et al.* (2004).

Table 7.1 Typical thermoreceptor agonists.

Chemical agonist	Botanical source	TRP channel
Capsaicin	Chili peppers	TRPV1
Piperine	Black pepper	TRPV1
[6] and [8] Gingerol	Ginger	TRPV1
alpha hydroxysanshool	Sichuan pepper	TRPV1, TRPA1
Allicin	Garlic	TRPV1, TRPA1
Camphor	*Cinnamomum camphora*	TRPV3, TRPV1
Cannabichromene	*Cannabis sativa*	TRPV2, TRPA1
(–) Menthol	Peppermint	TRPM8, TRPV3
1,8-Cineole	Eucalyptus	TRPM8
Carvacrol	Oregano oil	TRPV3,
WS-3	Synthetic	TRPM8
Cinnamaldehyde	Cinnamon, cassia	TRPA1, TRPV3
Allyl isothiocyanate	Mustard, horseradish	TRPA1

7.2 Molecular structure and physiological cooling

For a number of years, chemical compounds that occur in nature, such as menthol, menthone, and menthyl acetate as well as essential oils such as peppermint oil (which have high concentrations of these chemicals in their composition), have been used to provide a cooling effect both orally and topically. The use of these compounds in applications such as body wash, shampoo, and other personal care products continues. However, due to the relatively high dosage level required and the odor and substantial aroma associated with them, the majority of uses of these materials have been focused in the area of flavor application, such as chewing gum, mint confection, and oral care applications, for example, toothpaste and mouthwash.

These products, however, have certain drawbacks. Notably L-menthol, which continues to be the most widely used physiological coolant, has a limited dosage range beyond which both the peppermint odor and associated bitterness and burn become undesirable. For over four decades there has been considerable research into alternative compounds that impart a physiological cooling effect without imparting significant bitterness or undesirable odor. The work of Watson (1977) and his coworkers at the British company Wilkinson Sword were focused primarily on the corporate goal of creating products that could be sold along with their other consumer products, and were therefore aimed mainly towards cosmetic applications (Watson *et al.*, 1978).

A number of published physiological studies suggested that the interaction of menthol with relevant receptors could be interpreted to give a reasonable correlation between chemical structure and activity (Hansch and Fujita, 1964; Hansch and Steward, 1964; Fujuta *et al.*, 1964; Hansch *et al.*, 1965; Leo *et al.*, 1971; Hansch and Dunn, 1972; Hansch and Clayton, 1973; Fleischer *et al.*, 1976). Around 1200 compounds were synthesized by the Wilkinson Sword team. Among the outcomes of this work was the belief that a number of important factors needed to be met for a compound to have effective cooling activity. These were:

- functionality within the molecule capable of hydrogen bonding;
- a compact hydrocarbon skeleton;
- a hydrophilic/hydrophobic balance (log *P*) within a certain range (1.5–5.0);
- molecular weight in the range 150–350.

In general, chemical structure–activity relationships in physiological coolants are not fully understood. Carboxamide materials, as a rule of thumb, will provide a more rapid onset of cooling than those compounds not containing nitrogen in their structure. It is easier, moving forward, to look at physiological coolants from a structural perspective in two separate classes. These are: (1) menthol and its derivatives, including menthol esters and ethers, and menthane carboxamides; and (2) non-menthol derivatives.

7.2.1 Menthol derivatives

7.2.1.1 Menthol

Before looking at molecules that take their starting point from menthol, we should first take a look at menthol and its enantiomers, as well as molecules with structural similarity to menthol. In terms of cooling, molecules derived from L-menthol (**1**) have stronger cooling properties than those derived from D-menthol (**2**), with L-menthol being around 45 times more active than D-menthol (Fig. 7.2). (L-)-Menthol can be obtained from natural and synthetic sources. It has the FEMA (Flavor and Extract Manufacturers Association) number 2665 and is on the EFSA (European Food Safety Authority) list of European flavor ingredients (2.218).

Menthol can be obtained from a variety of botanical species, and is prevalent in various mint oils such as peppermint oil (*Mentha piperita*) and cornmint oil (*Mentha arvensis*). Steam distilled oils from these species contain approximately 75 and 50% L-menthol, respectively, although this can vary. Synthetic menthol can be produced by a number of routes from different starting materials, such as thymol (Fleischer *et al.*, 1976), myrcene (Akutagawa, 1997), beta-phellandrene (Kane *et al.*, 1977), and citral via (–)-isopulegol, and (+)-(*R*)-citronellol (Rauls *et al.*, 2008; Heydrich, 2009). More recently Takasago patented a route that outlined the production of menthol from geraniol (Hisanri *et al.*, 2009).

Leffingwell has written an excellent white paper, available on his personal website (http://leffingwell.com/menthol1/menthol1.htm), which contains a comprehensive review of the enantiomers of menthol (**3–10**) and their sensory properties. Some of

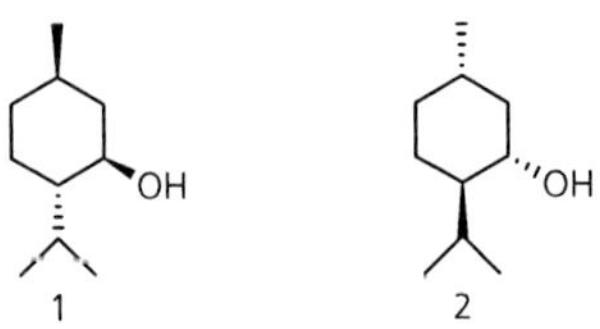

Fig. 7.2 Structures of L-menthol and D-menthol.

Table 7.2 Cooling and taste thresholds (ppm) of menthol enantiomers.

Isomer	Cooling threshold (ppm)	Taste threshold
(–)-Neoisomenthol	6	1
(+)-Neoisomenthol	28	0.2
(–)-Neomenthol	25	0.6
(–)-Neomenthol	2.5	0.5
(–)-Isomenthol	30	0.6
(+)-Isomenthol	7	0.7
(+)-Menthol	3	0.3
(–)-Menthol	0.8	0.4

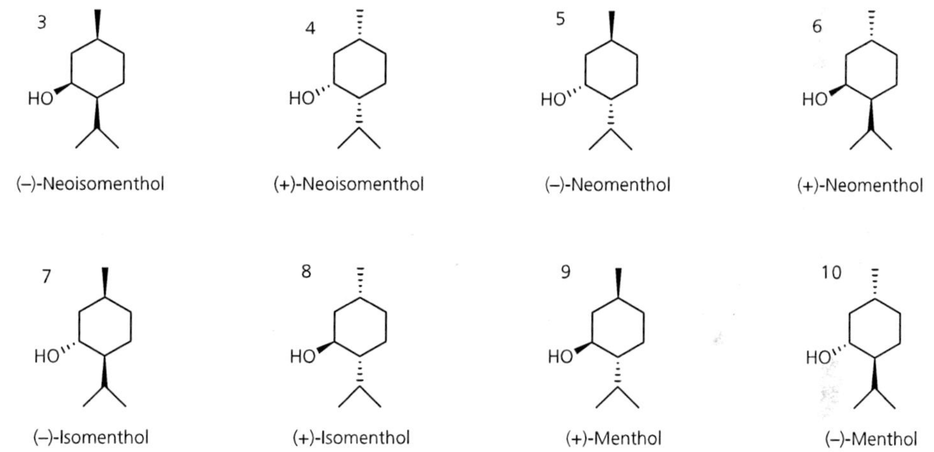

Fig. 7.3 Enantiomeric structures of menthol.

the data contained in that article are summarized in Table 7.2 and the stereospecific structures can be seen in Fig. 7.3.

7.2.1.2 (L)-Menthone

Menthone (**11**) is listed on the EFSA list (7.059) and has a FEMA number of 2667. It has a mentholic and minty odor, and at a 10% dilution in dipropylene glycol (DPG) it has a distinctive peppermint, herbal, and camphor note. It displays a slight cooling effect, however when compared with L-menthol this is extremely weak and is secondary to its olfactive properties.

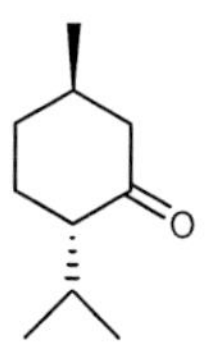

11

7.2.1.3 Menthol esters and ethers

There are a number of different ways in which to further derivatize menthol, such as those shown in compounds **12–16**. The simplest is to esterify menthol at the –OH group. Materials such as L-menthyl acetate (**12**), and L-menthyl lactate (**13**) are easy to prepare. Other derivatives such as L-menthol ethylene glycol carbonate (**14**), and L-menthol propylene glycol carbonate (**15**) are slightly more complex to synthesize but are still relatively straightforward. These materials are sold under the names Frescolat® MGC and Frescolat® MPC by Symrise and were the subject of a German patent (Pelzer *et al.*, 1994) that has recently expired.

12

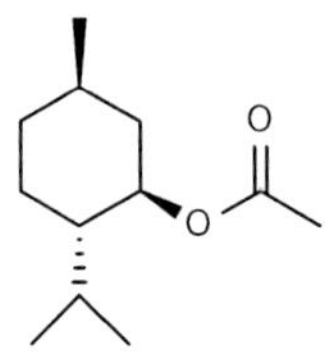

13

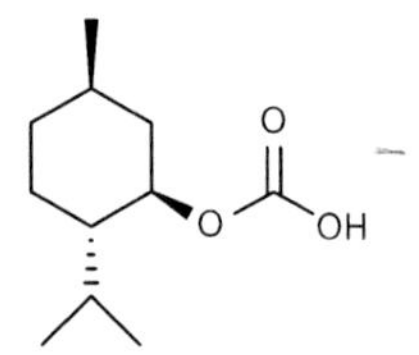

14

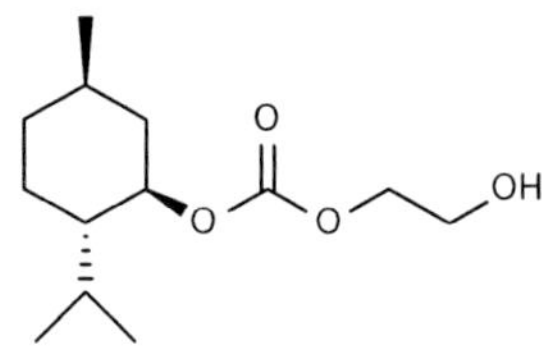

15

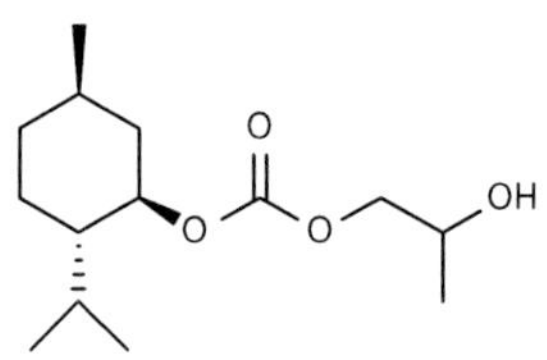

16

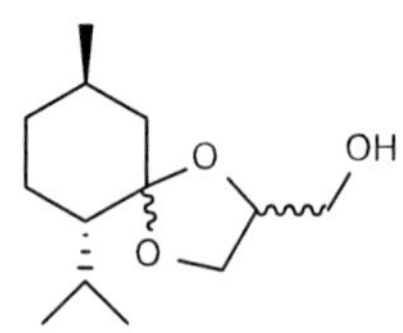

Menthone 1,2-glycerol ketal (**16**) is also sold by Symrise as Frescolat® MGA. There have been a number of articles published that propose its use in various edible and topical applications. There are a couple of articles specific to this compound (Greenberg, 1999; Jacobs and Johncock, 1999), however, much of the literature describes its usage with other physiological agents. The racemate L-isomer of Frescolat MGA is FEMA 3807 and the DL-isomer is 3808, with the latter being the commercially available material.

Additional menthol esters such as L-monomenthyl succinate (**17**), L-monomenthyl glutarate (**18**) (Cooler 2 – IFF), (–)-menthyl acetoacetate (**19**), (–)-menthyl 3-hydroxy-butyrate (**20**), and L-menthylpyrrolidone carboxylate (**21**) (Questice – Givaudan), along with others, have also been commercialized. Further molecules have utilized the hydroxyl moeity of L-menthol and created menthol ethers. Products such as 3-(L-menthoxy)propane-1,2-diol (**22**) is an example of this type of compound.

17
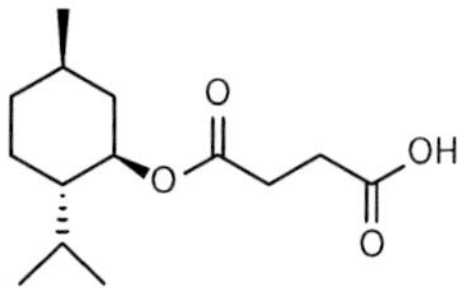

18
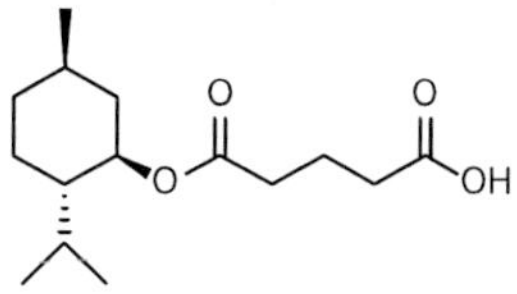

19
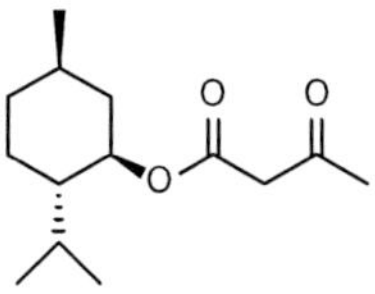

20
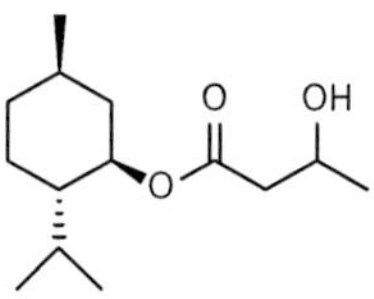

21
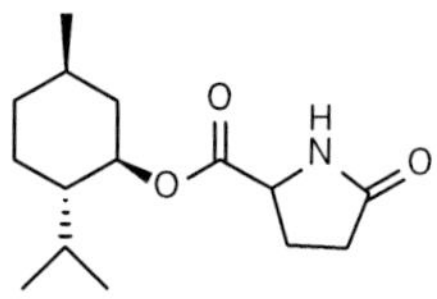

22
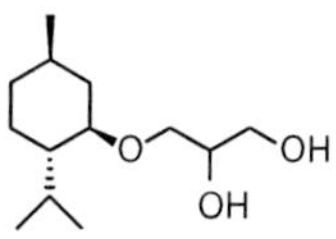

23
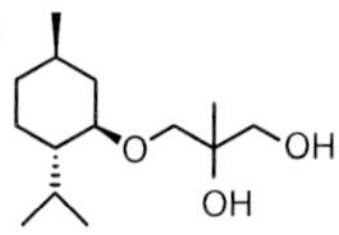

24
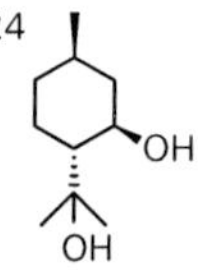

b
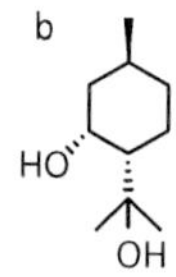

Table 7.3 shows a summary of the more important commercially available products and some of their basic properties. It should be noted that in Table 7.3, the cooling strength refers to that as a result of tasting rather than via another method of experiencing cooling, such as through topical application.

25
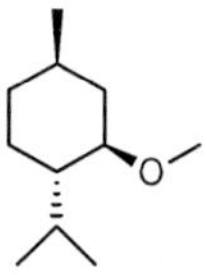

26
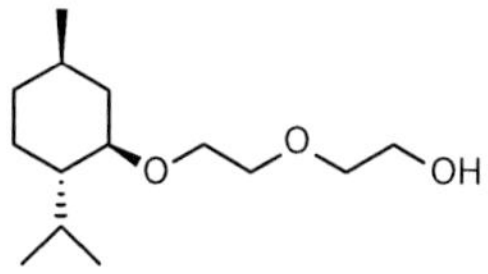

27
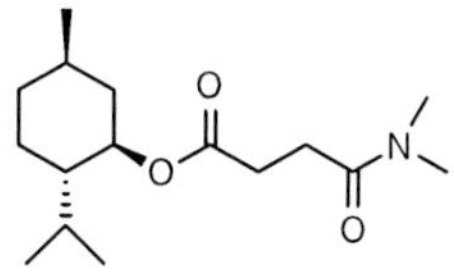

As can be seen in Table 7.3, the majority of commercially available coolants that utilize ester or ether linkages to functionalize menthol at the hydroxy group have cooling strengths that are below that of L-menthol. Considering that the cost of these materials is considerably above the market cost of L-menthol, we could question the use of these materials. However as previously mentioned, there are limitations to the use of L-menthol as a TRPM8 agonist, not least of these being the distinctive peppermint odor associated with the material, something which is undesirable in applications outside of purely peppermint flavors. Additionally, high load levels of menthol also induce the feeling of menthol "burn". In these instances, the effect of higher than normal levels of menthol induces an astringent, overpowering, and irritating effect, which detracts from the overall cooling sensation. Clearly

Table 7.3 Menthol derivative coolants.

Compound no.	Compound	Other names	FEMA	CAS	Flavis (EFSA)	Cooling strength*
(12)	L-Menthyl acetate			89-48-5	9.016	15
(13)	L-Menthyl lactate	Frescolat ML	3748	59259-38-0	9.551	40
(14)	Menthol ethylene glycol carbonate	Frescolat MGC	3805	156679-39-9	9.842	75
(15)	Menthol propylene glycol carbonate	Frescolat MPC	3806	30304-82-6	9.843	67.5
(16)	Menthone glycerol ketal	Frescolat MGA	3807/3808	63187-91-7		40
(17)	Monomenthyl succinate	Physcool	3810	77341-67-4	9.616	30
(18)	Monomenthyl glutarate	Cooler 2 (with dimenthly glutarate), Ultracool 2	4006	220621-22-7	9.929	45
(19)	Menthyl acetoacetate	Ultracool 7	4327	59557-05-0		200
(20)	Menthyl 3-hydroxybutyrate		4308	108766-16-1		150
(21)	Menthyl pyrrolidone carboxylate	Questice	4155	68127-22-0 (powder), 64519-44-4 (liquid)		130 (liquid), 105 (powder)
(22)	(–)-Menthoxypropane-1,2-diol	Coolact 10	3784	87061-04-9	2.224	40
(23)	3-(l-Menthoxy)-2-methylpropane-1,2-diol	TPG-1, Coolact 1	3849	195863-84-4	2.254	130
(24)	*cis* and *trans* *p*-Menthane-3,8-diols	Coolact 38D	4053	4822-86-6	2.246	10
(25)	Menthyl methyl ether	Optabreeze V	4054	2748-84-7	16.088	37.5
(26)	2-(*p*-Menthan-3-yloxy)ethanol	Coolact 5	4154	3618-23-4	2.247	150
(27)	(+/–)-*N*,*N*-Dimethyl menthyl succinamide		4230	544714-08-1		135
(28)	Menthone glycerol ketal		4285	831213-72-0	6.136	60
(29)	3*S*,3a*R*,3b*R*,4*S*,7*R*,7a*R*)-3,7-Dimethyl-4-(propan-2-yl) octahydro-1H-cyclopenta[1,3] cyclopropa[1,2]benzen-3-ol	Cubebol	4497	23445-02-5		15
(30)	Dimenthyl glutarate		4604	77304-30-4		0

*L-menthol =100.

then, one of the key considerations for the use of a physiological coolant is the ability of these compounds to interact and blend with menthol. In fact when using the material in a complex formulation, the key is to understand and evaluate how the physiological coolant works alongside menthol and the other aroma components in the entire system. Materials such as menthol glycerol ketal (**16**), menthol propylene glycol carbonate (**17**), and menthyl 3-hydroxybutyrate (**20**) work particularly well in this respect by blending well with the other aroma components such as L-carvone, anethole, and various mint oils. An interesting effect is observed when using monomenthyl glutarate (**18**). When used in flavor systems, and blended with menthol, a mild and pleasant cooling effect is experienced. When this material is used in conjunction with its very close relative dimenthyl glutarate (**30**), not only is there a noticeable increase in the cooling effect but also the effect of increased salivation is often perceived. This salivating experience often has the desirable effect of promoting or potentiating the flavor being experienced, which, depending on the consumer application, can give a more indulgent or effective taste experience.

28
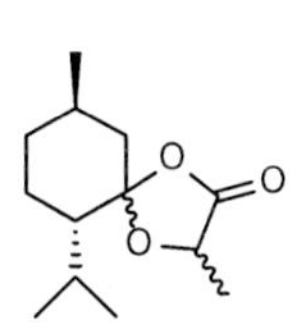
29
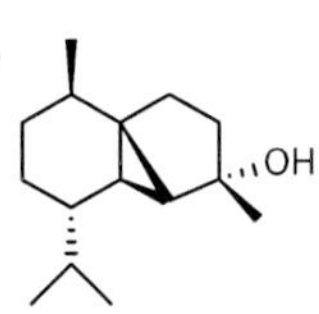
30
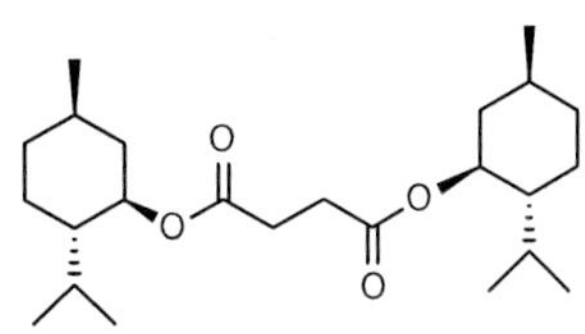

7.2.1.4 Menthane carboxamides

While there has been a wide number of structural variants based around derivatization of menthol at the hydroxyl (OH) moiety as described earlier and shown in Table 7.3, the majority of these materials have had limited commercial adoption. From a cost in use perspective, it is often found that their use can be prohibitive, given the levels of coolant needed in a formulation to achieve the desired level of cooling. In addition to these types of compounds, there is a separate menthol derived class of materials that have significantly improved cooling properties, making them more attractive to users. These materials are derivatized to remove the hydroxyl group of menthol, replacing it with an amide moiety. In general, these materials can be grouped together under the term menthane carboxamides. Probably the most widely used menthane carboxamides are *N*-ethyl-*p*-menthane-3-carboxamide or WS-3 (**31**) and ethyl-3-(*p*-menthane-3-carboxamido) acetate or WS-5 (**32**). More recently (1*R*,2*S*,5*R*)-*N*-(4-methoxyphenyl)-*p*-menthanecarboxamide or WS-12 (**33**) has been commercially introduced due to its longer lasting cooling properties.

31
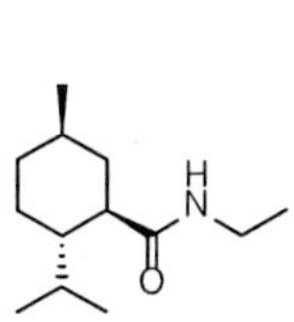
32
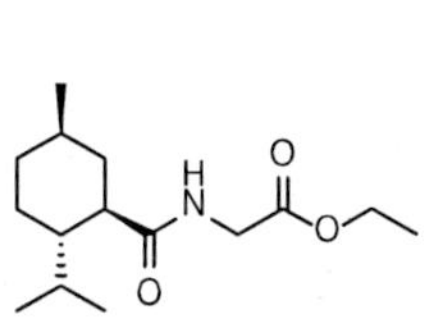
33
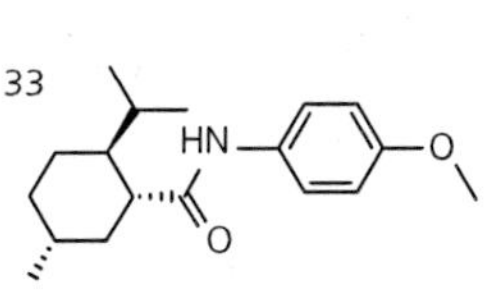

From the many compounds prepared by Watson and coworkers on behalf of Wilkinson Sword (Watson, 1977), the WS annotation for these materials has become widely accepted and is often used to refer to them. A number of the

Table 7.4 Non-FEMA GRAS WS menthane carboxamides.

Designation	Nomenclature	Relative cooling strength*
WS-63	*N*-(3-Hydroxy-4-methylphenyl)-*p*-menthane-3-carboxamide	300
WS-61	*N*-(3,4-Dimethylphenyl)-*p*-menthane-3-carboxamide	300
WS-35	*N*-(4-Methylphenyl)-*p*-menthane-3-carboxamide	100
WS-14	*N*-tert-Butyl-*p*-menthane-3-carboxamide	75
WS-11	*N*-(1,1-Dimethyl-2-hydroxyethyl)-*p*-menthane-3-carboxamide	67
WS-10	*N*-Isopropyl-*p*-menthane-3-carboxamide	67
WS-138	*N*-(3-Pyridyl)-*p*-menthane-3-carboxamide	60
WS-134	*N*-Cyclopentyl-*p*-menthane-3-carboxamide	60
WS-56	*N*-(5-Hydroxy-n-pentyl)-*p*-menthane-3-carboxamide	60
WS-34	*N*-sec-Butyl-*p*-menthane-3-carboxamide	43
WS-33	*N*-n-Propyl-*p*-menthane-3-carboxamide	38
WS-78	*N*-Cyclohexyl-*p*-menthane-3-carboxamide	30

*L-methol = 100.

materials synthesized were menthane carboxamides, and some of these materials are listed in Table 7.4, along with their approximate cooling strength relative to L-menthol (with menthol being designated 100). For a detailed comprehensive review of the work undertaken by Wilkinson Sword, the interested reader is referred to Leffingwell and Rowsell (2014).

Even though most of the materials in Table 7.4 are not used commercially due to their lack of regulatory approval through FEMA, EFSA or other regulatory bodies, this does not mean that there have not been materials with menthane carboxamide structures beyond WS-3 and WS-5 that have have been approved. Some of these are summarized in Table 7.5.

34
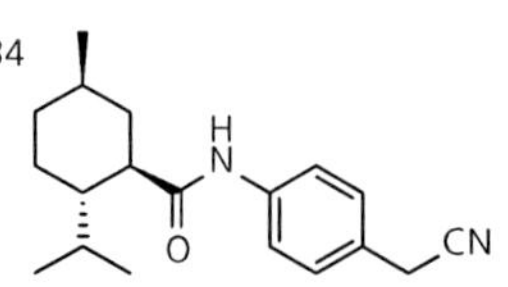

35
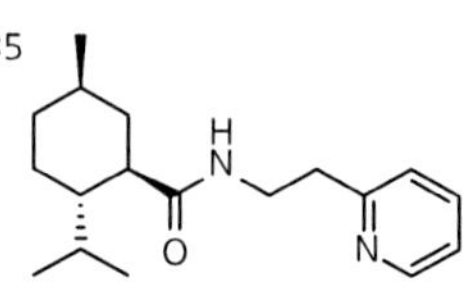

36
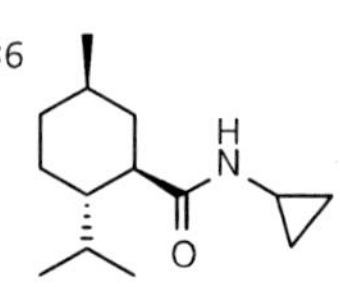

It can be seen from Table 7.5 that one of the reasons for the acceptance of menthane carboxamides and their adoption, especially in applications that are experienced orally by the consumer, such as those in confectionary and oral care, is their relatively high cooling strength when compared with L-menthol. Additionally, their ease of use when blending with L-menthol, mint oils, L-carvone, and other typical flavor compounds used to create mint flavor applications make them particularly attractive. From a cost in use standpoint, as they have strengths greater than menthol, they can be used to boost or increase the perception of freshness in mint

Table 7.5 Selected FEMA GRAS menthane carboxamides.

Compound	Name	CAS	FEMA	Flavis	Other names	Coolant strength*
31	*N*-Ethyl-*p*-menthane-3-carboxamide	39711-79-0	3455	16.013	WS-3	150
32	Ethyl-3-(*p*-menthane-3-carboxamido) acetate	68489-14-5	4309	16.111	WS-5	400
33	(1R,2S,5R)-*N*-(4-Methoxyphenyl)-*p*-menthanecarboxamide	68489-09-8	4681	16.123	WS-12	370
34	*N*-(4-cyanomethylphenyl)-*p*-menthanecarboxamide	852379-28-3	4496	16.117	Evercool 180	750
35	*N*-(2-(Pyridin-2-yl)ethyl)-3-*p*-menthanecarboxamide	847565-09-7	4549	16.118		600
36	*N*-Cyclopropyl-5-methyl-2-isopropylcyclohexanecarboxamide	73435-61-7	4692	16.124		100

*L-Menthol=100.

flavors and allow formulators to achieve higher efficacy in this area without the previously mentioned negative impacts of menthol alone, such as an overpowering peppermint note, or increased astringency.

While these materials have interesting and attractive cooling properties, one of the potential hurdles to be overcome was the more complex synthetic route to manufacturing these materials. Take for example WS-3, where preparation by conversion of menthol into menthyl chloride, followed by the corresponding Grignard reaction, and treatment with carbon dioxide allows the acid to be isolated. The acid can then be converted into the amide via acid chloride or similar methods. The challenge is then producing the material in sufficient purity such that any impurities do not create or contribute to an undesired olfactive or taste effect. This was easily solved in the case of WS-3 and hence the historical adoption of this material in formulations. However, with materials such as WS-5 the problem was not so easily solved, and it was not until Erman and colleagues discovered a way of sufficiently purifying this product that it became commercially available in a form that was acceptable for use and able to be used widely (Erman and Whelan, 2002). Prior to this discovery, WS-5 contained impurities that caused undesirable astringency to be experienced by the consumer, impairing its use by formulators. Since this discovery, and due to its increased cooling strength and less complex application patent landscape, more formulators are turning to WS-5. Care must still be taken, however, as material supplied outside of the method patented by Millenium Specialty Chemicals (now Renessenz LLC) can still have an undesirable taste profile.

The solubility of these materials is also a reason why compounds such as WS-3 (**31**) and WS-5 (**32**) have been adopted as being the materials of choice in flavor applications. As can be seen in Table 7.6, they can be easily dissolved in typical solvents used in flavor formulation and product development.

Other menthane carboxamides can be more difficult to use, and the requirement for increased longevity has led to the creation of compounds with chemical structures that do not lend themselves to be easily soluble. Even though materials such as (1*R*,2*S*,5*R*)-*N*-(4-methoxyphenyl)-*p*-menthanecarboxamide (WS-12, **33**) and *N*-(4-cyanomethylphenyl)-*p*-menthanecarboxamide (**34**) have some formulation issues, they have been introduced to the FEMA GRAS and EFSA lists because of other features of their cooling ability. Consumer demand for longer cooling effects in applications such as oral care has led to the development of materials with these properties. Figure 7.4 shows cooling longevity at 20 ppm in ethanol/water mixtures for established coolant products such as menthyl lactate, WS-3, WS-5, and WS-12.

Different hypotheses exist as to why materials such as WS-12 (**33**) exhibit longer cooling profiles. The first one states that this is simply a function of their increased initial strength. However, when comparing WS-5 versus WS-12 this hypothesis does not hold. In a second hypothesis, it is thought that the longer cooling effect is due to the reactivity at the receptor with these molecules being particularly slow to react, meaning that there remain unreacted molecules over a period of time, which imparts longevity. A further hypothesis is related to the solubility of these materials in organic and aqueous phases, and that in order for these materials to reach the relevant receptor and act as an agonist, they do not have sufficient

Table 7.6 Solubility of selected menthane carboxamide coolants.

Compound	Name	Propylene glycol (g/100 ml)	96% Ethanol g/100 ml	60% Ethanol g/100 ml	40% Ethanol g/100 ml
31	N-Ethyl-*p*-menthane-3-carboxamide	10.7	186	14.8	0.2
32	Ethyl-3-(*p*-menthane-3-carboxamido) acetate	11.7	79.4	50	1.9
33	(1R,2S,5R)-N-(4-Methoxyphenyl)-*p*-menthanecarboxamide	0.05	0.9	0.015	Practically insoluble

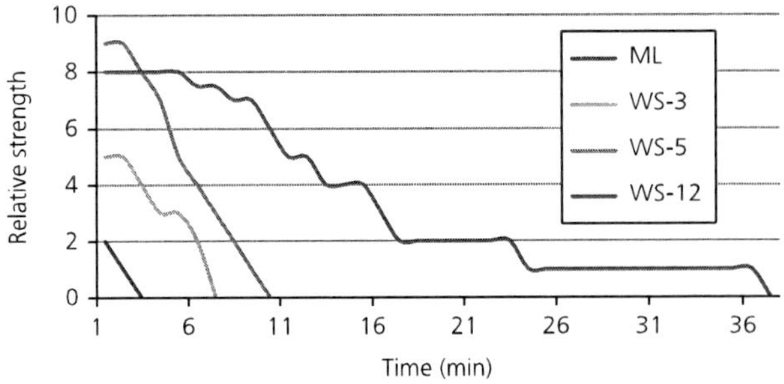

Fig. 7.4 Cooling longevity at 20 ppm in ethanol/water; left to right: ML, WS-3, WS-5, and WS-12.

solubility in either phase to create a rapid enough reaction at the receptor and instead sit in the oral cavity until they are slowly solubilized. Further work is needed in this area to determine the true mechanism of their longevity.

7.2.2 Non-menthol derived coolants

It is well known that in 1965 the first material with cooling properties added to the FEMA GRAS list was isopulegol (**37**). Since then, several other materials have been discovered and introduced as physiological coolants. Some of those, which are commercially available and/or have been added to the FEMA GRAS lists, are detailed in Table 7.7.

37

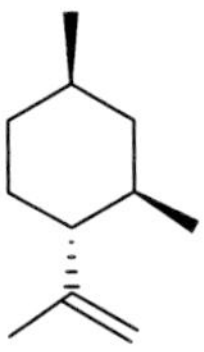

38

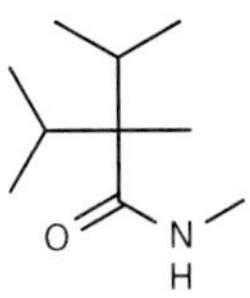

39

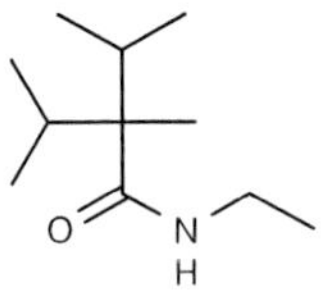

40

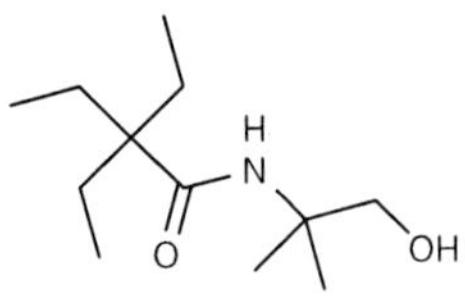

41

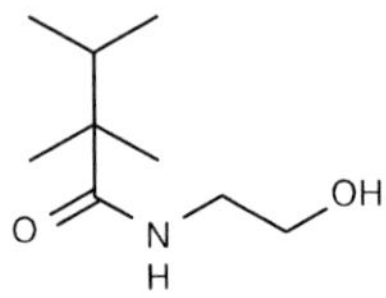

2-Isopropyl-*N*-2,3-trimethylbutyramide (WS-23) (**38**) gained acceptance with users some time ago and was added to the FEMA GRAS list in 1996. Since then it has become, along with WS-3, one of the widest used physiological coolants and finds use predominantly in flavor applications.

Research continues into discovering other structures that can act as TRPM8 receptor agonists. The search amongst chemical structures unrelated to menthol has been widened due to the emergence of technology that mimics TRP receptors and allows for high throughput screening of a variety of molecules. The US patent

Table 7.7 Selection of non-menthol derived coolants added to FEMA GRAS lists.

Compound no.	Name	Other names	CAS	FEMA	Flavis	Cooling intensity*
37	Isopulegol	Coolact® P	89-79-2	2962	02.067	25
38	2-Isopropyl-*N*,23-trimethylbutyramide	WS-23	51115-67-4	3804	16.053	75
39	*N*-Ethyl-2,2-diisopropylbutanamide	WS-27	51115-70-9	4557		150
40	*N*-(1,1-Dimethyl-2-hydroxyethyl)-2,2-diethylbutanamide	WS-116	51115-77-6	4603		15
41	*N*-(2-hydroxyethyl)-2,3-dimethyl-2-isopropylbutanamide	ICE 6000	883215-02-9	4602		20

*Menthol = 100.

application by Priest *et al.* (2013), and assigned to Senomyx, describes an extensive list of potential new molecules with physiological cooling properties. This technology is particularly interesting as it allows for rapid screening and evaluation of cooling strength of chemical structures that previously may have been ignored. The molecule **42** is an example of the compounds that appears to have strong cooling potency based on the EC_{50} (μM) values. Note the EC_{50} value is half the maximal effective concentration of a compound that induces a response halfway between the baseline and maximum after a specified exposure time when examined using a fluorimetric imaging plate reader assay (FLIPR™).

42

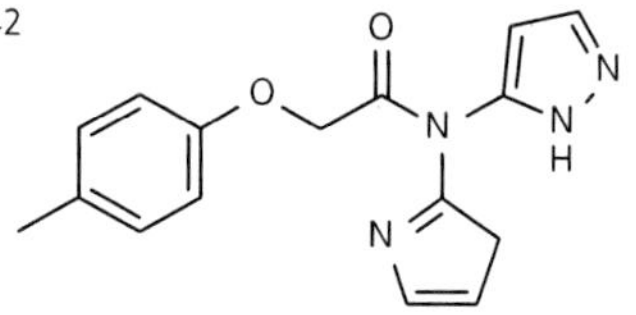

Other molecules such as 4-methyl-3-(1-pyrrolidinyl)-2[5H]-furanone (**43**) have been reported (Ottinger *et al.*, 2001). This compound, according to studies, is reported to have a cooling strength 35 times stronger than that of menthol. Other recent patents also report the structures **44–46** as having TRPM8 cooling efficacy similar to that of menthol (Krohn and Zinke, 2013).

43

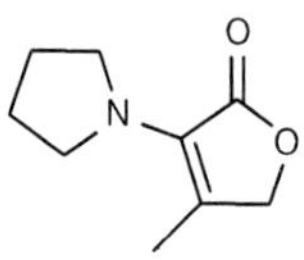

44

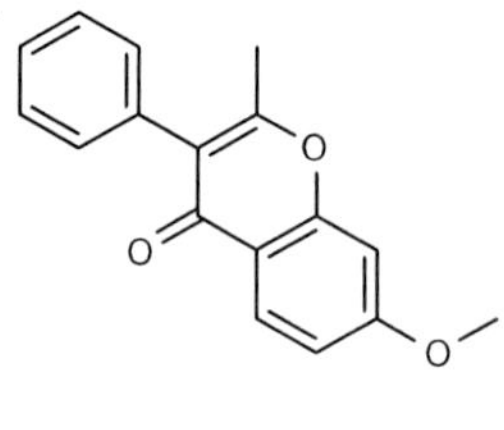

45

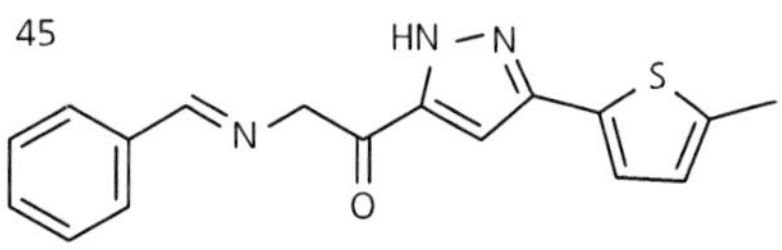

46

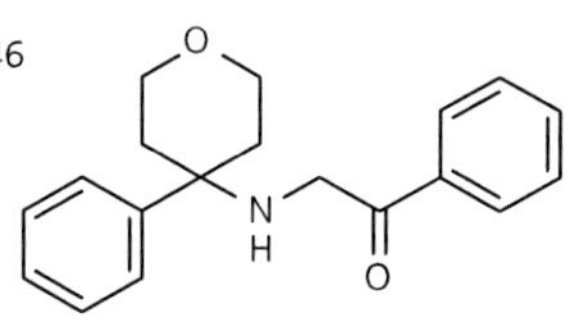

Materials such as **47–49** have been reported by Shirai *et al.* as a series of lignin type derivatives. These are disclosed as TRPM8 activators. Additionally, 1-(2-hydroxyphenyl)-4-(2-nitrophenyl)-1,2,3,6-tetrahydropyrimide-2-one or icilin (**50**) was discovered as far back as 1983 to produce coldness when in contact with mucous membranes when ingested (Wei and Seid, 1983). It is claimed that icilin is a super agonist being 200-fold more potent than L-menthol, however Andersson *et al.* suggest that the correlation between icilin and cold involves a different mechanism than activation by menthol (Andersson *et al.*, 2004).

47
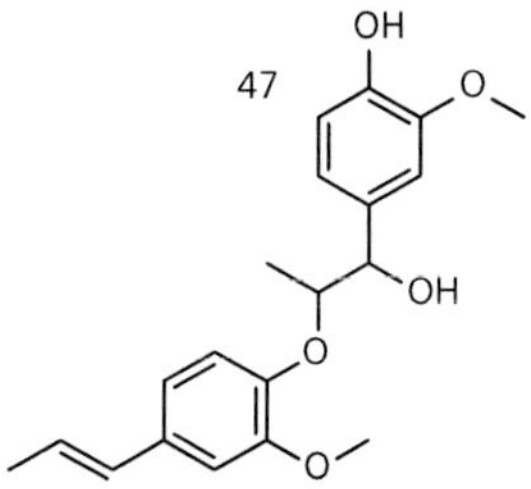

48
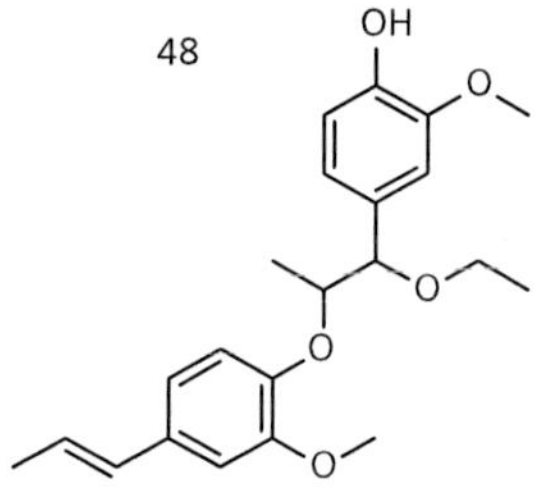

49
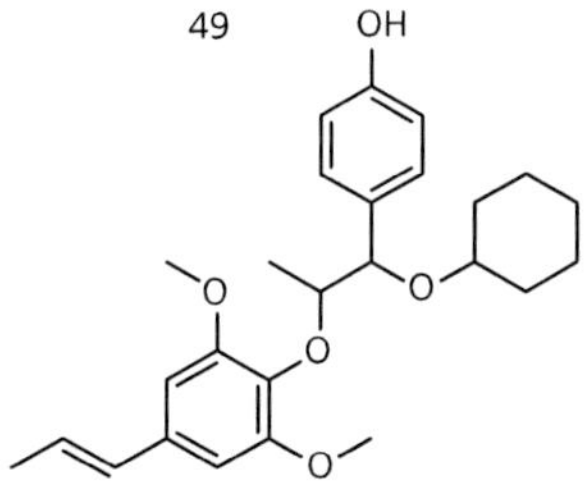

50
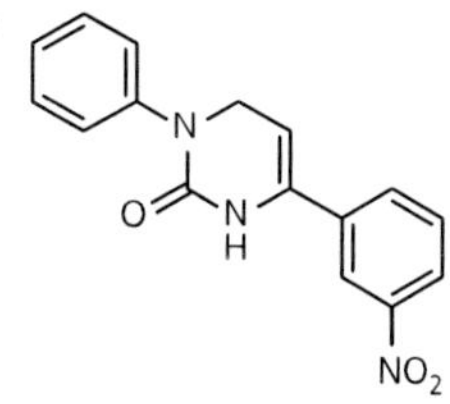

7.3 Physiological cooling outside of the oral cavity

Although the major applications in terms of quantity of use for physiological coolants are those where cooling is experienced in the oral cavity, there are still a wide number of potential uses in other areas. While menthol itself imparts a cooling sensation on the skin, one of the challenges in developing molecules to impart topical cooling is that available quantitative information to determine cooling strength and longevity is very fragmented. Different parts of the body vary in their sensitivity to cooling agents, however an attempt has been made to give an order of relative sensitivity (Watson *et al.*, 1977). This study gives the order indicated in Fig. 7.5. According to this study the order follows the general order of increasing thickness of the *stratum corneum*, and suggests that the main factor that affects sensitivity is the ability of the cooling agent to penetrate the skin barrier. An additional factor is the areas within the skin where there are higher concentrations of the relevant receptors to interpret heat and cooling, these being the face, scalp/forehead, armpits, thighs, and groin, while areas with fewer receptors include the arms and hands (Jacobs and Johncock, 1999). There is general acceptance that the vehicle in which the cooling agent is carried through the skin is another critical factor.

When a formulation which comprises an active ingredient such as a physiological coolant is applied to the skin, three processes must occur in order for cooling to be perceived, these being:

1 cooling molecules transfer from the vehicle and penetrate the *stratum corneum*;
2 molecules diffuse through the skin;
3 molecules reach and interact with the TRPM8 receptor.

From the work published by Watson *et al.* (1978) polarity is the most important factor in the success of the vehicle. After testing of nine potential cooling materials in 0.25% solutions in eight vehicles with varying polarities ranging from olive oil to aqueous ethanol,

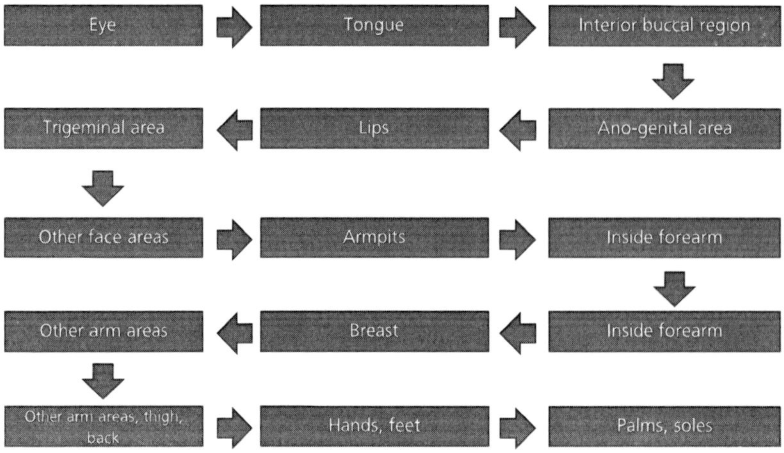

Fig. 7.5 Order of relative sensitivity to cooling of various skin areas.

all nine cooling agents gave stronger cooling in the more polar vehicles, with no cooling observed in the vehicles with lowest polarity (olive oil and hexyl laurate). As the coolants are relatively hydrophobic (log $P > 0$) they have a tendency not to migrate from less polar vehicles to the skin, while increasing the polarity allows the coolant to migrate. Once the coolant has been released into the *stratum corneum* the next step is the ability of the molecule to travel to reach a receptor. This would of course be dependent on the size of the molecule and it would therefore seem that the requirements of a molecule to exhibit cooling on the skin would be similar to those set out earlier in this chapter. However, it appears that this is not necessarily the case. The need for a compact hydrocarbon skeleton does not fit in with the cooling effect observed by molecules such as icilin (**50**) or indeed any of compounds **44–49**, which also display cooling. In addition, along with molecules such as WS-5 and WS-12, these also indicate that a low molecular weight is not necessary with these molecules becoming increasingly heavier.

The question of whether we could use the efficacy of cooling compounds within the oral cavity as an indication of their efficiency on the skin has also been addressed. It has been reported that oral thresholds are of limited use for prediction of topical effects (Watson *et al.*, 1978), and this was confirmed by experimentation at Millenium Specialty Chemicals (now Renessenz LLC). This experimentation was performed by applying several coolants on to the forearms of panelists at a level of 1% in PG with PG being used as a control. It was found that coolants were effective in the order of menthyl lactate >> WS-3 > WS-23 > Frescolat MGA = PG, the anomaly here being the comparatively high cooling strength of menthyl lactate when compared with the other coolants being studied. More recent experimentation at Renessenz used isopropyl palmitate as the vehicle and found that in terms of cooling strength WS-23 > WS-3 >> WS-5 = Questice liquid = menthyl lactate. This again gives results for cooling strength that differ from those obtained consistently when the same materials are evaluated through the oral cavity.

Other reported work on the use of physiological coolants in topical applications has been reported (Wei, 2006, 2007). This work shows that several materials related to WS-5 possess stronger levels of cooling and higher levels of longevity than WS-5 when

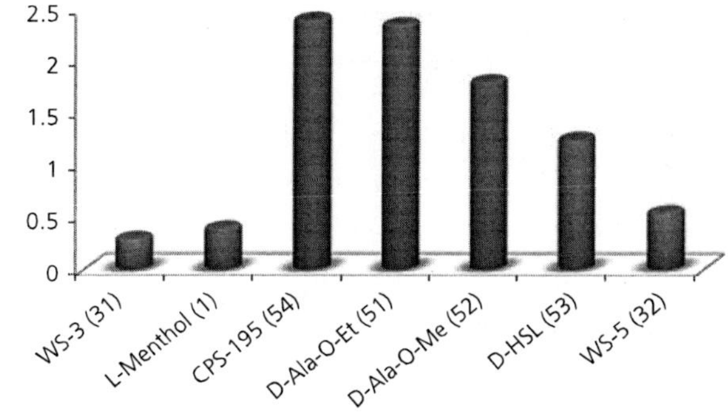

Fig. 7.6 Topical cooling duration (hours) 1% ointment.

applied in 1% petroleum based ointment to the skin. It appears that by making small structural changes to the molecule more effective cooling can be achieved (Fig. 7.6).

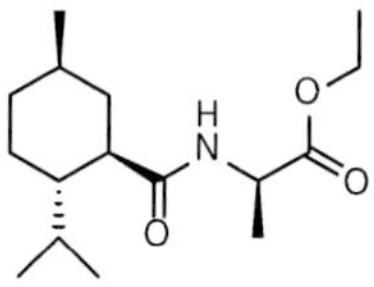

51

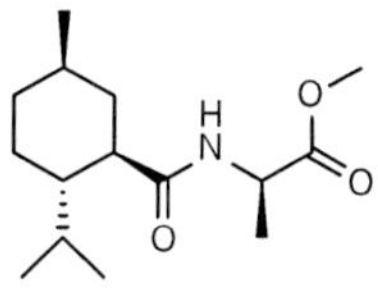

52

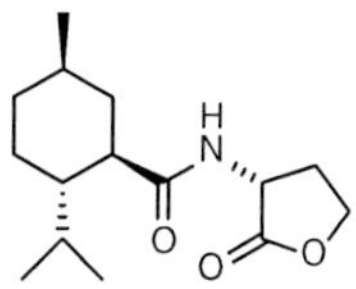

53

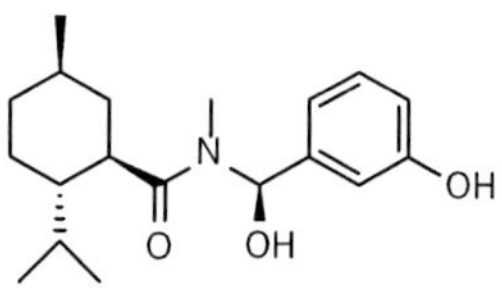

54

Many patents relating to consumer products in personal care applications disclose the use of physiological coolants. Menthyl lactate has been mentioned in patents (Deckner *et al.*, 1997; Lin and Baier, 1998; Fowler *et al.*, 1998) in compositions for artificial tanning, compositions for treating acne, and skin cleansing components. Other physiological coolants have also appeared in patents. For example Frescolat MGA may be added to powder compositions for reducing skin irritation (Nitikhunkasem *et al.*, 2002), menthyl lactate, Frescolat MGA, and WS-3 can be used in topical applications containing oxa-acids (Ptchelintsev *et al.*, 2003), and combinations of WS-3 and WS-23 were shown as being useful for cooling compositions with reduced stinging (Luke, 2001).

Searches also reveal some interesting studies in the field of pharmaceutical research. Work patented by Wei (2005, 2006, 2007, 2008a, 2008b, 20010a, 2010b, 2011a, 2011b, 2012, 2013a, 2013b) shows the modeling of a series of therapeutic analgesics consisting of a number of Wilkinson Sword coolants and closely related compounds as pain suppressant compounds for the treatment of non-keratinized

Table 7.8 Relative cooling versus menthol of various WS compounds.

Compound	Name	Cooling	Duration (hours)
WS-109	*N*-(*p*-Menthane-3-carbonyl)-D-alanine ethyl ester	0.1	2.4
WS-108	*N*-(*p*-Menthane-3-carbonyl)-L-alanine ethyl ester	1	0.7
WS-12	(1*R*,2*S*,5*R*)-*N*-(4-methoxyphenyl)-*p*-menthane carboxamide	1.25	0.4
WS-5	Ethyl 3-(*p*-menthane-3-carboxamido)acetate	1.5	0.5
WS-14	*N*-*t*-butyl-*p*-menthane-3-carboxamide	0.75	0.1
	Menthol	1	0.3

Table 7.9 Studies of WS-3 versus DEET as an insect repellent.

Compound	Days following treatment	Treated section present (%)	Untreated section present (%)	Dead present (%)
WS-3	1	1.5	98	0.5
WS-3	2	0.6	99	0.5
WS-3	3	4.5	95	0.5
WS-3	7	20	79	1
DEET	1	3	96.5	0.5
DEET	2	2.5	96	1.5

stratified epithelium (NKSE) around the larynx, eyes, trachea, and/or bronchi. In particular Wei sought coolants with high longevity with *N*-(*p*-menthane-3-carbonyl)-D-alanine ester (WS-109) being a preferred compound (Wei, 2006). Wei also discovered that the 4-ethoxy substituted analogue of WS-12 showed increased levels of cooling strength and longevity when compared with WS-12 (Wei, 2005). An outline of the cooling strengths and longevity versus menthol from Wei's work is summarized in Table 7.8.

Other novel applications for physiological coolants continue to be sought and there have been a number of reports of physiological coolants being used as insect repellents. Gautschi and Blondeau of Givaudan have discovered that WS-3 and related *N*-substituted *p*-menthane carboxamides have insect repelling capability. They have measured the activity in comparison with diethyl-*m*-toluamide (DEET), which is widely used as an insect repellent in many consumer products (Gautschi, 2006). Results of their study in repelling cockroaches are shown in Table 7.9.

Questice has also been patented as an insect repellent (Watkins *et al.*, 2002), and there have been other patents that claim the use of menthyl lactate or Frescolat MGA for repelling mites and other insects (Kalbe and Nentwig, 2000).

7.4 Usage and consumer perception

Typically, and as mentioned earlier, chemesthesis, when aligned to physiological cooling through application via a cavity flavor system, mainly has the effect of imparting freshness. Freshness, as a sensory profile, is strongly linked to mint

Table 7.10 Pressed mint formulations.

Ingredient	Control	Coolant enhanced pressed mint
Sorbitol	96.85%	98.088%
Sucralose	1.85%	0.060%
Magnesium stearate	0.30%	1.250%
Peppermint flavor	1.00%	0.400%
WS-5		0.020%
WS-12		0.002%

applications, which in turn are prevalent in oral care and confectionary applications. Mint flavors in these applications are, in the main, spearmint and peppermint, but other variations can occur. Presentations given by Renessenz have shown that by removing the physiological coolant in a mint flavor used in the pressed mint candies (formulations in Table 7.10), the perception is of a "weaker, lower quality flavor". In addition, physiological coolant added to a fruit flavor at or above the level of cooling gives the perception and feedback that the candy is perceived as "minty", despite there being no actual aroma chemicals associated with mint flavor present.

7.4.1 Physiological coolants in applications beyond cooling

How chemesthesis and physiological materials can play an important role in consumer perception of a taste experience has been alluded to earlier. It has been reported by Di Salle *et al.* (2013) that physiological effects such as carbonation can reduce the perception of sweetness in beverages (we all know flat soda tastes much sweeter too). Experiments by Bartoshuk *et al.* (1982) as well as by Green and Frankmann (1987), show that temperature can have a marked effect on the perceived intensity of taste. Furthermore, work by Fry *et al.* (2011) shows that Rebaudioside A (Stevia) is perceived as sweeter at lower temperatures, meaning that it could be more effective in applications that are served cold (beverages, ice creams) or those where there is a physiological cooling such as toothpaste, mint vconfectionary, and chewing gum.

If physically changing the temperature affects the way receptors interpret the multitude of signals being sent during the experience of tasting something, then can physiological changes in temperature have the same effect? Work by Renessenz presented at the American Association of Candy Technologists meeting (Campbell, 2014) suggests physiological coolants can also have an effect on perceived sweetness of a range of high-potency sweeteners (HPS). Figure 7.7 shows that in all instances, with the exception of Ace-K, addition of a level of WS-5 below that which imparts cooling increases the perception of sweetness. Sweetness was inhibited when dosed with WS-5 alone. However, when a 50 : 50 mixture of Ace-K and aspartame was dosed with the same level of WS-5, a significant increase in sweetness perception was observed. In addition, combining coolant with sucralose enabled the off-notes of the sweetener to be masked or inhibited.

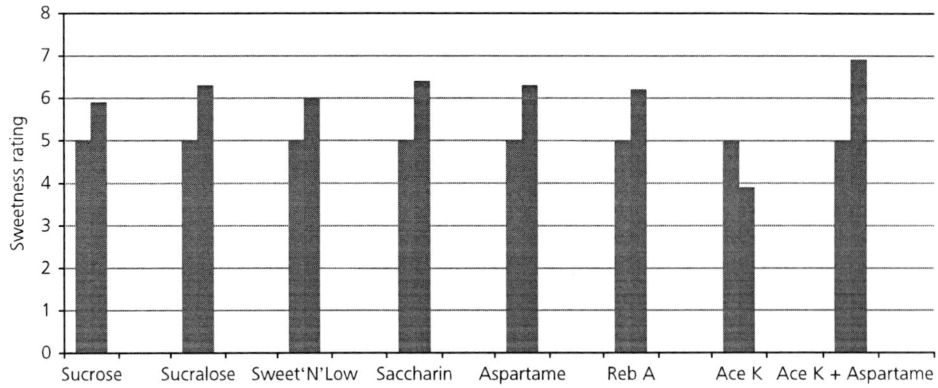

Fig. 7.7 Impact of WS-5 on the perceived sweetness of selected HPS.

7.4.2 Physiological cooling and flavor enhancement

Work by Petit *et al.* (2007) has shown a link between the use of coolants and flavor intensity. This work studied the effect of color–aroma–cooling interactions in beverages with congruent and incongruent formulations as well as the impact of exposure on flavor and cooling perception in an incongruent formulation. Results indicated that solutions with various levels of WS-3 and melon flavor in a green colored beverage (the color green being associated with melon) enhanced the perception of fruit flavor intensity. This is supported by work carried out by Barcelon *et al.* (1998). This work, in chewing gum, showed that the character of the flavor is generally found to be sharper or brighter, punctuated, more defined, and longer-lasting when compared with compositions containing the same flavoring agent without *N*-ethyl-*p*-menthane-3-carboxamide. Citrus flavors were found to be more citrusy and had a better defined and better perceived acid character. Cinnamon spice character was better defined with an intensified burn sensation.

Renessenz (Campbell, 2014) also shows that this effect can be seen beyond chewing gums. Table 7.11 shows three formula for strawberry gummy candies.

The control and two additional prototypes were sampled to a sensory panel and a series of forced response questions completed. Figure 7.8 shows a selection of the results presented by Campbell (2014) and clearly indicates that addition of a physiological coolant at levels below which cooling is perceived has a marked effect on several areas. The perceived increase in flavor intensity could be taken into account for the creation of more indulgent products or allow developers to reduce flavor load and decrease cost. Of note was the perception that the fruit flavor became more "juicy" and the belief that the character was more representative of the natural fruit. Overall liking of the product also increased by improving the flavor characteristics. The effect is not only applicable to strawberry flavors but is also relevant for the majority of fruit flavored products, which is of critical importance to developers in the confectionary and beverage categories as they continue to improve product perception and consumer acceptance of new and established brands.

It has been discussed earlier in the chapter that small structural variations can have effects on the chemesthetic properties of physiological cooling agents. This

Table 7.11 Strawberry gummy candy formulae.

Composition	Control	P1	P2
Strawberry flavor	0.40%	0.40%	0.40%
Citric acid – 50% solution	1.50%	1.50%	1.50%
Red 40–10% solution	0.05%	0.05%	0.05%
Gummy base	98.05%	98.0495%	98.0495%
WS-5	—	5 ppm	10 ppm

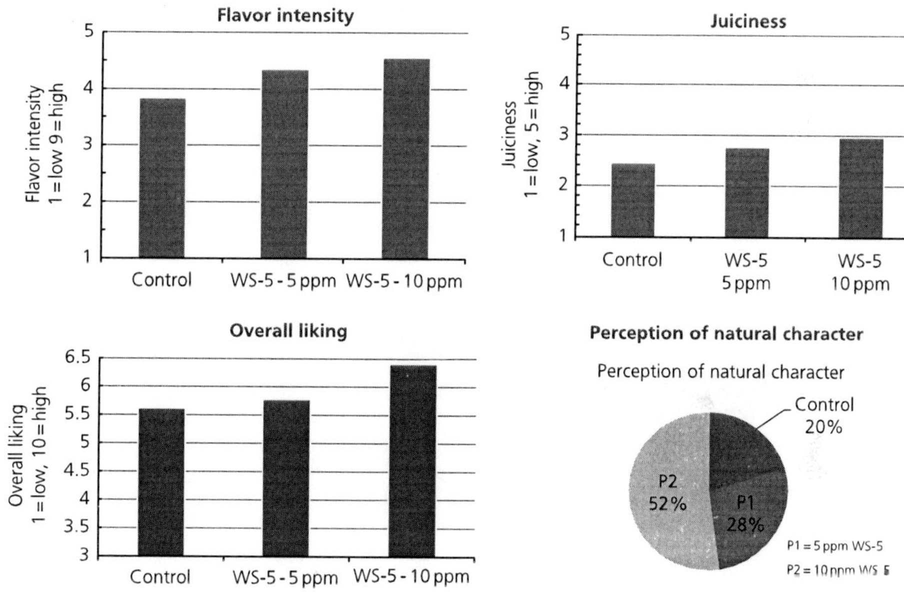

Fig. 7.8 Effect of addition of WS-5 on character of strawberry gummy candy.

is clearly demonstrated when we look at *N*-cyclopropyl-5-methyl-2-isopropyl-cyclohexanecarboxamide, FEMA 4693 (**36**). This material has been shown to have a cooling strength similar to L-menthol. The structurally related compound *N*-isopropyl-5-methylcyclohexyl cyclopropanecarboxamide, FEMA 4558 (**55**) is reported as having no cooling effect at all and instead has an umami character, with the neo menthol configuration being most effective (Looft *et al.*, 62). It has also been reported that physiological coolants themselves are capable of acting as salt enhancers at thresholds below cooling (Gray *et al.*, 2008). This obviously means that, at threshold levels below cooling, there are a wide array of potential applications for chemesthetic cooling materials to influence flavor potentiation and perception.

As we have indicated earlier there is also an interaction between physiological cooling materials and high-potency sweeteners. If these two materials combined in the right ratio, one of which has a cooling effect and the other acts as a sweetener,

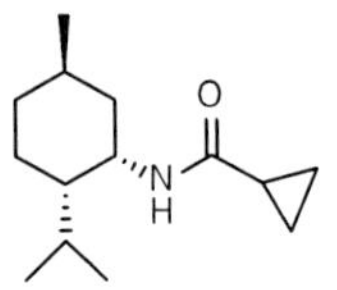

55

can increase the perception of sweetness, then it could be assumed that there is a synergistic effect between all TRP agonists on each other. Adding a small amount of cooling compound to a bitter material would give the perception of enhanced bitterness, or adding a small amount of an agonist that activates the TRPV1 (tingling) receptor to a cooling compound and creating a mixture that is predominantly a TRPM8 activator would increase the perception of cooling. Recently, combinations of various TRP agonists have been developed and reported. For example, Nakatsu *et al.* (2004) have used vanillin or a close derivative in combination with a physiological coolant as well as capsaicin, zingerone or other natural extracts. The composition is claimed as being a warming composition for fragrance and cosmetic applications (Kumamoto *et al.*, 2002). Other applications show the use of physiological combinations with other non-chemesthetic materials. For example, Haskel *et al.* showed the use of two chemesthetic compounds with a cationic polymeric compound for use in oral and personal care applications (Haskel *et al.*, 2011).

Other examples of trigeminal combinations have been reported for use in food applications, such as those reported by Johnson and Greenberg of the W.M. Wrigley company (Johnson and Greenberg, 2003; Greenberg and Johnson, 2002). These examples use not only cooling agents, tingling agents, and warming agents but they also claim that the addition of a hydrophobic sweetener can prolong the release of the flavor and sensory ingredients.

7.5 Cooling compounds – the next steps

The search for compounds that impart physiological cooling will continue. Access to high throughput screening of compounds that act as TRP agonists will allow greater understanding of structures outside of menthol derivatives that contribute to the cooling sensation, and then further derivatization of these molecules will allow for optimization of the cooling properties. There are already documented collaborations between companies such as Firmenich, Takasago, and others with Senomyx and Chromocell to name a few, showing that the interest in new molecule discovery is high, along with the rewards.

Application of these ingredients is likely to lead to even higher levels of activity. As developers explore the use of physiological coolants outside of applications that simply involve cooling, and as the need for different types and alternative cooling properties evolve to meet customer needs, so will the creativity of developers allow for solutions to be devised. Use of materials at thresholds below cooling as flavor modifiers or modulators to enhance or alter the perception of off-notes, as well as to enhance the perception of sweetness, alcohol levels, and other relevant factors that affect consumer experience, is an exciting area for exploration.

Future developments in sensory combinations have so far been limited due to the availability and properties of chemesthetic compounds outside of cooling, where odor is not imparted. However, in conjunction with new molecule discoveries, this is an area where there is potential for significant future development in product and product effectiveness and availability, as well as application and expansion into non-traditional uses for these materials.

References

S. Akutagawa. 1997. Enantioselective isomerization of allyamine to enamine; practical asymmetric synthesis of (–)-menthol bu Rh–BINAP catalysts. *Top. Catal.*, **4** (1997), 271–274.

D.A. Andersson, H.W.N. Chase, and S. Bevan. 2004. TRPM8 activation by menthol, icilin, and cold is differentially modulated by intracellular pH. *J. Neurosci.*, **24** (23), 5364–5369.

S.A. Barcelon, J.J. Keifer, H. Olaya, and S.J. Luo. 1998. Enhanced flavoring compositions containing *N*-ethyl-*p*-menthan-3-carboxamide and method of making and using same, EP 1 003 383 B1.

L.M. Bartoshuk, K. Rennert, J. Rodin, and J.C. Stevens. 1982. Effects of temperature on the perceived sweetness of sucrose. *Physiol. Behav.*, **28**, 905–910.

B. Bryant and W.L. Silver. 2000. Chemesthesis, in *The Neurobiology of Taste and Smell* (eds T.E. Finger, W.L. Silver, and D. Restrepo), New York: Wiley-Liss Inc., pp. 73–100.

I. S. Campbell. 2014. Sensory ingredients use in confectionery. American Association of Candy Technologists, Technical Seminar, Lincolnshire, IL.

G.E. Deckner, F.A. Pichardo, N.C. Alban, and M.C. Sills. 1997. Compositions for imparting an artificial tan and protecting the skin from ultra-violet radiation, US 5700452.

E. Demole and H. Wuest. 1967. Synthèses stéréosélectives de deux trioxydes C18H30O3 stéréo-isomères, d'ambréinolide et de sclaréol-lactone à partir de dérivés du (+)-manool. *Helv. Chem. Acta*, **50**, 1314–1327.

F. DiSalle, E. Cantone, M.F. Savarese, *et al.* 2013. Effect of carbonation on brain processing of sweet stimuli in humans. *Gastroenterology*, **145** (3), 537–539.

M.B. Erman and P.J. Whelan. 2002. Physiological cooling compositions containing highly purified ethyl ester of *N*-[[5-methyl-2-(1-methyethyl) cyclohexyl]carbonyl] glycine. US Patent 7,189760.

J. Fleischer, K. Bauer, and R. Hopp. 1976. Separating optically pure d-l-isomers of menthol, neomenthol and isomenthol, US Patent 3, 943,181 A.

T.J. Fowler, R.L. McManus, and E. Deckner. 1998. Personal cleaning compositions, US 5753245.

J.C. Fry, N. Yurttas, and K.L. Biermann. 2011. Sweetness concentration-response behavior of rebiana at room and refrigerator temperatures. *J. Food Sci.*, **76** (9), 545–548.

T. Fujita, J. Iwasa, and C. Hansch 1964. A new substituent constant, p, derived from partition coefficients. *J. Am. Chem. Soc.*, **86** (23), 5175–5180.

M. Gautschi. 2006. Insect repellents, US Patent Application 20060063764.

B.G. Green and S.P. Frankmann. 1987. The effect of cooling the tongue on the perceived intensity of taste. *Chem. Senses*, **12** (4), 609–619.

M.J. Greenberg. 1999. Preparation of menthone ketals for use as flavor enhancers. European Patent Application, EP485170 A1 19920513.

M.J. Greenberg and S.S. Johnson. 2002. A chewing gum having prolonged sensory benefits, International Patent Application, WO 02091849.

K. Gray, L.G.L. Yep, and R.G. Eilerman. 2008. Salt enhancement, US Patent 200880311266.

C. Hansch and T. Fujita. 1964. Analysis. A method for the correlation of biological activity and chemical structure. *J. Am. Chem. Soc.*, **86** (8), 1616–1626.

C. Hansch and A.R. Steward. 1964. The use of substituent constants in the analysis of the structure-activity relationship in penicillin derivatives. *J. Med. Chem.*, **7**, 691–694.

C. Hansch, A.R. Steward, J. Iwasa, and E.W. Deutsch. 1965. The use of a hydrophobic bonding constant for structure-activity correlations. *Mol. Pharmacol.*, **1** (3) 205–213.

C. Hansch and W.J. Dunn. 1972. Linear relationships between lipophilic character and biological activity of drugs. *J. Pharm. Sci.*, **61** (1), 1–19.

C. Hansch and J.M. Clayton. 1973. Lipophilic character and biological activity of drugs II: The parabolic case. *J. Pharm. Sci.*, **62** (1), 1–21.
A. Haskel, S. Harmalker, and J. Mattai. 2011. Compositions comprising combinations of sensates, US Patent 8084050.
G. Heydrich, G. Gralla, K. Ebel, W. Krause, and N. Kashani-shirazi, 2009. Continuous process for preparing menthol, in pure or enriched form, WIPO Patent Application WO/2009/033870.
I. Hisanri, H. Yojio, M. Hiroyuki, M. Kazuhiko, and M. Takaji, 2014. Method for producing optically active isopulegol and optically active menthol, WIPO Patent, WO2014077323.
P. Jacobs and W. Johncock. 1999. Some like it cool, menthol derivatives as effective cosmetic cooling agent. Results of a worldwide study. *Parfuem. Kosmet.*, **80**, 28–31.
S.S. Johnson and M.J. Greenberg. 2003. Hydrophobic sweetener containing chewing gum having prolonged sensory benefits, US Patent Application, US 2003072842.
S.-E. Jordt, D.D. McKemy, and D. Julius. 2003. Lessons from peppers and peppermint: the molecular logic of thermosensation. *Curr. Opin. Neurobiol.*, **13** (4), 487–492.
J. Kalbe and G. Nentwig. 2000. Non-toxic, inexpensive, arthropodicide comprises diacid ester, *N*-methyl-pyrrolidone, caprolactone or glycerol or menthol derivative, effective against insects, mites and ticks, especially dust mites, German patent 19840321.
B.J. Kane, K.E. Irving, J.O. Bledsoe Jr, and L.A. Canova. 1997. Hydrogenation of cyclic unsaturated compounds, US Patent, 4,058572.
M. Krohn and H. Zinke. 2013. Small molecule modulators of the cold and menthol receptor TRPM8. US Patent Application US 20130345200 A1.
H. Kumamoto and H. Ohta. 2002. Warming composition, US Patent Application 2002/0142015 A1.
J. Leffingwell and D. Rowsell. 2014. Wilkinson Sword Compounds: From beginning to now. *Perfum. Flavor.*, **39**, 34–44.
A. Leo, C. Hansch, and D. Elkins. 1971. Partition coefficients and their uses. *Chem. Rev.*, **71** (6), 525–616.
S.G. Lin and K.G. Baier. 1998. Method for treating skin wrinkles, US 5710141.
D.D. McKemy, W.M. Neuhausser, and D. Julius. 2004. Identification of a cold receptor reveals a general role for TRP channels in thermosensation. *Nature*, **427**, 260–265.
J. Looft, T. Vossing, J. Ley, M. Backes, and M. Blings. 2008. Susbtituted cyclopropanecarboxylic acid (3-methyl-cyclohexyl)amide as flavoring substance. US Patent Application 20080292763.W.M.
Luke. 2001. Cooling compositions with reduced stinging, US6277385.
T. Nakatsu, P.J. Mazeiko, and A.T. Lupo, *et al.* 2004. Sensate composition imparting initial sensation upon contact, US Patent 6780443 B1.
A. Nitikhunkasem, A.V. Khaiat, and J. Hopkins. 2002. Powder compositions comprising skin irritation reducing agent, US 6426092.
G. Ohloff. 1994. *Scent and Fragrances*, , Berlin Heidelberg: Springer-Verlag.
H. Ottinger, T. Soldo, and T. Hofmann. 2001. Systematic studies on structure and physiological activity of cyclic alpha-keto enamines, a novel class of "cooling" compounds. *J. Agric. Food Chem.*, **49**, 5383–5390.
G. H. Parker. 1912. The relations of smell, taste and the common chemical sense in vertebrates. *J. Acad. Nat. Sci. Phila.*, **14**, 221–234.
R. Pelzer, H. Surburg, and R. Hopp. 1994. Preparation of compounds, particularly menthyl carbonate derivatives, having a cooling sensation, DE 4226043 A1 19940201.
C.E.F. Petit, T.A. Hollowood, F. Wulfert, and J. Hort. 2007. Colour-coolant-aroma interactions and the impact of congruency and exposure on flavour perception. *Food Qual. Pref.*, **18** (6), 880–889.
C. Priest, A. Noncovich, A. Patron, and J. Ung. 2013. Compounds useful as modulators of TRPM8, US Patent Application 20130324557 A1.
D. Ptchelintsev, N. Scanarella, and R. Kalafinsky. 2003. Oxa acids and related compounds for treating skin conditions, US 6534543.
M. Rauls, C. Jakel, N. Kashani-shirazi, and K. Ebel. 2008. Method for the production of enriched isopulegol, US Patent Application 20080214877.
S. Watkins, M.J. Hills, and R.A. Birch. 2002. Use of menthyl-2 pyrrolidone-5-carboxylate as an insect repellent, US Patent 6451844.
H.R. Watson. 1977. Flavor characteristics of synthetic cooling compound, in *Flavor: Its Chemical Behavioural & Commercial Aspects* (ed. C.M. Apt), Boulder: Westview Press, pp. 31–50.

H.R Watson, R. Hems, D.G. Roswell, and D.J. Spring. 1978. New compounds with the menthol cooling effect, *J. Soc. Cosmetic. Chem.*, **29**, 185–200.

E.T. Wei and D.A. Seid. 1983. AG-3–5 (a chemical producing sensations of cold). *J. Pharm. Pharmacol.*, **35**, 110–112.

E.T. Wei. 2005. N-Arylsalkyl-carboxamide compositions and methods, US Patent Application 20050187211.

E.T. Wei. 2006. *N*-alkylcarbonylamino acid ester and *N*-alkylcarobonyl-amino lactone compounds and their use. WIPO Patent Application WO/2006/103401.

E.T. Wei. 2007. *N*-Arylshydroxyalkylidene-carboxamide compositions and methods. US Patent Application 20070155755.

E.T. Wei. 2008a. Aryl substituted derivatives of cycloalkyl and branched chain alkyl carboxylic acids used as antinociceptive drugs for peripheral targets. US Patent 7417048.

E.T. Wei. 2008b. *N*-Alkylcarbonyl-amino acid ester and *N*-alkylcarbonyl-amino lactone compounds and their use. US Patent Application 20080227857.

E.T Wei. 2010a *N*-Aryl-hydroxyalkylidene-carboxamide compounds and their use. US Patent Application 20100041757.

E.T. Wei. 2010b. *p*-Menthane-3-carboxylic acid esters to treat airways diseases, US Patent Application, 20100292325.

E.T. Wei, 2011a *N*-Alkylcarbonyl-D-amino hydroxyalkyl ester compounds and their use, US Patent Application 20110039812.

E.T. Wei, 2011b. *N*-Alkylcarbonyl-amino acid ester compounds and their use for skin irritation, itch and pain, US Patent Application 20110152366.

E.T. Wei. 2012. *N*-Alkylcarbonyl amino lactone compounds and their use, US Patent 8258320.

E.T. Wei. 2013a. [((1R, 2S, 5R)-2-Isopropyl-5-methyl-cyclohexanecarbonyl)-amino]-acetic acid isopropyl ester and related compounds and their use in therapy, US Patent 8426463.

E.T. Wei. 2013b. *N*-Alkylcarbonyl-amino acid ester compounds and their use for cough and pharyngitis. US Patent 8476317.

CHAPTER 8

Types of chemesthesis III. Tingling and numbing

Christopher T. Simons
Department of Food Science and Technology, The Ohio State University, Columbus, OH, USA

8.1 Introduction

Although most chemesthetic sensations are described using painful (e.g., irritant, pungent), thermal (e.g., cooling, warming, burning, hot) or pruritic (itch) qualities, some compounds elicit sensations that are best defined as tingling and/or numbing. Tingle can be evoked by a number of compounds including alkylamides such as α-hydroxy sanshool (Fig. 8.1a) from Szechuan pepper (*Xanthoxylum piperitum*), spilanthol (Fig. 8.1b) from jambu flowers or fruit (*Acmella sp.*), or the synthetically derived *N*-isobutyl (2*E*,4*E*,8*Z*)-unadeca-2,4,8-trienamide (isobutylalkylamide (IBA); Fig. 8.1c). These compounds are all conjugated, medium chain, fatty acid amides with fairly pronounced lipophilicity, giving them ready access to somatosensory nerve endings found in mucosal tissues. Carbon dioxide from carbonated beverages is also described as tingling (Cometto-Muñiz and Noriega, 1985; Dessirier, *et al.*, 2000, 2001; Green, 1992; Simons *et al.*, 1999; Wise and Bryant, 2014; Wise *et al.*, 2013; Yau and McDaniel, 1990, 1992) although the sensation and underlying mechanism are quite different from that of the alkylamides (see Section 8.2 on tingle mechanisms). In contrast to tingle, some compounds evoke a numbing sensation when applied to mucosal tissue. This localized anesthetic effect is different from adaptation or desensitization, both of which occur following prolonged or repeated application (see Chapter 6). Eugenol, from clove oil, is, perhaps, the most well known compound having numbing properties (Peters, 1971; Pramod *et al.*, 2010). However, many alkylamides also evoke a numbing sensation that becomes noticeable when the tingle subsides (Albin and Simons, 2010).

8.1.1 Historical use of tingling and numbing compounds

The use of tingle and numbing compounds has had longstanding and widespread use in cooking and as traditional medicaments in the control of pain. In particular, sanshool and carbon dioxide are used extensively to impart tingling sensations in Asian cuisines and beverages, respectively. Szechuan pepper (hua jiao) is one of the original compounds found in the traditional Chinese five-spice powder (Norman, 2002). Additionally, hua jiao is a key ingredient for the Szechuan version

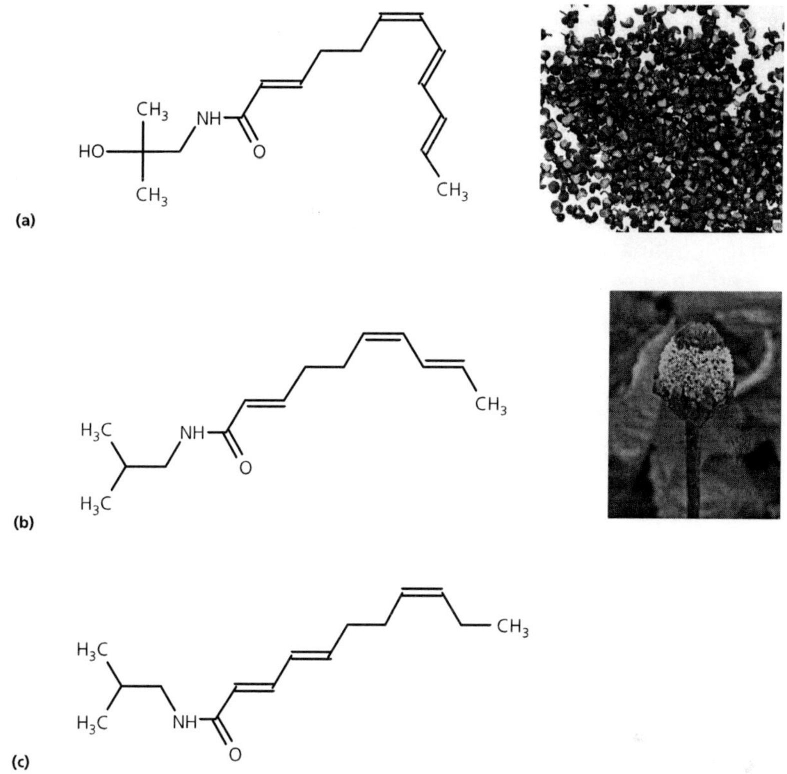

Fig. 8.1 Tingling alkylamides. Natural (a and b) and artificial (c) alkylamides used in sensory studies to elicit tingle sensations: (a) α-hydroxy sanshool from Szechuan pepper; (b) spilanthol from *Acmella oleracea* flower; and (c) *N*-isobutyl (2*E*,4*E*,8*Z*)-unadeca-2,4,8-trienamide (isobutylalkylamide; IBA), a synthetically derived, light and temperature stable tingle compound (Galopin *et al.*, 2004). Pictures: Szechuan pepper by N. Dulex (upper) (Source: FreeImages.com/Nathalie Dulex) and *Acmella oleracea* by H. Zell (lower) (Source: H. Zell. Reproduced under GNU Free Documentation License, http://www.gnu.org/copyleft/fdl.html).

of hot pot – má là hot pot. In addition to providing the unique tingling, the subsequent numbing sensations are thought to reduce the perceived pungency associated with this spicy dish. Similarly, carbonation is used to impart tingling sensations in a variety of beverages including soft drinks, beer, and champagne. Carbonation is said to impart "refreshing" characteristics (Beer Judge Certification Program, 2008) and is a key attribute associated with these products. More recently, carbonation has been reported in unconventional products including yogurt (Go-Gurt Fizzix, General Mills, Minneapolis, MN).

In addition to their use in foods and beverages, tingling and numbing compounds have been used extensively as medicaments and health aids. Alkylamides have recently been utilized in cosmetics to provide anti-wrinkle benefits to skin (Artaria *et al.*, 2011). Sanshool-containing extracts from *Xanthoxylum piperitum* have also been compounded into creams to aid in sexual performance, presumably due to its numbing effect (Tezuka and Tezuka, 1994). Spilanthol and eugenol have historical use in dentistry. Spilanthol, like sanshool, evokes a tingle sensation that subsides giving rise to a numbing sensation (Simons and Ward, 2005). Indigenous people from the Peruvian rainforest convert the spilanthol-containing jambu fruit (Fig. 8.1b) into a poultice used for toothaches and they affectionately refer to this plant (*Acmella oleracea*) as the "toothache tree." Recently, topical anesthetic gels have been formulated that have completed phase II clinical trials (Stovall, 2012). Similarly, eugenol is used extensively in dentistry as a topical anesthetic (Gurney, 1968; Markowitz *et al.*, 1992) and in combination with zinc oxide to form a paste used to cover the gingival tissues or extraction sockets (Freedman and Stassen, 2013) and for managing post-operative pain (Burgoyne *et al.*, 2010; Jorkjend and Skoglund, 1990).

8.2 Tingle mechanisms

To date there are two known specific mechanisms subserving chemically-evoked tingle sensations. Antagonism of two-pore K^+ channels underpins alkylamide tingle whereas activation of transient receptor potential (TRP) channel A1 by carbonic acid is responsible for the carbonation sensation. Each of these mechanisms are discussed in detail below.

8.2.1 Two-pore K^+ channels

Recent work has linked the tingle sensation evoked by α-hydroxy sanshool to the inhibition of two-pore K^+ channels expressed in mechanosensitive and TRPV1-expressing nociceptive neurons (Bautista *et al.*, 2008; Lennertz *et al.*, 2010). In particular, sanshool has been shown to inhibit specific members of the two-pore K^+ channel sub-family K, including KCNK3 (TASK-1), KCNK9 (TASK-3), and KCNK18 (TRESK; Bautista *et al.*, 2008). These receptors are expressed in neurons where they contribute to the resting membrane potential by allowing K^+ to leak across the plasma membrane as governed by the electrochemical gradient (reviewed in Lotshaw, 2007). As such, they are a key determinant of neuronal excitability, and inhibition by alkylamides results in the depolarization of the neuron. This mechanism is in contrast to that underpinning most chemesthetic compounds that activate excitatory TRP channels (see Chapter 4). However, in addition to two-pore K^+ channels, the alkylamides have been purported to activate TRPV1 and TRPA1, known receptor channels involved in pungent sensations (Sugai *et al.*, 2005; Koo *et al.*, 2007, although see Bautista *et al.*, 2008). The pungent sensations associated with TRP channel activation are described elsewhere in this book (see Chapter 4) but are not thought to contribute to the alkylamide-evoked tingle sensation *per se*.

8.2.2 Carbonic anhydrase/TRPA1

Although both described as tingling, the sensation elicited by carbonated drinks is qualitatively different from that evoked by the alkylamides. It is, therefore, not surprising that the molecular mechanism underpinning carbonation tingle differs from that subserving alklyamide tingle. Carbonation of beverages occurs when CO_2 is solubilized in water under pressure resulting in a supersaturated solution. In solution, the CO_2 reacts with water to form weak solutions of carbonic acid. However, the equilibrium constant for this reaction is quite low ($\sim 1.7\times10^{-3}$;

Housecroft and Sharpe, 2005) and, consequently, the majority of CO_2 is not converted into carbonic acid. Opening a can or bottle of a carbonated beverage releases the pressure and allows the CO_2 to come out of solution. Carbon dioxide bubbles are formed at nucleation sites on the container surface and, when released, migrate to the beverage surface.

Previously, investigators hypothesized that carbonation tingle was due to the excitation of mechanoreceptors in the oral epithelium resulting from the bursting of carbon dioxide bubbles (Yau and McDaniel, 1990, 1992). However, more recent studies have identified the mechanism as resulting from the acidification of tissue and activation of nociceptive nerve endings due to the enzymatic conversion of CO_2 and water into carbonic acid by carbonic anhydrase (Dessirier *et al.*, 2000, 2001; Komai and Bryant, 1993; Simons *et al.*, 1999; Wise and Bryant, 2014; Wise *et al.*, 2013; Wang *et al.*, 2010).

The carbonic anhydrase family of enzymes is large and contains multiple members (Supuran, 2007). Most carbonic anhydrase isoforms are cytoplasmic where they help to maintain cellular acid–base homeostasis (Supuran, 2007). Carbon dioxide passes freely through the plasma membrane of cells resulting in intracellular acidification via cytoplasmic carbonic anhydrase catalysis (Wang *et al.*, 2010). In contrast, carbonic anhydrase VI is the only secretory isozyme of the mammalian carbonic acid family (Kivela *et al.*, 1999a). This isozyme is found in secretory granules of the acinar cells of the parotid and submandibular glands (Parkkila *et al.*, 1990) and secreted into the saliva where it is thought to aid in the prevention of dental caries (Kivela *et al.*, 1999a,b) and the regulation of pH homeostasis on epithelial surfaces of the oral cavity and the upper alimentary canal (Parkkila *et al.*, 1997).

When ingested, the conversion of CO_2 and water into carbonic acid is catalyzed by carbonic anhydrase both intra- and extracellularly. Several potential candidate receptors could mediate trigeminal nociceptor responses of CO_2. Acid-sensing ion channels, a subfamily of the ENaC/Deg superfamily of ion channels (Waldmann *et al.*, 1997b), are proton-gated ion channels that respond to extracellular acidification (Lingueglia *et al.*, 1997; Waldmann *et al.*, 1997a). Similarly, the capsaicin receptor, TRPV1 (see Chapter 4), also responds to extracellular protons (Caterina *et al.*, 1997; Tominaga *et al.*, 1998). However, recent evidence suggests that intracellular acidification and activation of TRPA1 is the primary mechanism underpinning carbonation tingle (Wang *et al.*, 2010). Using ratiometric calcium imaging, CO_2 elicited a robust response in cultured wild-type trigeminal ganglion cells that was absent in cells cultured from TRPA1 knockouts. Similarly, in subsequent patch-clamp experiments, TRPA1-expressing, but not TRPV1-expressing, HEK cells responded to CO_2. Interestingly, TRPA1-expressing HEK cells responded to CO_2 (in HEPES-buffered saline at pH 5.3) but not to extracellular acidification alone (MES buffer, pH 5.3). In fact, TRPA1 mediated responses to cinnamic aldehyde were actually blocked by extracellular protons suggesting that extracellular acidification inhibits this receptor. Thus, to demonstrate that intracellular protons gate TRPA1, inside-out excised patches from transfected HEK cells were exposed to a cytoplasmic solution adjusted to pH 5 that robustly activated these receptors. In aggregate, these studies confirm that TRPA1 is both necessary and sufficient to mediate responses of trigeminal nociceptors to CO_2 via a mechanism requiring intracellular acidification.

8.3 Numbing (anaesthetic) mechanisms

Although relatively rare, some compounds found in foods elicit a numbing sensation. Mechanisms of anesthetic action are well studied and much is known regarding the molecular events subserving numbing anesthesia (for a review, see Barkin, 2013). As it relates to the alkylamides and eugenol, there are several mechanisms mediating the anesthetic action of these compounds.

8.3.1 Alkylamides and two-pore K^+ channels

To date it is unclear specifically how alkylamides elicit the numbing effect. Recent evidence suggests that a tingling sensation is elicited by alkylamides containing at least one cis-configured double bond whereas only numbing is evoked by all-trans-configured amides (Bader *et al.*, 2014). The different sensations evoked by the presence of a *cis*- versus a *trans*-double bond suggests that these compounds can be used to probe the specific mechanisms underpinning tingle versus anesthesia. There appears to be two potential mechanisms underpinning alkylamide numbing. The first mechanism is likely to involve the same receptors responsible for mediating the tingle sensation associated with alkylamides. KCNK channels contribute to the resting membrane potential of neurons (Lotshaw, 2007) by providing a pathway for K^+ ions to leak through the cellular membrane. Therefore, potentiating the flow of K^+ ions across the membrane will hyperpolarize the neuron, resulting in decreased sensitivity. Anesthetic compounds, such as halothane, have been found to increase the open probability of these channels (reviewed in Lotshaw, 2007) thus establishing a potential mechanism by which alkylamides cause numbing. As the structure–activity relationship suggested earlier (Bader *et al.*, 2014), simple modification of cis- to a trans-configuration likely changes an alkylamide from an antagonist of the KCNK channels to an agonist. How sanshool and other tingling compounds first evoke tingling followed subsequently by numbing is currently unknown, but possibly reflects the ability of these compounds to simultaneously inhibit and potentiate KCNK channels.

8.3.2 Alkylamides and voltage-gated Na^+ channels

A second mechanism by which alkylamides induce numbing has recently been shown to involve the inhibition of voltage-gated Na^+ channels expressed in a unique population of thinly myelinated mechano-nociceptive neurons (Tsunozaki *et al.*, 2013). Voltage-gated Na^+ channels are ubiquitous in nerve cells and, as suggested by their name, open in response to voltage changes across the cellular membrane, and subsequently allow for the rapid passage of Na^+ ions into the cell (reviewed in Peters and Ruben, 2014). The voltage dependency of these channels allows them to respond to, and produce, electrical signals within the body. As such, voltage-gated Na^+ channels are responsible for the excitation and conduction of action potentials in sensory neurons (Peters and Ruben, 2014). Voltage-gated Na^+ channels can be classified by their sensitivity to the classical channel blocker tetrodotoxin (TTX). TTX-sensitive (TTX-s) channels, such as Nav1.1, Nav1.3, Nav1.6, and Nav1.7, are preferentially expressed in medium- and large-diameter mechanosensory neurons (Chahine and O'Leary, 2014; Tsunozaki *et al.*, 2013). In contrast, small-diameter nociceptive cells express a combination of TTX-s (Nav1.7) and TTX resistant (TTX-r) channels, including Nav1.8 and Nav1.9 (Chahine and O'Leary, 2014).

Interestingly, the unique population of mechano-nociceptive cells inhibited by sanshool expresses a distinctive combination of voltage-gated Na^+ channels that differentiates them from small-diameter nociceptive cells or the population of mechanosensitive cells that are activated by sanshool (Tsunozaki *et al.*, 2013). In contrast to other large diameter mechanosensory neurons, those cells inhibited by sanshool express Nav1.7 to a much greater extent and also express the TTX-r channel Nav 1.8, similar to small-diameter nociceptive cells (Tsunozaki *et al.*, 2013). *In vitro* studies using these unique mechano-nociceptive cells dissociated from dorsal root ganglia have shown that sanshool inhibits action potential generation by blocking the voltage-gated sodium current (Tsunozaki *et al.*, 2013). Follow-up studies in which various isoforms of human voltage-gated Na^+ channels were heterologously expressed in Chinese hamster ovary cells indicated that all subtypes were inhibited to some extent by sanshool, however, Nav1.7 appeared to be most sensitive (Tsunozaki *et al.*, 2013). Combined, these results suggest that alkylamide induced anesthesia is facilitated, at least partially, via the inhibition of multiple voltage-gated Na^+ channel isoforms expressed in a unique population of large diameter mechanonociceptors.

8.3.3 Eugenol and voltage-gated sodium (Na^+) channels

Eugenol is used extensively in dentistry due to its ability to allay tooth pain (Markowitz *et al.*, 1992). The mechanisms by which eugenol reduce pain have been fairly well characterized. Application of eugenol has been shown to act on trigeminal ganglion neurons by inhibiting voltage-gated Na^+ channels (Park *et al.*, 2006, 2009). In whole-cell patch-clamp experiments on trigeminal ganglion cells, application of eugenol was found to inhibit both TTX-s and TTX-r channels, and the subsequent generation of action potentials (Park *et al.*, 2006). These results provide compelling evidence that eugenol anesthesia is, at least partially, mediated by the inability of mechanosensory and nociceptive neurons to generate action potentials in response to tactile or painful stimuli, respectively.

8.3.4 Eugenol and voltage-gated calcium (Ca^{2+}) channels

In addition to its inhibitory action against voltage-gated Na^+ channels, eugenol has also been shown to block voltage-gated Ca^{2+} channels. Voltage-gated Ca^{2+} channels transduce changes in membrane potential into intracellular Ca^{2+} transients and, as such, are key to converting the electrical signal of the action potential into an intracellular chemical signal. Intracellular calcium levels are tightly controlled and a sudden increase initiates many physiological events, including secretion, enzyme regulation, protein phosphorylation/dephosphorylation, gene transcription, and, critically, as it relates to sensation, synaptic transmission (reviewed in Catterall, 2011). There are multiple types of voltage-gated Ca^{2+} channels and the pattern of expression and function of these channels is dependent on the neuronal sub-type (reviewed in Altier and Zamponi, 2004). In particular, N-type channels function at neuronal synapses where they trigger neurotransmitter release and facilitate signaling up the neuroaxis (Altier and Zamponi, 2004). They have an established role in pain signaling (Altier and Zamponi, 2004) and are the target for therapeutic drug discovery (Lee, 2014). In contrast, T-type voltage-sensitive Ca^{2+} channels are expressed in high density in low-threshold mechanoreceptors (Cardenas *et al.*, 1995) – in particular D-hair cells (Heppenstall and Lewin, 2006) – as well as a unique population of neurons that express both mechanosensory and nociceptive behavior (Perez-Reyes *et al.*, 1998). Thus, voltage-gated Ca^{2+} channels are found in sensory neurons that process both tactile and nociceptive information.

Whole-cell patch-clamp studies have shown that eugenol can inhibit N-type voltage-gated Ca^{2+} currents in dissociated rat trigeminal ganglion cells as well as C2D7 cells stably expressing human N-type Ca^{2+} channels (encoded by the $\alpha_{1B\text{-}a}$, β_{1b}, $\alpha_{2b}\delta$ subunits; Lee *et al.*, 2005). Interestingly, the percent inhibition of Ca^{2+} current in the C2D7 cell line – that does not endogenously express voltage-gated Ca^{2+} channels – was greater than that observed in trigeminal ganglion cells (Lee *et al.*, 2005), which express multiple voltage-gated Ca^{2+} channel subtypes. This finding suggests that eugenol has higher affinity for the N-type Ca^{2+} channel compared with other subtypes. However, more recent data have shown that eugenol is a strong antagonist of T-type Ca^{2+} channels, which also includes $Ca_v3.1$, $Ca_v3.2$, $Ca_v3.3$ (Seo *et al.*, 2013). Whole-cell patch-clamp studies in HEK293 cells stably expressing $Ca_v3.1$, $Ca_v3.2$, and $Ca_v3.3$ indicate that eugenol inhibits the T-type current with IC_{50} values of 463 ± 12, 486 ± 12, and 708 ± 11 μM, respectively. The inhibitory effect of eugenol on T-type currents was confirmed in rat trigeminal ganglion cells natively expressing these channels (Seo *et al.*, 2013). Taken together, these data provide a mechanism that underpins the psychophysical finding of eugenol anesthesia. Voltage-gated Na^+ and Ca^{2+} channels expressed in trigeminal mechanosensory and nociceptive neurons are inhibited by eugenol, thus blocking the transmission of tactile and pain information to higher centers of the brain.

It should be noted that many irritants, including capsaicin, eugenol, mustard oil, and nicotine, evoke tachyphylaxis (Carstens *et al.*, 2002; Klein *et al.*, 2013; Patacchini *et al.*, 1990), a phenomenon that is characterized by decreased responsiveness of the neuron to repeated compound exposure. Tachyphylaxis has been shown in trigeminal ganglion cells to be a receptor-mediated event requiring extracellular Ca^{2+} (Liu and Simon 1996, 1998) and is believed to underpin the psychophysical phenomenon of desensitization (Carstens *et al.*, 2002). However, tachyphylaxis (and desensitization; see Chapter 6) is fundamentally different from anesthesia, in that neuronal excitability is decreased only for nociceptive cells expressing receptors for the desensitizing compound (e.g., TRPV1).

8.4 Tingle/numbing neural processing

When ingested, alkylamides and carbonation evoke tingle sensations arising from the oral cavity. These stimuli excite mechanosensory and nociceptive nerve endings from neurons comprising the trigeminal nerve. The trigeminal nerve carries somatosensory (e.g., tactile, nociceptive, proprioceptive, and thermal) information from the face and controls motor functions such as chewing (for review see Weaver, 2014). The trigeminal nerve is comprised of three branches including the ophthalmic nerve, the maxillary nerve, and the mandibular nerve. The ophthalmic and maxillary nerves convey sensory information whereas the mandibular nerve has dual sensory and motor functions. Of particular interest as it relates to oral tingle sensations, the lingual nerve is a branch of the mandibular division that innervates the anterior two-thirds of the tongue and carries afferent somatosensory information from this region. Somatosensory information from the posterior third of the tongue and the pharyngeal mucosa are carried by the glossopharyngeal (IX cranial nerve) and vagus (X cranial nerve) nerves, respectively. Following activation of lingual tactile and nociceptive neurons by tingle compounds and carbonation, the neural signal is conveyed to the trigeminal complex in the brainstem. Nociceptive information is carried by unmyelinated C-fibers or thinly myelinated Aδ-fibers, whereas mechanosensory information is conveyed via Aδ-fibers – in particular, D-hair fibers – and the larger diameter, myelinated Aβ-fibers (Brown and Iggio, 1967; McGlone *et al.*, 2014). C-fibers project to the brainstem trigeminal complex where they form synapses with second-order neurons in superficial laminae (reviewed in Braz *et al.*, 2014). Aδ-fibers project to both superficial and deeper layers, whereas Aβ-fibers project only to deeper layers of the brainstem (Braz *et al.*, 2014). This functional separation of pain and mechanosensory information allows for targeted investigations of nociceptive and tactile processing.

8.4.1 Activation of peripheral and central mechanosensory fibers by alkylamides

Initial studies indicated that alkylamides activate thermosensitive, nociceptive, and mechanosensitive fibers in the rat lingual nerve (Bryant and Mezine, 1999). Following application of sanshool to the tongue, increased neural activity was elicited in low- and high-threshold cold fibers, as well as in low-threshold mechanoreceptors, which lasted between 20 and 45 minutes. In calcium-imaging studies, a sanshool-derivative was found to activate small and large diameter rat dorsal root ganglion cells, some of which also responded to menthol, cinnamic aldehyde and/or capsaicin (Klein *et al.*, 2011). Cells that only responded to the alkylamide were of large diameter and thought to be mechanosensory. With repeated application, the alkylamide elicited tachyphylaxis in all cell types. However, pre-exposure to capsaicin did not elicit cross-desensitization.

Mechanistically, these data suggest that the tingle sensation could be the result of simultaneous input from many fiber types including mechanoreceptors, thermoreceptors, and nociceptors, all of which have been shown to be activated alkylamides. Alternatively, tingle may result strictly from activity in mechanosensory pathways with activation of nociceptive and thermosensitive pathways responsible for minor sensations of pungency and cooling, which have also been described (Albin and Simons, 2010; Sugai *et al.*, 2005).

More recent studies using an *ex vivo* rat skin preparation have shown that cutaneous application of sanshool activates specific subtypes of Aδ-, Aβ-, and C-fibers (Lennertz *et al.*, 2010). Among Aδ-fibers, D-hair mechanosensory afferents were robustly excited by sanshool, whereas high-threshold nociceptors and mechanically insensitive neurons were completely unresponsive (Lennertz *et al.*, 2010). D-hair fibers are the most sensitive of all mechanoreceptors, rapidly adapt to sustained mechanical force, and are instrumental to normal tactile acuity and movement detection (Brown and Iggo, 1967; Wetzel *et al.*, 2007). Similarly, differential activity was displayed by Aβ-fibers with rapidly adapting afferents responding strongly to sanshool, whereas slowly adapting fibers responded weakly (Lennertz *et al.*, 2010). Rapidly-adapting Aβ-fibers encode movement by guard hair follicles and slowly adapting afferents are thought to encode sustained pressure (SA-I) or stretch (SA-II) (LaMotte *et al.*, 1998; Srinivasan *et al.*, 1990).

Taken together, these data indicate that sanshool preferentially activates fibers that rapidly adapt to sustained mechanical force. Although hair follicle receptors do not exist in the tongue or oral mucosa, candidate mechanoreceptors mediating tingle sensation from oral tissues include the superficial low-threshold rapidly adapting Meissner cells as well as deeper slowly adapting Merkel (SA-1) and Ruffini (SA-II) mechanoreceptors (Biedenbach and Chan, 1971; Trulsson and Essick, 1997). Weak activity was also elicited from a unique population of slow-conducting C-fibers that was not correlated to capsaicin or mustard oil sensitivity (Lennertz *et al.*, 2010). Currently, it is unclear whether these fibers encode the irritant sensations that accompany sanshool application (Albin and Simons, 2010; Bryant and Mezine, 1999; Sugai *et al.*, 2005) or are a unique population of C-fiber that convey sensations of touch (Löken *et al.*, 2009). Interestingly, across all fiber types activated by sanshool, a subset of neurons responded with a unique bursting pattern of action potentials (Lennertz *et al.*, 2010). Bursting activity in mechanosensory neurons has been associated with tingling paresthesia (Ochoa and Torebjörk, 1980) and is thought to potentially underpin the tingling sensation associated with administration of sanshool and other alkylamides (Lennertz *et al.*, 2010).

Recordings from central neurons in the rat dorsal horn are consistent with findings obtained in peripheral recordings. Alkylamides were found to excite wide-dynamic range (WDR) and possibly low-threshold mechanosensitive (LTM) spinal neurons when injected into the hind paw (Sawyer *et al.*, 2009); although the classification of cells as LTM was confounded because following alkylamide administration cells also responded to noxious chemicals. Wide-dynamic range neurons show graded responses to all somatosensory sub-modalities (mechanical, chemical, and thermal) and are activated by noxious as well as non-noxious stimuli. Low-threshold mechanosensitive neurons, however, respond to non-noxious mechanical stimuli. Dorsal horn WDR and LTM responses to alkylamide were concentration dependent and displayed tachyphylaxis or desensitization to repeated injection, similar to that observed in peripheral recordings (Sawyer *et al.*, 2009). Additionally, nearly all cells also responded to the pungent compounds allyl isothiocyanate (mustard oil) and/or capsaicin. Currently, it is unclear how central neurons convey sensations of tingle. Two potential mechanisms have been put forward, including the asynchronous activity in multiple primary afferents having different conduction velocities, which might be interpreted as tingle, or, alternatively, a population code involving sanshool excitation of WDR neurons that mediate pain and co-excitation of mechanoreceptor afferents (Sawyer *et al.*, 2009). Further experiments examining central processing of tingle sensation are needed.

8.4.2 Activation of peripheral and central nociceptive fibers by carbonation

Several studies have investigated the peripheral and central processing of carbonation sensation. Multi- and single-unit responses of the rat lingual nerve responded to carbonated water solutions applied to the tongue with latencies of 3–6 seconds (Komai and Bryant, 1993). These lingual nerve fibers typically responded to a variety of other stimuli including noxious temperature (cold and heat) and acidic chemicals (NH_4Cl and HCl), although in some cases, chemical sensitivity was specific to CO_2. Systemic administration of the carbonic anhydrase inhibitor acetazolamide blocked lingual nerve responses to carbonated water, but not thermal or other acidic stimuli, confirming the role of this enzyme in the transduction of the carbonation sensation. Similar findings were observed in central recordings of rat trigeminal caudalis neurons. Carbonated water evoked activity in WDR neurons that also responded to tactile stimuli and noxious heat (54 °C; Simons *et al.*, 1999). Responses to CO_2 were inhibited by lingual pretreatment with the carbonic anhydrase inhibitor dorzolamide. Complementary c-fos immunohistochemical studies were also completed (Simons *et al.*, 1999; Dessirier *et al.*, 2000) and showed that expression of the immediate-early gene protein c-fos, an indicator of neuronal activation, was upregulated in trigeminal caudalis neurons following lingual application of carbonated water, but not when administration was preceded by dorzolamide or acetazolamide pretreatment of the tongue. Taken together, peripheral and central neurophysiological studies confirm that carbonic anhydrase is essential for sensing carbon dioxide and that the tingle sensation associated with carbonated beverages is processed through the activation of nociceptive pathways.

8.4.3 Inhibition of peripheral fibers by alkylamides and eugenol

In addition to the tingle sensation, sanshool administration results in a numbing anesthesia. To differentiate the neural processing of tingle from numbing anesthesia, cutaneous nerve endings were exposed to sanshool using the *ex vivo* skin preparation in rat (Tsunozaki *et al.*, 2013). Sanshool was shown to inhibit mechanically evoked activity in nearly all Aδ-mechanosensitive fibers as well as a subset of slowly adapting Aβ-fibers, but not mechanically sensitive C-fibers. The inhibitory action of sanshool is specific to slowly adapting, myelinated fibers and is in contrast to the rapidly adapting fibers that are activated by sanshool as described earlier (see Section 8.4.1). Inhibition of mechanically evoked activity by sanshool was concentration dependent and, in some cases, could last in excess of 30 minutes (Tsunozaki *et al.*, 2013).

The ability of eugenol to reduce tooth pain underpins its widespread use in dentistry (Markowitz *et al.*, 1992). When applied to the frog sciatic nerve, eugenol concentrations as low as 0.05% irreversibly blocked compound action potentials (Kozam, 1977). Similar results were observed in rat vagus and phrenic nerve preparations (Brodin, 1982; Brodin and Røed, 1984) with the caveat that at low concentrations (<2.44 mM), the nerve block was reversible (Brodin and Røed, 1984). The mechanism of the reversible nerve block is probably due to the ability of eugenol to inhibit voltage-gated sodium (Park *et al.*, 2006, 2009) and calcium (Lee *et al.*, 2005; Seo *et al.*, 2013) channels, whereas the irreversible block may be due to cytotoxic effects (Markowitz *et al.*, 1992).

8.5 Psychophysical evaluations of tingle

In contrast to other chemesthetic sensations, tingle has been relatively understudied psychophysically. This dearth of information is surprising given that tingle contributes positively to the hedonic experience associated with the consumption of foods and beverages containing alkylamides or carbonation. Recent psychophysical investigations into the sensations evoked by alkylamides and carbonation have provided significant insight linking perceived tingle with underlying neurophysiological mechanisms.

8.5.1 Alkylamide tingle: temporal phenomena

Initial orosensory evaluations with α- and ε-hydroxy sanshool clearly indicated the alkylamide-evoked sensation was different from the pungency elicited by capsaicin and other chemoirritants (Bryant and Mezine, 1999). After a 30–90 second delay, oral application of low concentrations (25–50 μg) of α- and ε-hydroxy sanshool elicited a sensation characterized as tingling or "carbonation-like" (Bryant and Mezine, 1999). At higher concentrations (>100 μg), these compounds were described as painful (Bryant and Mezine, 1999). A subsequent study identified the recognition thresholds of a number of naturally occurring alkylamides to range between 3.5 and 8 nmol cm^{-2} of tongue surface stimulated (Bader *et al.*, 2014). The absence of a cis-double bond, as found for β-hydroxy sanshool, rendered the compound inactive in human taste trials (Bryant and Mezine, 1999) – even at high concentrations – a finding that was confirmed in a more recent study (Bader *et al.*, 2014). The temporal dynamics of alkylamide-evoked sensation was studied more specifically using a discrete time intensity method (Albin and Simons, 2010). When painted on the tongue, a synthetic alkylamide (IBA; 0.52%) evoked an intense tingle sensation that remained constant for the first 10 minutes and slowly decayed thereafter; by 30 minutes, most subjects reported the tingle as absent. Interestingly, subjects also reported the presence of burning, cooling, and numbing sensations that appeared temporally in that order.

Similar to other chemesthetic stimuli, repeated application of IBA evoked a sensation that either increased or decreased in intensity depending upon the interstimulus interval. At short interstimulus intervals of 30 seconds, IBA applied to the same region of the tongue elicited tingling that grew progressively, such that the sensation evoked by each application was perceived as more intense than preceding applications (Albin and Simons, 2010). Activation of previously quiescent alkylamide-sensitive neurons as the compound diffuses through the lingual epithelium (spatial summation) is thought to underpin this sensitizing pattern of tingle. In contrast, self-desensitization was observed when IBA was applied to the tongue at longer interstimulus intervals (Albin and Simons, 2010). Following a 30 minute hiatus, the perceived tingle intensity of IBA when re-applied to an area of the tongue previously treated with the same compound was significantly less than the tingle intensity perceived on untreated tissue. Although unknown, the mechanism underpinning alkylamide self-desensitization probably reflects the anesthetic action of these compounds on two-pore potassium channels and/or voltage-gated Na^+ channels (Bautista *et al.*, 2008; Tsunozaki *et al.*, 2013).

8.5.2 Alkylamide tingle: mechanosensory sensitivity

Physiological studies indicate that alkylamides activate mechanosensory pathways (Bautista *et al.*, 2008, Lennertz *et al.*, 2010) and several psychophysical investigations have produced data consistent with this hypothesis. Pretreating the tongue with capsaicin or mustard oil activates lingual nociceptors expressing TRPV1 and TRPA1, respectively, and, following a brief delay of 3 or more minutes, induces desensitization (see Chapter 4). If IBA is applied during this desensitized state, the tingling sensation intensity is not reduced, indicating a lack of cross-desensitization (Albin and Simons, 2010). This finding suggests that peripheral transduction and processing of alkylamide-evoked tingle does not involve capsaicin- or mustard oil-sensitive pathways.

Further studies investigated the relationship between alkylamide-evoked tingle and physical, tactile stimuli. Using a sensitive signal detection methodology, lingual tactile sensitivity was assessed following pretreatment with IBA or capsaicin (Albin and Simons, 2010). If alkylamides activated mechanosensitive pathways, IBA administration would be speculated to increase background neural "noise" in these channels, and tactile sensitivity would be hypothesized to decrease. If, on the other hand, alkylamides activated nociceptive pathways, the effect of IBA on tactile thresholds would be expected to be similar to that observed following lingual capsaicin pretreatment. Indeed, immediately following IBA pretreatment, when the tingling sensation was at its most intense, lingual tactile sensitivity was significantly impacted and subjects had a difficult time identifying weak punctate stimuli applied to the tongue. The same effect was not observed following lingual capsaicin pretreatment.

More recently, a ground Szechuan pepper slurry was used to identify the frequency of sanshool-evoked tingle on the lip that matched the frequency of a vibratory stimulus applied to the index finger (Hagura *et al.*, 2013). Subjects consistently matched the perceived sanshool tingle to a 50 Hz vibration – a frequency corresponding to the range of activity of RA-I fibers (Toma and Nakajima, 1995). Moreover, the perceived frequency of sanshool-evoked tingle was significantly reduced following adaptation to a mechanical vibration stimulus previously applied to the lip (Hagura *et al.*, 2013). Cross-adaptation of the vibratory stimulus to the chemical stimulus suggests that the tingle sensation is conveyed through activation of mechanosensory pathways including RA-I fibers (see Chapter 6 for a discussion of the doctrine of specific nerve energies).

8.5.3 Alkylamide tingle: effect of temperature

Initial electrophysiological studies indicated that alkylamides activate cold-sensitive fibers in the rat lingual nerve (Bryant and Mezine, 1999). Subsequent psychophysical studies investigated alkylamide-evoked cooling and identified tingle-thermal interactions. When applied to the tongue, IBA was found to evoke a low-intensity cooling sensation that became notable around 12 minutes post-administration and was identified by about 50% of the test subjects (Albin and Simons, 2010). In contrast, the percentage of panelists identifying cooling as a descriptor when ground Szechuan pepper was applied to the lip was not different from the control condition when the solvent was applied (Hagura *et al.*, 2013). However, these data are confounded by the relative impurity of the ground pepper stimulus and the presence of ethanol as the solvent. Additional studies specifically investigated tingle-thermal interactions. Cold (0 °C) and cool (21 °C) temperatures were found to enhance the IBA-evoked sensation whereas warm (37 °C) and hot (41 °C) temperatures had no effect (Albin and Simons, 2010). To date, the specific mechanism of this thermal–tingle interaction is unknown. However, in addition to mechanosensory neurons, various KCNK channels are expressed in thermally sensitive neurons (Madrid *et al.*, 2009; Noel *et al.*, 2009). Recently, the sanshool-sensitive KCNK 9 (TASK-3) channel was found to be enriched in a sub-population of thermosensitive cells that co-express the TRPM8 cold receptor (Morenilla-Palao *et al.*, 2014). Interestingly, the threshold temperature needed to activate these cells was increased from 28.7 to 30.6 °C when testedinthepresenceof1-(1-(6-(biphenyl-4-ylcarbonyl)-5,6,7,8-tetrahydropyrido[4,3-d]-pyrimidin-4-yl)piperidin-4-yl) (propan-1-one) (Morenilla-Palao *et al.*, 2014), a recently identified KCNK 9 channel blocker (Coburn *et al.*, 2012). Alkylamides may similarly modulate thermosensitivity and thus provide a mechanism that underpins these tingle-temperature interactions.

8.5.4 CO_2 tingle: concentration and tastant effects

As many consumers can attest when sampling a "flat" carbonated beverage, perceived tingle intensity is dependent on the aqueous CO_2 concentration. Empirical data exist confirming these anecdotal observations and suggest that for the oral cavity, perceived tingle and/or pungency increases as a function of CO_2 concentration (Commetto-Muñiz and Noriega, 1985; Yau and McDaniel, 1990; Green, 1992; Harper and McDaniel, 1993; Hewson *et al.*, 2009; Wise and Bryant, 2014). Indeed, perceived carbonation sensation increases as a power function of CO_2 concentration with exponents ranging from ~1.1 to ~2.4 (Commetto-Muñiz and Noriega, 1985; Yau and McDaniel, 1990; Green, 1992; Wise and Bryant, 2014). The difference in the exponent of the power function obtained in these studies was attributed to the level of panelist training (Green, 1992) and the diverse sensitivities, personalities and backgrounds of the subjects, wherein different criteria for "pungency" may have influenced the perception of carbonation (Yau and McDaniel, 1990). The presence of some tastant compounds influenced CO_2 pungency. Sucrose (Yau and McDaniel, 1992; although see Commetto-Muñiz *et al.*, 1987), glucose (Hewson *et al.*, 2009), and quinine (Commetto-Muñiz *et al.*, 1987), but not fructose (Hewson *et al.*, 2009) or phosphoric acid (Yau and McDaniel, 1992), reduced the perceived carbonation intensity, whereas sodium chloride (Commetto-Muñiz *et al.*, 1987) and tartaric acid (Commetto-Muñiz *et al.*, 1987) enhanced the carbonation pungency. Whether tastants impact perceived carbonation intensity via changes in solution viscosity (in particular for sweeteners) or through peripheral or central mechanisms is not known.

8.5.5 CO_2 tingle: impact of carbonic anhydrase blockers

Anecdotal reports indicate mountain climbers using carbonic anhydrase blockers to prevent acute mountain sickness experience the "champagne blues" in which celebratory carbonated beverages were perceived as "flat" and without fizz (Graber and Kelleher, 1988). These findings suggest the carbonation sensation is, at least partially, of chemogenic origin. Controlled studies assessing the role of carbonic anhydrase in carbonation tingle showed that pretreating the dorsal lingual surface with dorzolamide (Simons *et al.*, 1999) or acetazolamide (Dessirier *et al.*, 2000) reduced, but did not eliminate, the sensation. Similarly, when carbonated water was flowed over the tongue for 15 seconds verses 5 seconds, the effect of dorzolamide was lost (Simons *et al.*, 1999). Together, these data indicate that carbonation tingle has chemogenic origins, however, other sensory mechanisms, including mechanical activation of tactile pathways may also contribute.

8.5.6 CO_2 tingle: impact of bubbles

To directly assess the role of mechanosensory pathways in carbonation tingle, carbonated beverages were assessed in a hyperbaric chamber where atmospheric pressure could be regulated (Wise *et al.*, 2013) to influence bubble formation. Ratings of carbonation bite assessed at normal atmospheric pressure (where bubbles formed) did not differ from the ratings obtained at 2.0 atm, which prevented bubble formation (Wise *et al.*, 2013). These findings indicate that bubbles are not required to elicit carbonation tingle. However, the presence of bubbles may well modulate perceived carbonation. When air bubbles generated in a mildly carbonated solution are directed at the tongue, perceived carbonation bite is enhanced (Wise *et al.*, 2013). Accordingly, bubbles may augment the carbonation sensation by stirring the liquid layer to facilitate a higher concentration gradient of CO_2 at the surface of the tongue or through pathophysiological conditions such as mechanical allodynia (Wise *et al.*, 2013).

8.5.7 CO_2 tingle: self-desensitization and cross-desensitization by capsaicin

As with many chemoirritants, prolonged application of a carbonated water stimulus to the tongue elicits a temporally dynamic pattern of irritation. In one study, holding a carbonated water solution in the oral cavity for 5–10 seconds evoked a carbonation sensation that became painful, suggesting over this short time frame that sensitization occurs (Green, 1992). Subsequent experiments showed that longer stimulation periods of 15 or 60 seconds elicited a carbonation sensation of reduced intensity (Dessirier *et al.*, 2001). Indeed, the perceived carbonation bite was found to be significantly more intense at 5 seconds compared with that evoked at either 15 or 60 seconds and the 60 second stimulus evoked the least intense sensation (Dessirier *et al.*, 2001). This is consistent with a self-desensitizing pattern of irritation. However, a continuous flow of carbonated water is equivalent to the shortest possible interstimulus interval, 0 seconds, and should be expected to evoke sensitization. That the perceived carbonation intensity is greatest between 5 and 10 seconds implies that CO_2 diffusion through the lingual tissue peaks during this short time frame, such that spatial summation and sensitization is maximal. Thus, at longer stimulus durations of 15 or 60 seconds, the impact of spatial summation is offset by the activation of desensitization mechanisms and perceived irritation declines. A similar phenomenon was observed with mustard oil (Simons *et al.*, 2003).

Consistent with cross-desensitization, pretreatment of the tongue with low (5 ppm; Dessirier *et al.*, 2000) or high (55 ppm; Dessirier *et al.*, 2001) concentrations of capsaicin decreased the perceived intensity of carbonation bite. This was true independent of the duration of carbonated water stimulation (5, 15 or 60 seconds; Dessirier *et al.*, 2001). These results suggest that the oral sensation evoked by carbonated water is mediated, at least partly, by capsaicin-sensitive nociceptive pathways. Given the sufficiency of TRPA1 to mediate CO_2 responses *in vitro* (Wang *et al.*, 2010; see Section 8.2.2), the capsaicin cross-desensitization data suggest a mechanism whereby CO_2 activates a subclass of nociceptors that co-express TRPV1 and TRPA1. Alternatively, CO_2 could excite peripheral nociceptors that express only TRPA1, but synapse on central neurons that receive and summate input from both CO_2- and capsaicin-sensitive fibers. In this latter mechanism, capsaicin desensitization may reduce spontaneous activity in capsaicin-sensitive pathways resulting in a decreased integrated signal.

8.5.8 CO_2 tingle: effect of temperature

Temperature has a modulating effect on the intensity of the carbonation sensation that is independent of the ability of cold solutions to dissolve higher amounts of CO_2 (Green, 1992; Wise and Bryant, 2014). As carbonated solutions are cooled, the perceived intensity of the carbonation bite increases (Yau and McDaniel, 1992; Green, 1992; Wise and Bryant, 2014). This phenomenon has been observed for multiple temperatures and CO_2 concentrations. Interestingly, when pretreating the tongue with a menthol solution that elicits a strong cooling sensation (1.28 mM), no enhancement of carbonation bite was observed (Wise and Bryant, 2014). Similarly, warming carbonated solutions appears to have no effect on the carbonation pungency (Wise and Bryant, 2014). Indeed, the carbonation bite of solutions served between 30 and 45 °C did not differ (Wise and Bryant, 2014). Taken together, these results suggest cold-sensitive fibers play a potentiating role in the perception of the carbonation sensation albeit without the involvement of the cold/menthol receptor TRPM8 (McKemy *et al.*, 2002). Recent data have shown that in mice, many cold-sensitive neurons do not express TRPM8 (Munns *et al.*, 2007). Perhaps a similar population of neurons exists in humans that underpin this effect.

8.6 Psychophysical evaluations of numbing

In general, controlled psychophysical evaluations of the numbing properties of food spices are not well documented. Recently, several studies have investigated the anesthetic properties associated with alkylamides (Albin and Simons, 2010; Bader *et al.*, 2014). The psychophysical assessment of the numbing properties associated with eugenol is better documented owing in large part to this compound's historical use as an anesthetic and analgesic in dentistry (for a review, see Markowitz *et al.*, 1992).

8.6.1 Alkylamide numbing

Oral alkylamide administration elicits a complex sensation that is initially defined as having tingle, irritant, and cooling sub-modalities followed later by numbing (Albin and Simons, 2010). The time course by which the numbing sensation appears is dependent, to some extent, upon the presence or absence of a cis-bond in the alkyl chain (Bader *et al.*, 2014). In particular, all trans-conformations of alkylamide compounds confer only numbing properties without the concomitant tingling or pungent sensations. In time–intensity studies, numbing was reported as occurring within approximately 10–15 minutes following lingual application of 0.52% IBA (Albin and Simons, 2010). The anesthetic duration lasted for at least 30 minutes as subsequent lingual applications of IBA elicited only very weak tingling sensations (Albin and Simons, 2010). Further studies are needed to determine the full time course of anesthetic action following topical administration.

8.6.2 Eugenol numbing

In contrast to many other well-studied chemesthetic compounds, the oral sensation evoked by eugenol was found to be characterized primarily by a numbing quality (Cliff and Heymann, 1992). This description is consistent with the use of eugenol in dentistry to treat post-operative pain. Following gingivectomy, periodontal dressings containing eugenol significantly reduced the pain experienced in the immediate 12 hours post-surgery compared with dressings containing no eugenol (Jorkend and Skoglund, 1990). More recently, repeated application of eugenol to the tongue (1 minute interstimulus intervals) was shown to elicit self-desensitization (Klein *et al.*, 2013). On initial stimulation, eugenol evoked a strong irritant sensation, however, by the third application, desensitization was complete and subsequent eugenol administration evoked no sensation. This rapid decrease in perceived sensation is mirrored by the temporal onset of a numbing sensation that is identified by nearly 80% of panelists within 2 minutes following lingual eugenol administration (Klein *et al.*, 2013). Pretreatment with eugenol also reduced the burn associated with capsaicin administration and the ability to detect weak tactile stimuli on the lingual surface (Klein *et al.*, 2013). These data are commensurate with an anesthetic mechanism whereby eugenol application reduces sensitivity to other somatosensory stimuli. However, in these same studies, eugenol was found to enhance perceived warmth and heat pain while having no effect on cold sensitivity (Klein *et al.*, 2013), thus arguing against local anesthetic action. Further studies are needed to assess the unique sensory effects of eugenol and the mechanisms that underpin them.

8.7 Summary

The tingle and numbing sensations evoked by certain chemesthetic compounds contribute to the pleasure associated with the consumption of various foods and beverages. Significant insight into the mechanistic underpinnings have been elucidated in recent years, however, there still remains a number of questions to be addressed. The development of novel chemical probes, animal models, and psychophysical techniques will undoubtedly further our understanding of these distinctive sensations. In the meantime, chefs, culinary experts, and food scientists will continue to experiment with the use of these compounds in the development of novel food offerings. As scientists and foodies, let's enjoy the unique creations.

References

Albin KC and Simons CT. 2010. Psychophysical evaluation of a sanshool derivative (alkylamide) and the elucidation of mechanisms subserving tingle. *PLoS One*, **5** (3), e9520.

Altier C and Zamponi GW. 2004. Targeting Ca^{2+} channels to treat pain: T-type versus N-type. *Trends in Pharmacological Science*, **25**, 465–470.

Artaria C, Maramaldi G, Bonfigli A, Rigano L, and Appendino G. 2011. Lifting properties of the alkamide fraction from the fruit husks of Zanthoxylum bungeanum. *International Journal of Cosmetic Science*, **33** (4), 328–333.

Bader M, Stark TD, Dawid C, Lösch S, and Hofmann T. 2014. All-trans-configuration in Zanthoxylum alkylamides swaps the tingling with a numbing sensation and diminishes salivation. *Journal of Agricultural and Food Chemistry*, **62**, 2479–2488.

Barkin RL. 2013. The pharmacology of topical analgesics. *Postgraduate Medicine*, **125** (4), 7–18.

Bautista DM, Sigal YM, Milstein AD, *et al.* 2008. Pungent agents from Szechuan peppers excite sensory neurons by inhibiting two-pore potassium channels. *Nature Neuroscience*, **11** (7), 772–779.

Beer Judge Certification Program, Inc. 2008. BJCP *Style Guidelines for Beer, Mead, and Cider*.

Biedenbach MA and Chan KY. 1971. Tongue mechanoreceptors: comparison of afferent fibers in the lingual nerve and chorda tympani. *Brain Research*, **35** (2), 584–588.

Braz J, Solorzano C, Wang X, and Basbaum AI. 2014. Transmitting pain and itch messages: a contemporary view of the spinal cord circuits that generate gate control. *Neuron*, **82** (3), 522–536.

Burgoyne CC, Giglio JA, Reese SE, Sima AP, and Laskin DM. 2010. The efficacy of a topical anesthetic gel in the relief of pain associated with localized alveolar osteitis. *Journal of Oral Maxillofacial Surgery*, **68** (1), 144–148.

Brodin P. 1985. Differential inhibition of A, B and C fibres in the rat vagus nerve by lidocaine, eugenol and formaldehyde. *Archives of Oral Biology*, **30** (6), 477–480.

Brodin P and Røed A. 1984. Effects of eugenol on rat phrenic nerve and phrenic nerve-diaphragm preparations. *Archives of Oral Biology*, **29** (8), 611–615.

Brown AG and Iggo A. 1967. A quantitative study of cutaneous receptors and afferent fibres in the cat and rabbit. *Journal of Physiology*, **193** (3), 707–733.

Bryant BP and Mezine I. 1999. Alkylamides that produce tingling paresthesia activate tactile and thermal trigeminal neurons. *Brain Research*, **842** (2), 452–460.

Cardenas GG, Del Mar LP, and Scroggs RS. 1995. Variation in serotonergic inhibition of calcium channel currents in four types of rat sensory neurons differentiated by membrane properties. *Journal of Neurophysiology*, **74**, 1870–1879.

Carstens E, Iodi Carstens M, Dessirier JM, *et al.* 2002. It hurts so good: oral irritation by spices and carbonated drinks and the underlying neural mechanisms. *Food Quality and Preference*, **13** (7), 431–443.

Caterina MJ, Schumacher MA, Tominaga M, *et al.* 1997. The capsaicin receptor: a heat-activated ion channel in the pain pathway. *Nature*, **389**, 816–824.

Catterall WA. 2011. Voltage-gated calcium channels. *Cold Spring Harbor Perspectives in Biology*, **3** (8), a003947.

Chahine M and O'Leary ME. 2014. Regulation/modulation of sensory neuron sodium channels, in *Voltage Gated Sodium Channels* (ed. PC Ruben), Heidelberg: Springer, pp. 111–136.

Cliff M and Heymann H. 1992. Descriptive analysis of oral pungency. *Journal of Sensory Studies*, **7**, 279–290.

Coburn CA, Luo Y, Cui M, *et al.* 2012. Discovery of a pharmacologically active antagonist of the two-pore-domain potassium channel K2P9.1 (TASK-3). *ChemMedChem*, **7** (1), 123–133.

Cometto-Muñiz JE, García-Medina MR, Calviño AM, and Noriega G. 1987. Interactions between CO_2 oral pungency and taste. *Perception*, **16** (5), 629–640.

Cometto-Muñiz JE and Noriega G. 1985. Gender differences in the perception of pungency. *Physiology and Behavior*, **34** (3), 385–389.

Dessirier JM, Simons CT, Carstens MI, O'Mahony M, and Carstens E. 2000. Psychophysical and neurobiological evidence that the oral sensation elicited by carbonated water is of chemogenic origin. *Chemical Senses*, **25** (3), 277–284.

Dessirier JM, Simons CT, O'Mahony M, and Carstens E. 2001. The oral sensation of carbonated water: cross-desensitization by capsaicin and potentiation by amiloride. *Chemical Senses*, **26** (6), 639–643.

Freedman M and Strassman LFA. 2013. Commonly used topical oral wound dressing materials in dental and surgical practice – a literature review. *Journal of the Irish Dental Association*, **59** (4), 90–195.

Galopin CC, Furrer SM, and Goeke A. 2004. Pungent and tingling compounds in Asian cuisine, in *Challenges in Taste Chemistry and Biology, ACS Symposium Series*, vol. **867**, (eds. T Hofmann, CT Ho, and W Pickenhagen), Washington DC: American Chemical Society, pp. 139–152.

Graber M and Kelleher S. 1988. Side effects of acetazolamide: the champagne blues. *American Journal of Medicine*, **84** (5), 979–980.

Green BG. 1992. The effects of temperature and concentration on the perceived intensity and quality of carbonation. *Chemical Senses*, **17** (4), 435–450.

Gurney BF. 1968. Chemotherapy in dental practice. Compound topical anesthetics. *Dental Digest*, **74** (3), 129.

Hagura N, Barber H, and Haggard P. 2013. Food vibrations: Asian spice sets lips trembling. *Proceedings of the Royal Society B, Biological Sciences*, **280**, 20131680.

Harper SJ and McDaniel MR. 1993. Carbonated water lexicon: Temperature and CO_2 level influence on descriptive ratings. *Journal of Food Science*, **58** (4), 893–898.

Heppenstall PA and Lewin GR. 2006. A role for T-type Ca^{2+} channels in mechanosensation. *Cell Calcium*, **40** (2), 165–174.

Hewson L, Hollowood T, Chandra S, and Hort J. 2009. Gustatory, olfactory and trigeminal interactions in a model carbonated beverage. *Chemosensory Perception*, **2**, 94–107.

Housecroft CE and Sharpe AG. 2005. The group 14 elements, in *Inorganic Chemistry*, 2nd edn, Essex, UK: Pearson Education Limited, p. 368.

Jorkjend L and Skoglund L. 1990. Effect of non-eugenol and eugenol-containing periodontal dressings on the incidence and severity of pain after periodontal soft tissue surgery. *Journal of Clinical Periodontology*, **17** (6), 341–344

Kivela J, Parkkila S, Parkkila AK, Leinonen J, and Rajaniemi H. 1999a. Salivary carbonic anhydrase isoenzyme VI. *Journal of Physiology*, **520** (2), 315–320.

Kivela J, Parkkila S, Parkkila AK, and Rajaniemi H. 1999b. A low concentration of carbonic anhydrase isoenzyme VI in whole saliva is associated with caries prevalence. *Caries Research*, **33**, 178–184.

Klein AH, Carstens MI, and Carstens E. 2013. Eugenol and carvacrol induce temporally desensitizing patterns of oral irritation and enhance innocuous warmth and noxious heat sensation on the tongue. *Pain*, **154** (10), 2078–2087.

Klein AH, Sawyer CM, Zanotto KL, *et al.* 2011. A tingling sanshool derivative excites primary sensory neurons and elicits nocifensive behavior in rats. *Journal of Neurophysiology*, **105** (4), 1701–1710.

Koo JY, Jang Y, Cho H, *et al.* 2007. Hydroxy-alpha-sanshool activates TRPV1 and TRPA1 in sensory neurons. *European Journal of Neuroscience*, **26** (5), 1139–1147.

Komai M, Bryant BP. 1993. Acetazolamide specifically inhibits lingual trigeminal nerve responses to carbon dioxide. *Brain Research*, **612** (1–2), 122–129.

Kozam G. 1977. The effect of eugenol on nerve transmission. *Oral Surgery*, **44** (5), 799–805.

LaMotte RH, Friedman RM, Lu C, Khalsa PS, and Srinivasan MA. 1998. Raised object on a planar surface stroked across the fingerpad: responses of cutaneous mechanoreceptors to shape and orientation. *Journal of Neurophysiology*, **80**, 2446–2466.

Lee MH, Yeon KY, Park CK, *et al.* 2005. Eugenol inhibits calcium currents in dental afferent neurons. *Journal of Dental Research*, **84** (9), 848–851.

Lee MS. 2014. Recent progress in the discovery and development of N-type calcium channel modulators for the treatment of pain. *Progress in Medicinal Chemistry*, **53**, 147–186.

Lennertz RC, Tsunozaki M, Bautista DM, and Stucky CL. 2010. Physiological basis of tingling paresthesia evoked by hydroxy-alpha-sanshool. *Journal of Neuroscience*, **30** (12), 4353–4361.

Lingueglia E, de Weille JR, Bassilana F, *et al.* 1997. A modulatory subunit of acid sensing ion channels in brain and dorsal root ganglion cells. *Journal of Biological Chemistry*, **272**, 29778 –29783.

Liu L and Simon SA. 1996. Capsaicin-induced currents with distinct desensitization and Ca^{2+} dependence in rat trigeminal ganglion cells. *Journal of Neurophysiology*, **75** (4), 1503–1514.

Liu L and Simon SA. 1998. The influence of removing extracellular Ca^{2+} in the desensitization responses to capsaicin, zingerone and olvanil in rat trigeminal ganglion neurons. *Brain Research*, **809** (2), 246–252.

Löken LS, Wessberg J, Morrison I, McGlone F, and Olausson H. 2009. Coding of pleasant touch by unmyelinated afferents in humans. *Nature Neuroscience*, **12** (5), 547–548.

Lotshaw DP. 2007. Biophysical, pharmacological, and functional characteristics of cloned and native mammalian two-pore domain K^+ channels. *Cell Biochemistry and Biophysics*, **47** (2), 209–256.

Madrid R, de la Pena E, Donovan-Rodriguez T, Belmonte C, and Viana F. 2009. Variable threshold of trigeminal cold-thermosensitive neurons is determined by a balance between TRPM8 and Kv1 potassium channels. *Journal of Neuroscience*, **29**, 3120–3131.

Markowitz K, Moynihan M, Liu M, and Kim S. 1992. Biologic properties of eugenol and zinc oxide-eugenol. A clinically oriented review. *Oral Surgery, Oral Medicine, Oral Pathology*, **73**, 729–737.

McGlone F, Wessberg J, and Olausson H. 2014. Discriminative and affective touch: sensing and feeling. *Neuron*, **82** (4), 737–755.

McKemy DD, Neuhausser WM, and Julius D. 2002. Identification of a cold receptor reveals a general role for TRP channels in thermosensation. *Nature*, **416** (6876), 52–58.

Morenilla-Palao C, Luis E, Fernández-Peña C, *et al.* 2014. Ion channel profile of TRPM8 cold receptors reveals a role of TASK-3 potassium channels in thermosensation. *Cell Reports*, **8** (5), 1571–1582.

Munns C, Al Qatari M, and Koltzenburg M. 2007. Many cold sensitive peripheral neurons of the mouse do not express TRPM8 or TRPA1. *Cell Calcium*, **41** (4), 331–342.

Noel J, Zimmermann K, Busserolles J, *et al.* 2009. The mechano-activated K^+ channels TRAAK and TREK-1 control both warm and cold perception. *The EMBO Journal*, **28**, 1308–1318.

Norman J. 2002. *Herbs & Spices: The Cook's Reference*, New York: DK Publishing Inc.

Ochoa JL and Torebjörk HE. 1980. Paraesthesiae from ectopic impulse generation in human sensory nerves. *Brain*, **103** (4), 835–853.

Park CK, Kim K, Jung SJ, *et al.* 2009. Molecular mechanism for local anesthetic action of eugenol in the rat trigeminal system. *Pain*, **144**, 84–94.

Park CK, Li HY, Yeon KY, *et al.* 2006. Eugenol inhibits sodium currents in dental afferent neurons. *Journal of Dental Research*, **85** (10), 900–904.

Parkkila S, Parkkila AK, Lehtola J, *et al.* 1997. Salivary carbonic anhydrase protects gastroesophageal mucosa from acid injury. *Digestive Diseases and Sciences*, **42**, 1013–1019.

Parkkila S, Kaunisto K, Rajaniemi L, *et al.* 1990. Immunohistochemical localization of carbonic anhydrase isoenzymes VI, II and I inhuman parotid and submandibular glands. *Journal of Histochemistry and Cytochemistry*, **38**, 941–947.

Patacchini R, Maggi CA, and MeliA. 1990. Capsaicin-like activity of some natural pungent substances on peripheral endings of visceral primary afferents. *Naunyn-Schmiedeberg's Archives of Pharmacology*, **342** (1), 72–77.

Perez-Reyes E, Cribbs LL, Daud A, *et al.* 1998. Molecular characterization of a neuronal low-voltage-activated T-type calcium channel. *Nature*, **391**, 896–900.

Peters CH and Ruben PC. 2014. Introduction to sodium channels, in *Voltage Gated Sodium Channels* (ed. PC Ruben), Heidelberg: Springer, pp. 1–6.

Peters S. 1971. An old home remedy against toothache. *Die Quintessenz*, **22** (3), 33–34.

Pramod K, Ansari SH, and Ali J. 2010. Eugenol: a natural compound with versatile pharmacological actions. *Natural Products Communications*, **5** (12), 1999–2006.

Sawyer CM, Carstens MI, Simons CT, *et al.* 2009. Activation of lumbar spinal wide-dynamic range neurons by a sanshool derivative. *Journal of Neurophysiology*, **101** (4), 1742–1748.

Seo H, Li HY, Perez-Reyes E, and Lee JH. 2013. Effects of eugenol on T-type Ca^{2+} channel isoforms. *Journal of Pharmacology and Experimental Therapeutics*, **347** (2), 310–317.

Simons CT, Carstens MI, and Carstens E. 2003. Oral irritation by mustard oil: self-desensitization and cross-desensitization with capsaicin. *Chemical Senses*, **28** (6), 459–465.

Simons CT, Dessirier JM, Carstens MI, O'Mahony M, and CarstensE. 1999. Neurobiological and psychophysical mechanisms underlying the oral sensation produced by carbonated water. *Journal of Neuroscience*, **19** (18), 8134–8144.

Simons CT and Ward CD. 2005. Temporal characteristics and regional sensitivity of tingle compounds [Abstract]. *Chemical Senses*, **30**, A203.

Srinivasan MA, Whitehouse JM, LaMotte RH. 1990. Tactile detection of slip: surface microgeometry and peripheral neural codes. *Journal of Neurophysiology*, **63**, 1323–1332.

Stovall S. Rainforest remedy may tackle toothache, *The Wall Street Journal*, 12 March 2012.

Sugai E, Morimitsu Y, Iwasaki Y, *et al.* 2005. Pungent qualities of sanshool-related compounds evaluated by a sensory test and activation of rat TRPV1. *Bioscience, Biotechnology, and Biochemistry*, **69** (10), 1951–1957.

Supuran CT. 2007. Carbonic anhydrases as drug targets—an overview. *Current Topics in Medicinal Chemistry*, **7** (9), 825–833.

Tezuka S and TezukaG. 1994. Private part-coating cream containing Zanthoxylum piperitum and sexual hormone. Japanese Patent JPH06211676, filed Jan. 11, 1993 and issued Aug. 2, 1994.

Toma S and Nakajima Y. 1995. Response characteristics of cutaneous mechanoreceptors to vibratory stimuli in human glabrous skin. *Neuroscience Letters*, **195**, 61–63.

Tominaga M, Caterina MJ, Malmberg AB, *et al.* 1998. The cloned capsaicin receptor integrates multiple pain-producing stimuli. *Neuron*, **21** (3), 531–543.

Trulsson M and Essick GK. 1997. Low-threshold mechanoreceptive afferents in the human lingual nerve. *Journal of Neurophysiology*, **77** (2), 737–748.

Tsunozaki M, Lennertz RC, Vilceanu D, *et al.* 2013. A 'toothache tree' alkylamide inhibits Aδ mechanonociceptors to alleviate mechanical pain. *Journal of Physiology*, **591**, 3325–3340.

Waldmann R, Bassilana F, de WeilleJ, *et al.* 1997a. Molecular cloning of a non-inactivating proton-gated Na^+ channel specific for sensory neurons. *Journal of Biological Chemistry*, **272**, 20975–20978.

Waldmann R, Champigny G, Bassilana F, Heurteaux C, and Lazdunski M. 1997b. A proton-gated cation channel involved in acid-sensing. *Nature*, **386**, 173–177.

Wang YY, Chang RB, and Liman ER. 2010. TRPA1 is a component of the nociceptive response to CO_2. *Journal of Neuroscience*, **30** (39), 12958–12963.

Weaver F. 2014. The trigeminal nerve, in *Structures of the Head and Neck*, Philadelphia: F. A. Davis Company, pp. 250–265.

Wetzel C, Hu J, Riethmacher D, *et al.* 2007. A stomatin-domain protein essential for touch sensation in the mouse. *Nature*, **445**, 206–209.

Wise PM and Bryant B. 2014. The effect of temperature and menthol on carbonation bite. *Chemical Senses*, **39** (7), 571–582.

Wise PM, Wolf M, Thom SR, and Bryant B. 2013. The influence of bubbles on the perception carbonation bite. *PLoS One*, **8** (8), e71488.

Yau NJN and McDaniel MR. 1990. The power function of carbonation. *Journal of Sensory Studies*, **5**, 117–128.

Yau NJN and McDaniel MR. 1992. Carbonation interactions with sweetness and sourness. *Journal of Food Science*, **57**, 1412–1416.

CHAPTER 9

Interactions in chemesthesis: everything affects everything else

Brian Byrne
Natural Advantage LLC, Oakdale, LA, USA

9.1 Introduction

When we taste food it is a complex mixture of taste, mouth flavor, and nasal aromas all coming together to create an impression that the brain has been trained to interpret as a particular food, smell, or flavor.

Many models have appeared to explain the different aspects of taste. Earlier models were simple, but they have now become more complex. This has been necessary in order to explain the intricate inter-dependency of all the stimuli working in concert to create an impression.

9.2 Coolants

Throughout the early work on topical coolants one can clearly see the progression in the understanding of the theory of cooling, as it developed from simple to more complex. Seminal work by Wilkinson-Sword, Ltd shows that not only is *levo*-menthol cooling, but artificially created chemicals that are structurally similar to *levo*-menthol are frequently coolants as well (e.g., WS-3) (Watson *et al.*, 1977; Watson *et al.*, 1979b) (Fig. 9.1). It thus follows that structurally unrelated coolants can be created that contain similar regions in the molecule which are necessary for coolness. They require polar, non-polar, hydrophilic, and hydrophobic, and so on, conditions that must be met for coolness to be perceived (Watson *et al.*, 1979a; Watson *et al.*, 1979b).

These newly created and commercialized coolants have levels of coolness that are expressed relative to *levo*-menthol. When these were used at their recommended level of use in flavors, the expected results were obtained. However, when two different coolants were used together at their standard use levels, the end product was dramatically cooler than expected. This was attributed to an additive effect. Startling results occurred when two different coolants were used together at levels of half their recommended use levels. They were in fact cooler than either

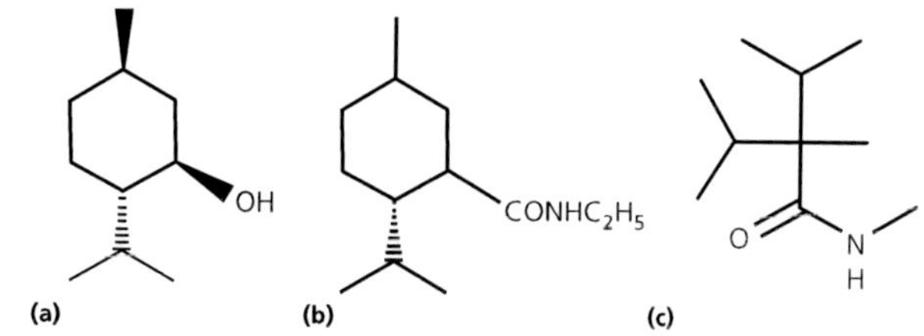

Fig. 9.1 Structurally similar and dissimilar coolants: (a) *levo*-menthol, (b) WS-3, and (c) WS-23.

Table 9.1 Comparing isopulegol, *levo*-menthyl lactate, and *levo*-menthol.

Material tasted in 5% sugar water*	Level (ppm)	Flavor observations	Cooling observations
levo-Isopulegol	10	No flavor	No cooling
	20	No flavor	No cooling
	30	No flavor	No cooling
levo-Menthyl lactate	10	No flavor	No cooling
	20	No flavor	No cooling
	30	No flavor	Faint cooling
levo-Menthol	10	Peppermint	No cooling
	20	Very peppermint	Slight cooling
	30	Strong peppermint	Cooling

*Control = 5% sugar water. All solutions are 1% dilutions in alcohol tasted in 5% sugar water.

one used separately, suggesting a synergy of the two coolants (Mane and Ponge, 1998). For example, findings in the author's laboratory demonstrate that separate selected coolants were not cool at use levels of 10 ppm (Table 9.1).

However, when the three coolants from Table 9.1 are used below their cooling level (10 ppm) but in conjunction with one other coolant, also at 10 ppm, the resulting cooling is stronger than expected (Table 9.2). The results support the theory that one coolant synergizes or positively modulates the other. Similar effects are observed in other areas, such as sweetness and mouthfeel.

These results suggest that cooling using *levo*-menthol could be achieved at reduced levels (below typical flavor use levels of 20–30 ppm), without the characteristic peppermint flavor, burning and bite. To further examine this theory, we used a beverage base (5% sugar water) to evaluate *levo*-menthol with non-coolant *levo*-carvone, which has a similar structure to that of menthol (Fig. 9.2). The most frequent use of *levo*-carvone is in chewing gum and oral care products to give the characteristic spearmint taste. levo-Carvone is not cool at 30 ppm, nor is it reported to be a coolant.

All of the coolants in the study have *no cooling effect* in 5% sugar water at *10 ppm*. However, all of the coolants have cooling effects at 30 ppm in sweetened water. Thus, the coolants were added at levels of 10 ppm to the 5% sugar water base, and the results are reported in Table 9.3 (S. Fontenot, personal communication, September, 2014).

Table 9.2 Comparing *levo*-menthol with *levo*-menthyl lactate and *levo*-isopulegol when used together with the other coolants at 10 ppm.

Material tasted in 5% sugar water	Level (ppm)	Flavor observations	Cooling observations
levo-Isopulegol	10	No flavor	No cooling
levo-Menthol	10	Slight peppermint	No cooling
levo-Isopulegol + *levo*-menthol	10 + 10	No peppermint	Very strong cooling
levo-Menthyl lactate + *levo*-menthol	10 + 10	No peppermint	Very strong cooling
levo-Isopulegol + *levo*-methyl lactate	10 + 10	No flavor	Slight tingle

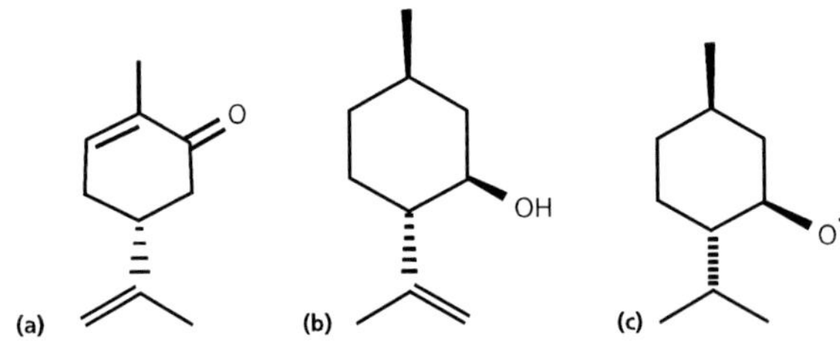

Fig. 9.2 Structural comparison of non-coolant with coolants: (a) *levo*-carvone, (b) *levo*-isopulegol, and (c) *levo*-menthyl lactate.

Table 9.3 Evaluation of several coolants in combination with the flavorant *levo*-carvone.

Material tasted in 5% sugar water	Level (ppm)	Flavor observations	Cooling observations
levo-Carvone only	10	Strong spearmint flavor	No cooling
levo-Carvone + *levo*-menthol	10 + 10	Same flavor as control	Mild cooling
levo-Carvone + *levo*-isopulegol	10 + 10	Stronger spearmint flavor with fresh-picked minty taste	Strong cooling
levo-Carvone + *levo*-menthyl lactate	10 + 10	Same flavor as control	Strong cooling

It was observed that *levo*-menthol at 30 ppm is cool, and has a peppermint flavor with a quick onset of cooling. At 10 ppm, *levo*-menthol has an almost imperceptible cooling and *no peppermint flavor*. However, a 10 ppm *levo*-menthol and 10 ppm *levo*-carvone solution (in 5% sugar water) does exhibit a perceptible cooling, but it tastes like the spearmint control. Therefore, a non-coolant can positively modulate a coolant, hence making it appear cooler in beverages.

Two other examples are worth noting. In the case of *levo*-carvone at 10 ppm (in 5% sugar water) with two other known coolants, *levo*-isopulegol and *levo*-menthyl lactate at 10 ppm (Table 9.3), even stronger cooling effects are caused by *levo*-carvone than by *levo*-menthol. Therefore, higher use levels gave consequently stronger cooling.

Evaluations of 10 ppm *levo*-carvone in 5% sugar water added to the sweetened and flavored base coolants are shown in Table 9.3. From the results, it appears that synergies occur with non-coolant materials, rather than just with coolant plus coolant.

9.3 Sweet

Much work has been done over the years to prepare and use alternatives to sugar for reasons of health and cost. Many of the newer sweeteners have negative tastes that thwart their use as direct sugar replacements. For example, stevia, the extract of *Stevia rebaudiana*, is made up of eight glycosides, with Rebaudioside-A (Reb-A) being the prominent component. Stevia Reb-A is used in food and beverages, however, its taste is marred by a slow onset of sweetness, prolonged after-sweetness, and a metallic finish. While Reb-A is doing well in the commercial market, people are continuing to isolate other Rebaudiosides with cleaner tastes. The scope of this chapter is not to cover the history or the commercially available natural or artificial sweeteners, but to present some data regarding the modulating of existing sweeteners. The mechanism of modulation is not understood to the level that allows the prediction of a successful sweetener candidate. However, several materials have been developed based on structural similarities, which is exemplified by the dipeptide sweeteners (i.e., aspartame and neotame, Fig. 9.3). Sweetness enhancing is similar to the findings for cooling in that adding one sweetener to another gives enhanced sweetness.

Synergy is one characteristic often associated with sweeteners when combined with other sweeteners. This refers to the total sweetness intensity of a mixture

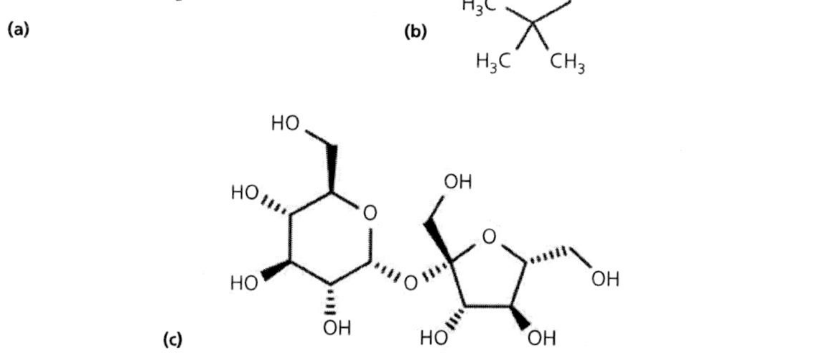

Fig. 9.3 Structural comparison of artificial dipeptide sweeteners to sugar: (a) aspartame, (b) neotame, and (c) sucrose.

Table 9.4 Tasting 5% sugar water solutions with enhancer.

Item	Base level	Observation
Esifurane, 0.01% in Alc	1 drop/100 ml 0.016 ppm	Sweeter than control
Mesifurane, 0.01%	1 drop 0.016 ppm	Sweeter than esifurane
Monk fruit, 0.01% in water	1 drop 0.025 ppm	Sweeter than mesifurane

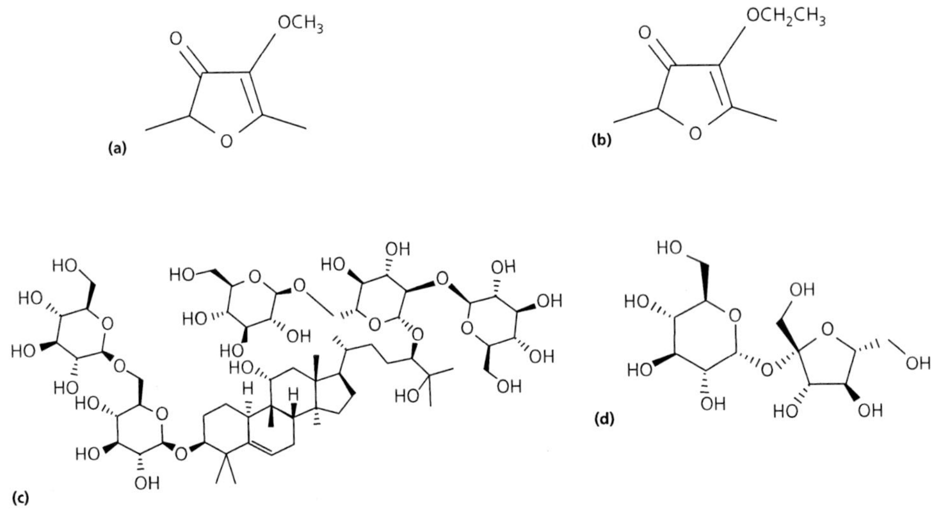

Fig. 9.4 Structural comparison of non-sweeteners and sweeteners: (a) mesifurane, (b) esifurane, (c) monk fruit (mogroside V), and (d) sucrose.

being greater than the theoretical sum of the intensities of the individual components. Another alternative is to utilize sweetness enhancers, which are themselves not sweet, to enhance the perception of the sweet taste and to reduce the quantity of sugar used in the food product. The ideal sweetness enhancer will have no intrinsic taste or aroma, but will increase the sweetness of a sweetener without imparting any negative effect on other flavor profiles. Work in the author's lab is consistent with the theory of synergy in sweeteners (S. Fontenot, personal communication, September, 2014).

A selection of natural sweeteners, monk-fruit powder, and two non-sweeteners, mesifurane and esifuran, which naturally occur in sherry, were evaluated in 5% sugar water (see Table 9.4 and Fig. 9.4). Monk fruit is the purified extract from the plant *Siraitia grosvenorii*, having a sweetness 300 times stronger than that of sugar. The sweetness components in monk fruit are known as mogrosides, and the predominant mogroside in monk fruit is mogroside-V.

The results show that monk fruit at 0.025 ppm in water is not sweet, as is expected, since it should be equivalent to 7.5 ppm of sugar. Sugar (sucrose) has an established taste threshold of 0.41% in water. This is equivalent to 4100 ppm. The conclusion is that monk fruit has a pronounced synergistic effect on sugar at 0.025 ppm in 5% sugar water.

In addition, Table 9.4 shows that the non-sweetener esifurane at 0.015 ppm is sweeter in 5% sugar water than the control (5% sugar water), and that mesifurane is even stronger than the esifurane. The aroma of esifurane is that of purified white granular sugar, whilst mesifurane has a more pronounced "sugar bowl" aroma than esifurane. Although speculative, the aroma of sugar may be contributing to the overall intensity of the sweetening synergies. Others have suggested that association of familiar smells and tastes may cause a learned response interaction with the brain (Labbe *et al.*, 2007). Such learned responses are discussed further in the Section 9.7 on aroma.

9.4 Salt

Salt (sodium chloride) is the penultimate target for salty taste, because our societies taste has already been trained to expect the exact taste of salt. Other similar salts such as potassium chloride taste salty but the wrong kind of salty. The most successful salt reduction products on the commercial market actually contain salt, NaCl. Many approaches to creating a salt replacement have been taken, with varying success. Approaches to salt reduction products have two common themes:

- to use other materials that are salty;
- to use salt enhancers, thereby increasing the perception of saltiness.

Thus far, no commercial product mix has equaled salt (NaCl) in saltiness. The most commonly used replacer is potassium chloride (KCl). While KCl is salty, it is a different type of saltiness, with a bitter, metallic taste and after-taste. The saltiness of KCl lacks the depth of saltiness of NaCl, which can be seen in the middle of the taste profile. Blends of NaCl and KCl do not have the intensity of salt, and still have the bitter metallic taste. In the next generation of salt replacers, blends of NaCl–KCl are positively modulated by another component to give mixtures that block the bitter metallic taste of KCl. Commonly used modulators are edible salts of acids, both mineral and carboxylic acids, for example, sodium acetate, ammonium acetate, sodium glutamate, potassium sulfate, calcium gluconate, sodium gluconate, calcium chloride, and so on (Bonorden *et al.*, 2003). Other positive modulators in use are based on fermentation. "Naturally fermented" autolyzed yeast and glutamate salts are successful in amplifying the saltiness of the NaCl–KCl mixture. The most notable defects are that autolyzed yeast can be "doctored up" by use of glutamate as a "nutrient" prior to harvesting the yeast. Added glutamates have to be declared on the label and have an umami taste. This has given autolyzed yeast a bad label claim and poor market acceptance. On the other hand, fermented soy sauce contains soy sauce flavor and more nucleotides than the traditional acid hydrolyzed soy sauce. The nucleotides in fermented soy sauce strongly enhance the saltiness of NaCl and NaCl–KCl mixtures. The last of the innovative approaches is to change the structured form of the salt crystal to achieve greater surface area, and therefore greater salt impression in topical applications only.

9.5 Mouthfeel

Mouthfeel is a more diverse topic than the previous sections in that it covers such a wide range of food and beverages, including meat products, alcoholic beverages, baked goods, salad dressing, and snacks, wine and so on (see Table 9.5). Basically, it covers everything that we eat and drink, all of which have their own language to describe the presence or absence of these particular tastes or mouthfeel effects. The common component of all these diverse areas is in the tasting. During the tasting process, there is a beginning, middle, end, and a finish to the product (after it is swallowed). The language descriptors are used to compare the sample versus the target, or an individual's personal perception of the ideal product.

In addition to making better tasting products, there is a large market for healthier, or good-for-you products. By understanding the importance that mouthfeel brings to the product, only then can a better product be created. For a successful product, all mouthfeel, taste, and aroma components have to be be met, and must be in balance with the target or ideal product.

Many tasty calorific products have achieved market success; however, when lower sugar, fat, salt, or less calorific starting ingredients are used, the product frequently lacks the flavor and texture that is expected. Thus building back the missing taste and mouthfeel are the challenges for the technologists. As an example, non-fat salad dressing lacks the emulsifying action and lubricity of oil, so natural gums and hydrocolloids are frequently used to give the texture, lubricity, and the mouthfeel of oil.

However, salad dressing frequently contains vinegar for flavor and apparently the vegetable oil blocks the sour receptors. Therefore, removal of the oil creates a product that is too sour, in turn causing market rejection of the product. In the evaluation of spices and essential oils that might suppress the sourness of vinegar, several candidates work well: lime oil (cold pressed), ylang-ylang oil, and black pepper oil (deodorized) are most effective at acid blocking, or positive modulation, at the 1–10 ppm level in the finished food product. Although the mechanism of action is unclear, it is believed that the receptors for sourness are hydrophilic in character, and they are surrounded by hydrophobic neighboring groups that are

Table 9.5 Some examples of mouthfeel.

Astringent	Cooling	Mouth-watering
Bitter	Refreshing	Dry
Creamy	Gummy	Smooth
Fatty	Watery	Smokey
Brittle, crisp	Umami	Buttery
Hard to chew	Metallic	Nutty
Softness	Prickly	Alcohol burn (tongue)
Sour	Hot, warming	Lubricity
Salty	Painful	Alcohol burn and bite
Sweet	Succulent	Soapy

blocked by lime oil, ylang-ylang oil, black pepper oil, and vegetable oil, preventing the acid from entering the ion channel (Medler and Kinnamon, 2004).

Similarly, sugar, the most commonly used sweetener, brings a sweetness profile, aroma, and mouth presence that is typically not met by high-intensity sweetness. One solution is to add low-caloric polyols, such as erythritol, to improve the mouth presence of stevia. Additionally, the additives help to dilute stevia to ready-to-use levels for the consumer.

Other modulators have structural similarities to that of sucrose (sugar) (see Fig. 9.4d), enhancing or having a positive modulation effect. The early modulators that enhanced the perceived sweetness of sugar, however, have their own characteristic tastes that prevent an exact duplication of the milder taste of sugar. In this category are Furaneol® (registered trademark of Firmenich™), cyclotene, maltol, ethyl maltol, and the like, and their characterizing tastes are that of cotton-candy (spun sugar) to maple flavor. However, earlier modulators give a different mouthfeel, taste, and aroma to that of sugar and replacing the older modulators with newer, clean-tasting modulators is the way of the future for better tasting sugar replacers.

9.6 Astringency and bitterness

Astringency and bitterness have been an area of study due to the health benefits of polyphenols. While they are good radical scavengers, protecting the human body from cancer, aging, and other degenerative maladies (Bors *et al.*, 1990), they can be too astringent and/or bitter to be consumed in food and beverages. Polyphenols are found in wine, pomegranates, cranberries, tea, chocolate, raspberries, strawberries, broccoli, cabbage, celery, onion, and parsley (Breton, 2008). In addition, they are present in tree bark, which is where they are extracted from (Araghi-Niknam *et al.*, 2000) and used in over-the-counter products to be consumed as vitamins. Processed proteins can have a bitterness associated with bitter peptides. While sugar, salt, and added flavors can increase the palatability of some products, the actual blocking of bitterness remains under study.

Strategies to improve the palatability of astringent and bitter materials in food goods can vary from a chef's approach to a scientist's approach. The chef's approach is to use tried and true solutions to bitterness. Among these are white vinegar added to a bitter tomato sauce to block the bitter taste, and too tangy or sour a sauce can be helped by a pinch of sugar or salt. Such anecdotal remedies frequently are based in truth and can be explained by science.

Scientifically, the ion channel receptors for salt and acidity are believed to be responsible for white vinegar (acetic acid) out-competing the bitter component for the receptor site. Alternatively, the vinegar may protonate the bitter component, changing its polarity, and therefore its taste. Similarly, vegetable oil blocks some receptors by binding to the hydrophobic groups surrounding the ion channel. Vegetable oil and coating-like oily substances have been demonstrated to block bitterness. Symrise (Ley *et al.*, 2005) scientists have shown that homoeriodictyol, found in Herba Santa (*Eriodictyon californicum*), being a polyphenolic flavone, is a phenolic acid that has powerful bitter blocking properties in its sodium salt form.

Sucralose, a chlorinated sucrose, has been found to be another bitterness blocker as well as a high-intensity sweetener. Sucralose is 8000 times sweeter than sugar, but has been found to block bitter receptor sites, possibly due to the non-polar chloro-group (Roy, 1997). However, in addition it has a water soluble end of the molecule making it able to fill the ion channel. Neotame, a potent dipeptide sweetener has a much larger non-polar neopentyl group, which is likely to bind to the hydrophobic entrance to the receptor channel while having a polar opposite end of the molecule. Neotame has been found to have a better blocking effect than aspartame, in addition to the intense sweetness from the molecule (Windschauer and Virgallito, 2012).

Adenosine monophosphate (AMP) has been found to inhibit the bitter taste of quinine and other bitter agents. This study was based on the neural response of mice to bitter materials with and without potential positive bitterness modulators (Ding *et al.*, 1999).

In the area of wines, particularly red wines, polyphenols from the skin of the grape and the seeds are present, which can lead to a bitterness or astringency of the wine. As red wine ages, the taste usually improves. Exposure of wine to air and light can greatly change the taste of the final product. However, as wine ages it develops ethyl esters of the acids that are found in the grapes, as the ethanol in the new wine is produced. The most prominent acid in new wine is tartaric acid, and as the aging process proceeds, diethyltartrate (DET) forms in the bottle, cask, or tank. DET has been linked to improved wine taste (Edwards *et al.*, 1985), it reduces the perceived acidity, in addition to lowering the perceived astringency and harshness, thus producing a more mellow wine. Although not present in wine at a level that would contribute to the taste (aroma), it does have a mellowing effect. The DET content, along with the alcohol content contribute the "legginess" of wine when swirled in a glass. This is a viscosity phenomenon, however, it is deemed to be important in better wines. In the author's laboratory, work has been carried out to isolate DET from wine. Evaluating the smoothness, perceived acidity, and astringency of low-quality wines, as well as high-quality wines without added DET, establishes a baseline for a trained tasting panel. After the addition of 2 ppm of DET to the wines, the lower quality wines are improved in all categories; however, it does not improve the taste of the higher quality wines. There is a noticeable lowering of the perceived acidity and astringency in the lower quality or young wines using 2 ppm DET.

9.7 Aroma (retronasal and orthonasal)

Aroma is an integral part of the "tasting" of edible products. A number of researchers have been able to link aroma to the overall impression of taste. Their work correlates well with the author's findings.

In the studies by Verhagen and Engelen (2006), Small and Prescott (2005), and Prescott (2004), a connection has been made between odor and taste integration in the orbitofrontal cortex of the brain. These converging data have been measured by neuroimaging (Rolls and Bayliss, 1994) of taste and odor in a unimodal stimulation (one stimulus at a time), and the data used to evaluate taste and odor in a bimodal (simultaneous) presentation of the same tastes and odors (Small and Prescott 2005; Djordjevic *et al.*, 2004). These findings confirm that odor and taste integration happens at the neural level. The brain connects all of the components of taste, odor, and mouthfeel to provide an image to our minds.

As discussed earlier, salt reduction generally causes consumer rejection. However, some studies (Girgis *et al.*, 2003) show that a lowering of salt in some products does not affect consumer acceptance or cause a loss in perceived saltiness. Therefore, it is a product–byproduct analysis that will provide low-salt products with odor-enhancement compensating for the lower salt content. In order to evaluate different groups that consistently give a positive odor–taste correlation, Lawrence *et al.* (2009) performed a multidimensional analysis using only water, and then with low salt water solutions, with and without commercially available flavors (not foods). Their data reveal saltiness differences between samples with flavor in the water solutions not containing salt. These findings show that odors perceived retronasally do not correlate well with the flavored salt-less solutions and the water control. However, if odor is introduced orthonasally, then perceived saltiness correlates well with the odors commonly associated with salt (fish, bacon, ham). Nevertheless, using a low salt content and the same flavors shows that the salt water is judged to be saltier with salt-associated products such as anchovy, bacon, smoked salmon, dry sausage, peanuts, bouillon cube, sardine, as opposed to the low salt ratings given for sweet flavored low salt water (citrus, chocolate, strawberry, and vanilla).

In general, aroma enhancement of one of the major taste groups (salt, sweet, sour, umami) is possible and has been studied (Cliff and Noble, 1990). Sourness has been studied in a similar manner by Saint-Eve *et al.* (2004) to establish taste–aroma interactions. Recent work by Bartoshuk *et al.* (2013) has been published as a patent application (US 20130280400-A1) showing that certain volatile components found in strawberries are responsible for a sweeter tasting strawberry. They state that perceived sweetness in consumer products is enhanced by the addition of two or more volatiles from the following group: neral, 4-carene, 3-methyl-1-butanol, 6-Methyl-5-hepten-2-ol, isovaleric acid, and geranial. According to the claims, this is applicable to sugar, natural sweeteners, and artificial sweeteners, although none of these materials are sweeteners, only flavorants that enhance sweetness.

These results are consistent with the author's finding that mesifurane and esifurane strongly enhance sweeteners to provide a greater sweetness. The data are presented in Table 9.4. The "sugar bowl" aroma that they create would work in the same way as sugar in combination with other sweeteners.

9.8 Conclusion

Taste, aroma, and mouthfeel work in concert. Our brain has learned to associate certain food products with salt, sweet, sour, and bitter. The simultaneous actions of all of these stimuli are responsible for enhancement or blocking and are therefore a combination of all the parts of the tasting experience. In other words, *Everything Affects Everything Else.*

References

Araghi-Niknam, M., Hosseini, S., Larson, D., Rohdewald, P., and Watson, R.R. 2000. Pine bark extract reduces platelet aggregation. *Integrative Medicine: Integrating Conventional and Alternative Medicine*, **2** (2), 73–77.

Bartoshuk, L., Colquhoun, T.A., Clark, *et al.* 2013. Compositions and methods for modifying perception of sweet taste, US Patent Application 20130280400, filed April 24, 2013 and issued Oct. 24, 2013.

Bonorden, W.R., Giordano, D.A., and Lee, B.L. 2003. Salt flavor enhancing compositions, food products including such compositions, and methods for preparing such products, US Patent 6,541,050 B1, filed Dec. 1, 1998 and issued Apr. 1, 2003.

Bors, W.H., Michel, C., and Saran, M. 1990. Flavanoids as antioxidants: determination of radical-scavenging efficiencies. *Methods in Enzymology*, **186**, 343–355.

Breton, F. 2008. Polyphenol antioxidants in red wine, [online], available: http://www.frenchscout.com/polyphenols (accessed 28 September 2015).

Cliff, M. and Noble, A.C. 1990. Time-intensity evaluation of sweetness and fruitness and their interaction in a model solution. *Journal of Food Science*, **55**, 450–454.

Ding, M., Ninomiya, Y., and Margolskee R.F. 1999. Blocking taste receptor activation of gustducin inhibits gustatory responses to bitter compounds. *Proceedings of the National Academy of Sciences*, **96** (17), 9903–9908.

Djordjevic, J., Zatorre, R.J., and Jones-Gotman, M. 2004. Effects of perceived and imagined odours on taste detection. *Chemical Senses*, **29**, 199–208.

Edwards, T.L., Singleton V.L., and Boulton, R. 1985. Formation of ethyl esters of tartaric acid during wine aging: chemical and sensory effects. *American Journal of Enology and Viticulture*, **36** (2), 118–124.

Girgis, S., Neal, B., Prescott, J., *et al.* 2003. A one-quarter reduction in the salt content of bread can be made without detection. *European Journal of Clinical Nutrition*, **57**, 616–620.

Labbe, D., Rytz, A., Morgenegg, S., and Martin, N. 2007. Subthreshold olfactory stimulation can enhance sweetness. *Chemical Senses*, **32** (3), 205–214.

Lawrence, G., Salles, C., Septier, C., Busch, J., and Thomas-Danquin, T. 2009. Odour-taste interactions: A way to enhance saltiness in low-salt content solutions. *Food Quality and Preference*, **20**, 241–248.

Ley, J.P., Krammer, G., Reinders, G., Gatfield, I.L., and Bertram, H.J. 2005. Evaluation of bitter masking flavanones from Herba Santa (*Erodictyon californicum* (H. and A.) Torr., Hydrophyllaceae. *Journal of Agricultural and Food Chemistry*, **53** (15), 6061–6066.

Mane, J.M., and Ponge, J.L. 1998. Coolant Compositions, US Patent 5,725,865, filed Aug. 29, 1995 and issued March 10, 1998.

Medler, K. and Kinnamon, S.C. 2004. Transduction mechanisms in taste cells, in *Transduction Channels in Sensory Cells* (eds. S. Frings and J. Bradley), Weinham: Wiley-VCH Verlag GmbH and Co. KGaA, pp. 153–157.

Prescott, J. 2004. Psychological process in flavor perception, in *Flavour Perception (*eds. A.J. Taylor and D.D. Roberts), Oxford: Blackwell Publishing, pp. 256–277.

Rolls, E.T. and Baylis, L.L. 1994. Gustatory, olfactory, and visual convergence within the primate orbitofrontal cortex. *Journal of Neuroscience*, **14**, 5437–5452.

Roy, G.M. (ed.) 1997. *Modifying Bitterness: Mechanism, Ingredients, and Applications*, Valhalla, NY: CRC Press.

Saint-Eve, A., Paci Kora, E., and Martin, N. 2004. Impact of the olfactory quality and chemical complexity of the flavouring agent on the texture of low fat stirred yogurts assessed by three different sensory methodologies. *Food Quality and Preference*, **15**, 655–668.

Small, D.M. and Prescott, J. 2005. Odour/taste integration and the perception of flavor. *Experimental Brain Research*, **166**, 345–357.

Verhagen, J.V. and Engelen, L. 2006. The neurocognitive bases of human multimodal food perception: Sensory integration. *Neuroscience and Biobehavioral Reviews*, **30**, 613–650.

Watson, H.R., Rowsell, D.G., and Browning J.H.D., 1977. Substituted p-menthanes, US Patent 4,033,994, filed July 8, 1974 and issued July 5, 1977.

Watson, H.R., Rowsell, D.G., and Spring, D.J., 1979a. p-Menthane carboxamides having a physiological cooling effect, US Patent 4,136,163, filed July 8, 1974 and issued Jan. 23, 1979.
Watson, H.R., Rowsell, D.G., and Spring, D.J., 1979b. N-Substituted paramenthane carboxamides, US Patent 4,150,052, filed May 16, 1977 and issued Apr. 17, 1979.
Windschauer, R.J. and Virgallito, T.T. 2012. Taste masking compositions and edible forms, US Patent Application US20120321727 A1, filed Apr. 25, 2012 and issued Dec. 20, 2012.

CHAPTER 10

Some like it hot! Sensory analysis of products containing chemesthetic compounds

Cindy Ward

Sensation Research, Maineville, OH, USA

10.1 Introduction

Can you imagine a soda without any fizz? Would you eat salsa that did not offer any heat? Would Bengay® be marketable if it did not provide cooling/burning to a sore back? Would anyone crave a bourbon that did not give a warm burn as it glides down your throat? These questions may not seem to have any bearing on each other but indeed they do. Products (and our environment) are experienced through our senses. That seems straightforward until you actually start trying to understand how those systems work, and how to measure sensation.

The trigeminal, glossopharyngeal, and vagus nerves are responsible for carrying the signals to the brain for *chemesthetic sensations*, which include touch, thermal sensation, and pain. Chemesthetic agents, sometimes called trigeminal compounds, are chemicals that elicit touch sensations (e.g., carbonation bubbly feelings), thermal sensations (e.g., warming from ginger and cooling from menthol), and pain (many compounds above threshold concentrations cause pain including those with touch and thermal aspects) (see Table 10.1). These chemesthetic sensations are critical to the experience of sensory pleasure as well as rejection of many products. Liking for chemesthetic sensations may be learned since children tend to reject products that evoke these feelings. As consumers age, these sensations often become associated with greater pleasure. Additionally, individuals who are less neophobic have greater liking for foods containing trigeminal compounds (Törnwall *et al.*, 2014).

As with most sensations, there is an intensity of a sensation, chemesthetic in this case, that is related to consumer liking as measured on a bipolar hedonic scale (Fig. 10.1) (Peryam and Pilgrim, 1957). For a specific consumer, the intensity associated with optimized liking (highest liking) would be shifted left if the consumer likes more mild intensities and right if the consumer prefers higher intensities of the attribute.

Table 10.1 Common trigeminal compounds found in consumer products.

Chemesthetic agent	Sensory character terms for intensity rating	Consumer term for liking or intensity	Common products
Acids (Dessirier *et al.*, 2000; Carstens *et al.*, 2002)	Sour, stinging/burning at high concentrations nasal (volatile acids), oral, skin	Sour or tart oral taste, burning on skin at high concentrations	Beverages, sauces, skin care, candies
Alliaceous (fresh: onion, garlic and shallots)	Burning, pungent, causing watering eyes	–	Fresh onion, garlic, shallots, savory dishes
Capsicum (and other capsaicinoids)	Burning, pepper heat	Burning, hot	Peppers, salsas, a variety of savory cuisines, beverages and candies, sports rubs, pain reliever creams, and etc.
Carbon dioxide incorporated in a liquid, carbonic acid (Carstens *et al.*, 2002)	Carbonation, size of bubbles	Bubbly, fizzy	Beverages – soft drinks, sparkling waters, juices, wines, beers and other alcohols
Cooling compounds (menthol, menthol-based compounds, and other cooling agents) (Eccles, 1994; Bharate and Bharate, 2012)	Cooling at low to moderate levels, cold-burning-pain at high levels in the oral and nasal cavity; warming/cooling on skin. Note: menthol has a bitterness side-taste at moderate levels that some other cooling compounds do not (Eccles, 1994)	Cooling, cold (nasal and oral); warming/ cooling skin	Mints, chewing gums, toothpastes, candies, pain reliever creams, sports rubs, shampoos
Cinnamon (Cliff and Heymann, 1992)	Cinnamon (aroma/ flavor), burning feeling	Red hot candy, cinnamon, burning	Candies, baked goods, fragrances, toothpaste
Ethyl alcohol (and other alcohols)	Burning and tingling (at high proofs in nasal, oral and throat tissues), Cooling (from evaporation on undamaged skin), painful stinging on damaged skin	Alcohol taste, burning, stinging, cooling	Alcohol beverages Personal care products – astringents, creams, lotions, gels, etc.
Eugenol (clove) (Cliff and Heymann, 1992; Klein *et al.*, 2013)	Clove (aroma/flavor), numbing feeling, warming	Numbing and warm	Cookies, cured meats, cough syrups, toothpastes
Galangal acetate, dihydrogalangal acetate	Burning heat (possibly higher on the tip of the tongue) (Yang *et al.*, 2009)	Burning	Thai and other Asian cuisines, possible alcohol enhancement (Yang *et al.*, 2009)

(*Continued*)

Table 10.1 (*Continued*)

Chemesthetic agent	Sensory character terms for intensity rating	Consumer term for liking or intensity	Common products
Ginger (zingerone) (Cliff and Heymann, 1992)	Ginger (aroma/flavor), burning and warming (throat and chest)	Ginger, warming, burn	Beverages, baked goods and savory products for examples Ginger ale, ginger bread, Asian dishes
Horseradish and Wasabi	Burning and stinging (occurring in nasal passages as well as oral cavity) eye watering	Hot, burning	Sauces, condiments
Mustard	Mustard (aroma/flavor), burning and stinging (Jacquot *et al.*, 2005)	Sharp, burning and stinging	Mustard condiments, sauces
Piperine (peppercorns) (Cliff and Heymann, 1992)	(Pepper aroma/flavor), Burning	Burning, hot	Widely used in seasonings (black pepper, white pepper, etc.) and savory products
Salts – high concentrations (Carstens *et al.*, 2002)	Oral cavity – salty, burning at high concentrations No nasal impression, Burning/stinging on damaged skin or in eyes No short-term effect on healthy skin	Salty, burning at high concentrations	Savory foods for higher concentration, all s
Szechuan pepper – Sanshools (huajiao) (Yang, 2008)	Tingling, burning–numbing	Burning, bubbly, electric, buzzing feeling	Szechuan cuisine such as Kung Pao chicken

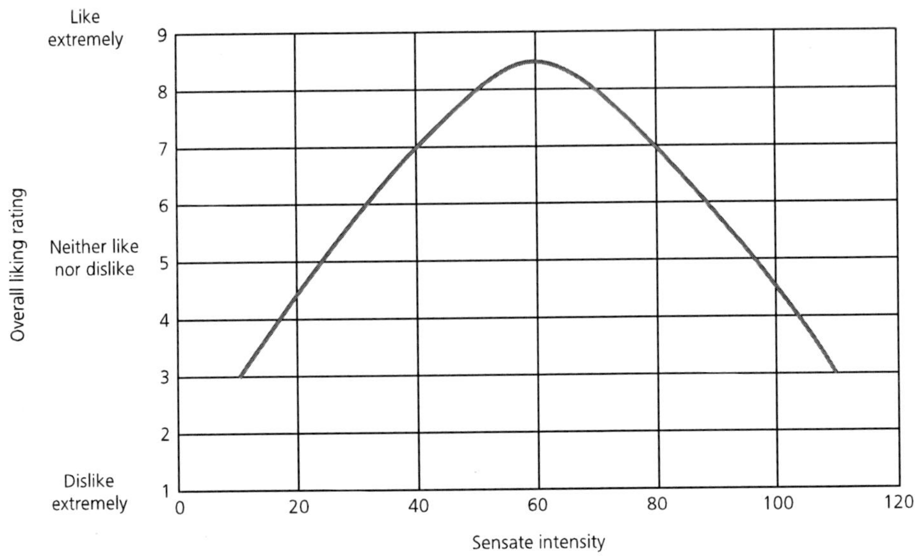

Fig. 10.1 Relationship of product liking versus sensory attribute intensity.

10.2 Overview of test approaches for sensory evaluation of chemesthetic compounds in consumer products

Characterizing and measuring the intensity of chemesthetic stimulants is complex. These compounds create pain responses that vary in level of intensity across human subjects. While age and gender show differences in the effect of a product experience, cultural and social factors have the most profound effect (Cardello and Wise, 2008). These intensity effects can be due to differing levels of sensitivity, experience, and tolerance to the sensations. Measurement of chemesthetic sensations commonly occurs in the consumer product industries with product development teams in pre-screening as well as by formalized testing in sensory experiments including: discrimination tests, intensity scaling experiments, profiling by descriptive panels, or opinions from consumers in either quantitative or qualitative studies. Questions answered in product research on chemesthetic agents include:

1 Discrimination tests, used to determine differences in samples that are similar. Some examples could be:
   (a) At what change in alcohol percentage can a consumer detect a difference in this ale?
   (b) If menthol level increases by 5%, does it make a difference to the chewing gum perception?
2 Descriptive and scaling tests, used to measure intensity in an analytical sensory method. While various methods are available (ASTM, 1996), generally, panelists are oriented to the scaling and descriptive task through experience with a range of samples, definitions, references, and project specific training. These techniques apply to a wide range of research questions including:
   (a) How much burning is attributed to this specific pepper?
   (b) How does an increase in alcohol change the aroma and flavor profile of my product?
   (c) How do combinations of chemesthetic agents influence the sensory profile?
   (d) Are there chemesthetic synergies or antagonists in these formulations?
3 Consumer tests, used to make determinations on liking such as:
   (a) How much capsicum should we add to this snack chip to optimize consumer acceptance?
   (b) What level of carbonation leads to more liked products?

10.3 The phenomena of sensitization and desensitization

Regardless of the human evaluation method selected, the impact of chemesthetic compounds is hard to measure. These compounds are considered "painful" at concentrations above threshold, yet people may learn to like them (see Chapter 2). The quality (or type) of pain associated with chemesthetic agents creates an effect whereby unfamiliar and untrained panelists may judge measures of intensity of the sensation to be very high. Additionally, chemesthetic compounds are fatiguing, and therefore can skew the accuracy of intensity ratings. Fatigue is amplified by several processes in chemesthetic agents including: sensitization and desensitization as well as cross-sensitization and cross-desensitization (Carstens *et al.*, 2002; Green, 1989). *Sensitization* is the process by which the intensity of the sensation increases in intensity. When only one chemesthetic agent is involved, then the process is called self-sensitization. Self-sensitization occurs when the intensity of a sensation climbs as the same compound at the same concentration is experienced several times, usually over a usually short period of time (seconds to a few minutes). The issue of sensitization is compounded in the presence of other chemesthetic agents. Cross-sensitization occurs where stimulation from one chemesthetic agent in the presence of another agent leads to a higher overall response than either agent would cause alone. *Desensitization* is when a chemesthetic sensation decreases in intensity, even when the concentration of the irritant remains constant (see Chapter 6). Similarly to sensitization, self-desensitization and cross-desensitization can occur. Self- desensitization usually occurs over longer time spans than self-sensitization. With capsaicin (the burning/hot compound in chilis) self-desensitization commences post evaluation around 5.5 minutes, but at less than 14.5 minutes (Green, 1991).

To further understand sensitization and desensitization, imagine this common scenario in a Mexican restaurant. A consumer is eating spicy salsa on a tortilla chip. The first bite is hot and burning a few seconds after tasting. As the consumer continues to quickly consume the chips and salsa, the intensity begins to increase (self-sensitization). However, if the chip and salsa consumer would wait 6 minutes between bites then the salsa would seem less spicy on each new bite than it did in the initial bite (self-desensitization). However, if that chip and salsa eater is savvy to the ways of the chili, he/she drinks an icy margarita with the chips, which seems to dull the pain. Assuming the pain dulling is not due to numerous margaritas, this decrease in pain is probably due to cross-desensitization where the cold from the ice has reduced the perceived thermal pain from the chilis. However, what if a more "macho" consumer decides to take a shot of tequila? In this situation, this "macho" may experience a rapidly increasing level of pain over what would be experienced by eating the chips and salsa alone. This is an example of cross-sensitization where alcohol burning increases the burning experienced by the chilis. While these processes seem rather straightforward when discussed in regards to a common experience, many sensory scientists, psychophysicists (scientists who quantitatively measure perception), and other scientists have spent years studying these sensations. In fact, this area of research is constantly growing as we learn more about sensation and receptors. All of these experiences complicate the challenge of reliably measuring chemesthetic sensations.

10.4 Testing products containing chemesthetic compounds

Appropriate measurement of chemesthetic sensations involves carefully limiting the number of samples and references for evaluation, while monitoring sensory fatigue, sensitization, cross-sensitization, desensitization, and cross-desensitization.

Failure to control these issues pertaining to chemesthetic agent evaluation ultimately reduces the sensory data reliability by introducing too much variation or noise into the data.

Various workers have investigated methods to reduce capsaicin fatigue of the chemesthetic burning sensation (Hutchinson *et al.*, 1990; Allison *et al.*, 1999). Water, crackers, and a resting time (8 minutes with water and crackers and 16 minutes without) between samples decrease the likelihood of sensitization, which happens at short intervals in time of less than 3.5 minutes, and desensitization, which happens at longer periods of more than 5.5 minutes but less than 14.5 minutes for capsaicin (Green, 1991; Allison *et al.*, 1999).

Several studies have investigated the ability of common food ingredients to diminish trigeminal sensations, particularly capsaicin (Nasrawi and Pangborn, 1989; Hutchinson *et al.*, 1990; Allison *et al.*, 1999; Lawless *et al.*, 2000). Sucrose decreases duration and intensity of pepper heat, while gums only retarded its detection (Nasrawi and Pangborn, 1989). Incorporating fat into a system containing capsicum dramatically reduces the perceived intensity of capsaicin burn; this effect is due to the high hydrophobicity of capsaicin whereby capsaicin remains in the oil of the food product rather than solving into saliva and interacting with the trigeminal receptor (Lawless *et al.*, 2000). While capsaicin has been studied by many workers, other chemesthetic compounds have been investigated less.

Sensory scientists should be cognizant that, given the well-documented complexities in capsaicin heat measurement, most other chemesthetic stimulants are likely to have similar effects. For example, when evaluating alcohol products, mints, and other chemesthetic compound containing products, it is important to incorporate long breaks between samples and use palate cleansers to decrease recovery time. Additionally, products with very high levels of chemesthetic agents or/and multiple trigeminal compounds (e.g., toothpastes) may need longer breaks (interstimulus intervals) between samples and lower numbers of total products evaluated per day (1 2). When sensory testing commercial toothpastes Allison *et al.* (2007) found that long delays of even 24 minutes had carry-over effects associated with sensitization.

Measuring chemesthetic agent intensity requires methodical approaches with interstimulus aids (palate cleansers) to help recovery. Most commonly, water and crackers are used for the oral system recovery, while sniffing of clean skin may assist in olfactory system recovery. Additionally, rest times (interstimulus intervals) of 8–15 minutes (or longer) are helpful, as indicated in the literature, due to the sensitization and desensitization processes (Green, 1991; Allison *et al.*, 1999; Allison *et al.*, 2007). Furthermore, each chemesthetic agent may have sensitization and desensitization processes which are unique to that given compound as well as interactions with other chemesthetic agents (cross-sensitization and cross-desensitization). Skin and eyes can take a very long time to recover. Additionally, various parts of the body may respond differently to a given chemesthetic agent, for example, menthol is cooling in the oral and nasal cavities, but warming and/or cooling on other areas of the skin (Eccles, 1994).

10.5 Discrimination testing with trigeminal compounds

Discrimination techniques include multiple test types such as triangle, duo–trio, *n*-alternative forced choice, and *r*-index. One commonality across these test types is that two or more samples (stimuli) are compared, and then a judgment is made. In practice these measurements are made in a short period of time (within seconds) to decrease the effects of memory. Owing to fatigue, sensitization, and desensitization problems associated with chemesthetic agents, these types of sensory tests are particularly difficult. One procedure utilized in skin evaluations is to place the sample and control on different sides of the body then judge the differences. Similarly, in the laboratory, it is possible to conduct measurements on opposite sides of the tongue by applying the test sample via a cotton swab or filter paper by half mouth tasting protocols for liquids; however, this testing process is not possible where products must be chewed or complex mixtures are evaluated. Forced breaks (interstimulus intervals) between samples must be enforced if accurate differences are to be determined between samples; however, these breaks may lead to a decreased ability to detect differences even in situations where the panelist is familiar with the stimuli (Almedia *et al.*, 1999). Descriptive analysis of the samples may be easier to conduct since character and intensity can be measured with adequate breaks between samples.

10.6 Rating of chemesthetic agent intensity

There are many methods for rating intensity of sensation. The most common methods utilized are magnitude estimation as well as using an intensity scale such as categorical scales of intensity, unstructured and structured line scales, and the labeled magnitude scale (Green *et al.*, 1996; Meilgaard *et al.*, 1991). Psychophysicists often utilize the labeled magnitude scale (Fig. 10.2), while sensory scientists working in industry (food, beverage, personal care, and home care) often use either categorical scales (where categories are 0 for none to 15 for strong, and half points are included) or line scales (either 0 to 15, 0 to 150, or 0 to 100) (Fig. 10.3). Any of these approaches can be valid depending on the goals of the test, and each has their own inherent issues. Note that with very high intensity sensations, such as chemesthetic sensations, the categorical and line scales may need to be expanded in length to accurately measure the intensity of the sensation. For a discussion of ceiling and compression effects, see Hayes, Allen and Bennett 2012.

10.7 Dose response

When studying a new chemesthetic agent, it is frequently necessary to understand its intensity rating over a large range of concentrations. New compounds are often compared with industry standards. For example, new cooling compounds are often discussed in relation to menthol. Additionally, there are ASTM standardized methods for

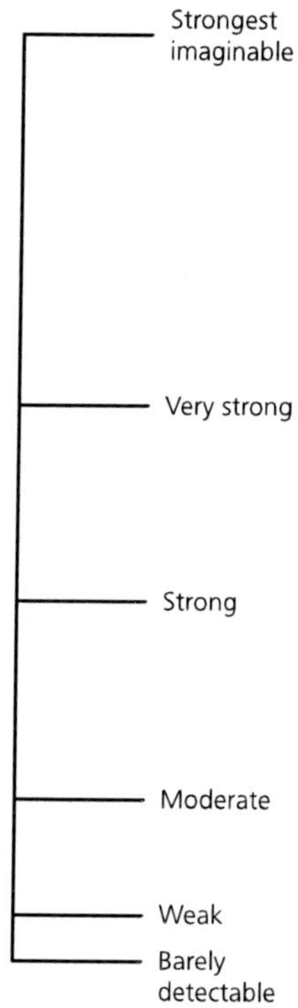

Fig. 10.2 Labeled magnitude scale (LMS). Green *et al.* 1996.

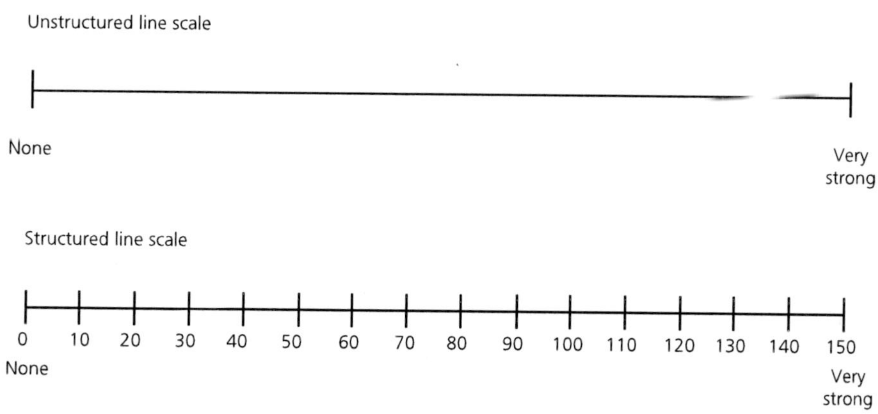

Fig. 10.3 Unstructured and structured line scales for intensity scaling.

evaluating chili heat (ASTM, 2011a, 2011b), which correlate with analytical chemistry and the Scoville heat units (standardized units for measuring the heat from chilis).

Dose response curves can be determined to understand the resultant intensity (using any of the aforementioned rating methods) for a given concentration as well as the relative behavior of a new compound as compared with an existing industry

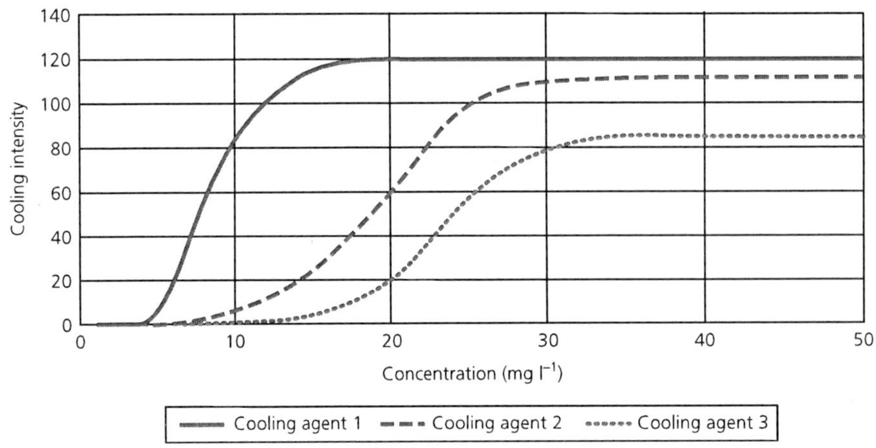

Fig. 10.4 Example dose response curves, left to right: Cooling Agent 1, Cooling Agent 2, and Cooling Agent 3.

standard. In Fig. 10.4 an example set of curves are depicted. Cooling Agent 1 is an example of a high intensity chemesthetic agent, since at low concentrations it has a high cooling intensity perception, while compound Cooling Agent 2 is moderate, and Cooling Agent 3 is a weaker cooling compound. Compared with Cooling Agent 1, the threshold concentration of Cooling Agent 2 is higher, the range of concentrations with a linear response is greater, and the maximum intensity is not as high.

10.8 Descriptive analysis of chemesthetic agents containing samples

Descriptive analysis can be a powerful technique to understand and characterize chemesthetic agents (Cliff and Heymann, 1992). The panelists must be trained and the methods need to be carefully developed and documented to increase the likelihood of reliable guidance. The number of samples for evaluation of samples containing chemesthetic compounds should be limited per day. When tasting foods and beverages, there should be limits of seven or less products per day for tasting, with a minimum of 8 minutes between samples with water and crackers (Allison *et al.*, 1999).

While references help trained panelists understand an attribute, it is important to remember that references with chemesthetic properties can be fatiguing themselves and need appropriate recovery time and palate cleansers. Therefore, panelists should try to memorize the sensations evoked by the references. Additional training time may be needed to make sure the panelists understand the attributes clearly, so that they have experience with multiple references and products. Owing to the processes of sensitization and desensitization, descriptive analysis training and evaluation takes longer than profiling of non-chemesthetic sensations. Furthermore, training tends to decrease the differences in ability

Box 10.1 Chemesthetic agent evaluation protocol considerations.

How is the product evaluated?
- Order of attributes on the ballot.
- Definition of attributes.
- How long after receiving or opening must the product be evaluated? This is very important in products served hot, carbonated products or products that dry out due to evaporation of either water or alcohol.

Are there differences in attribute evolution across the samples?
- If so, would a time intensity of the attribute be of interest to the project team?
- What is the duration of the evaluation?
- How much product to evaluate?
- Is it best to portion the whole product for evaluation by weight or volume?
- Alternatively, give panelists a spoon or syringe to portion their own product from larger containers of 3-digit coded product.

Product interactions with the panelist – document the exact protocols
- Personal care products.
  - Application procedure.
  - What are the test sites?
  - How is the test site prepared prior to product application?
  - Length of time on the skin prior to evaluation.
  - Are there multiple time points of interest?
- Food and beverages
  - Length of time in the mouth.
  - Number of chews or swishes before expectoration or swallowing (if necessary).
  - Is an aftertaste of interest? If so, how long?
- Sniffing procedures
  - Number of sniffs.
  - Use of clean skin, coffee beans (or other) for recovery.
  - Product temperature.
  - Ambient temperature and relative humidity.
  - Glass or plastic?
  - Sample cup - covered or uncovered?
  - Is it necessary to release built up aroma then set time to recover prior to sniffing? (Alcohol products may need this).
  - Liquids – swished or still?

to rate intensity differences reliably and decrease the data spread in intensity measurements, especially for chemesthetic agent evaluations. To help minimize the noise of intensity measurement within a sample, procedures for attribute intensity measurements should be carefully documented and applied to the evaluation (see Box 10.1; ASTM, 2014). For example, it may be appropriate to evaluate burning on the first sip or taste of a sample rather than at an ambiguous time point in the evaluation. Some considerations for documenting the methodology include: (a) how much sample is evaluated; (b) how should the product be applied or tasted; (c) at what

Box 10.2 Beverage evaluation in the oral cavity protocol.

- Panelists should refrain from using food and beverages with trigeminal stimulants 1–2 hours prior to the panel. Remember, toothpaste and mouthwashes have multiple chemesthetic properties (cooling, burning, and numbing) that can interfere with product evaluations of chemesthetic products as well as other sensations.
- Sip 5 ml of product, swish gently in the mouth for 5 seconds, expectorate, and immediately evaluate the product for cooling and burning. Evaluate again at 30 seconds and at 1 minute.
- Cooling mouthfeel definition – the amount to which the sample feels cold. Reference – menthol in solution.
- Burning/irritation mouthfeel definition – the amount to which the sample is perceived to burn or irritate the mouth. Reference – capsaicin in solution.

time past application should the chemesthetic agent intensity be evaluated; and (d) should the mouth be open or closed while breathing for evaluations (Box 10.2). Whichever methodology is deemed appropriate, the panelists should practice multiple times, receiving feedback or scores prior to blind data collection. If procedures are not carefully documented, then the resulting noise in the data may be too high to measure the true differences between the samples.

10.9 Alcohol burn case study

In this study, a complete analysis of aroma, flavor, and texture for a wide range of proofs for a single type of spirit was investigated. A broad variety of references were utilized to calibrate the panel and training for this large set of 20 spirits extended over two weeks prior to data collection. There were 20 minute breaks between samples where panelists used bottled water, soda crackers, carrots, string cheese, and apples to assist in recovery. Five samples were profiled per day over 3 hours. Data were collected from 12 trained descriptive panelists with 1500 hours of alcohol beverage experience. Sample evaluations were replicated using completely balanced and rotated serving order designs where each panelist had their own unique design. All samples were expectorated.

In this panel experience with this product, nasal burn (also described as stinging) recovered much more quickly than oral burn. After a brief resting phase and with a sniff of clean skin, products were ready to be sniffed again. Oral burn lasted around 6 minutes even with the interstimulus delays and rinses applied. Additionally, panelists were cautioned and trained to make decisions quickly to avoid holding the product in the mouth while making judgments. In training it was determined that five products a day were the maximum the panelists could handle due to sensory fatigue and sensitization.

The relationship between alcohol burn intensity and proof is shown for nasal burn (Fig. 10.5), the initial sip (Fig. 10.6), and for the 30 second aftertaste after multiple sips (Fig. 10.7). For each attribute an evaluation procedure was developed. Panelists practiced the procedures in training. The relationship

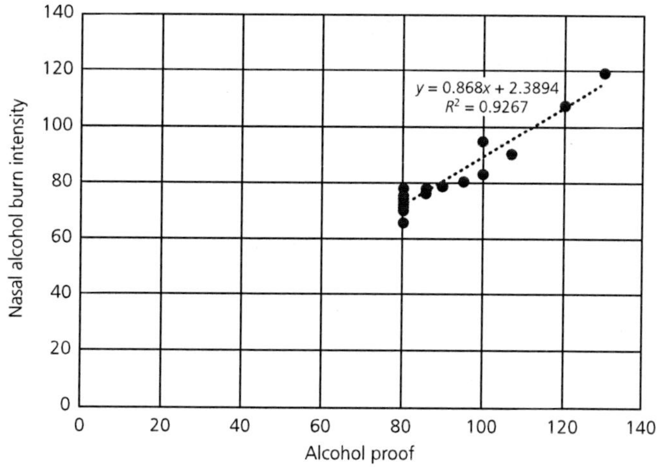

Fig. 10.5 Descriptive nasal burn intensity as a function of alcohol proof for spirits.

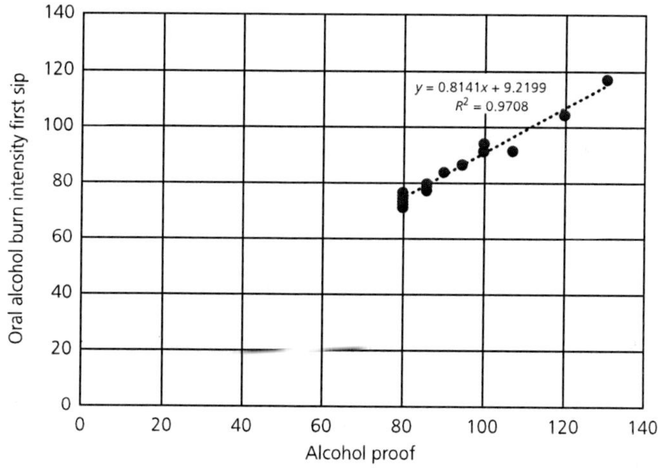

Fig. 10.6 Descriptive first sip alcohol burn intensity as a function of alcohol proof for spirits.

for nasal burn, first sip alcohol burn, and 30 second aftertaste burn is linear within the 80–140 proof range for this type of spirit. The intensity range for nasal burn is from 75 to 120. The intensity range for alcohol burn is from 75 to 120 on the first sip. Thirty seconds after completing the product profile, panelists rated the amount of burn remaining in the mouth. This range of burning sensation is from 60 to 100. Nasal intensity and oral burn are similar in intensity for these spirits. However, that is not the case for every spirit based on the experience of this panel with numerous spirits. Burning intensity is reduced 20 points on average from the first sip of the sample to the 30 seconds after full profile completion.

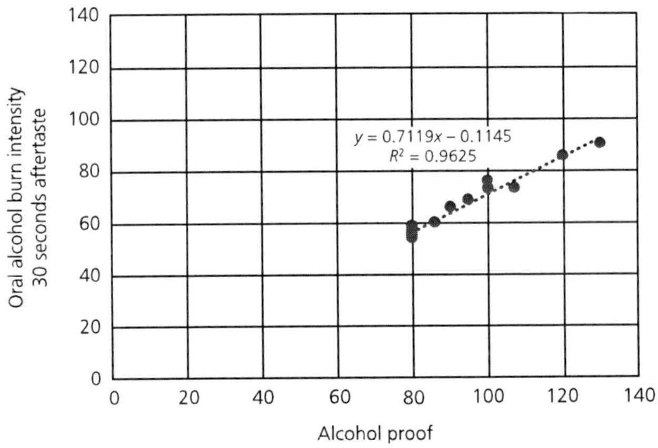

Fig. 10.7 Descriptive 30 second aftertaste alcohol burn intensity as a function of alcohol proof for spirits.

10.10 Time intensity

Time intensity can be particularly helpful in characterizing the differences between chemesthetic compounds. Time-intensity methods allow for the intensity of an attribute to be studied over time thus allowing for the temporal profile (attribute evolution over time) to be investigated. This is of interest because the character of different chemesthetic agents is not only in the overall intensity of that attribute but also in how the attribute evolves (ASTM, 2014). For example, when comparing different toothpaste formulations, how do the cooling attributes differ among the samples? Does the toothpaste quickly provide cooling or is the maximum cooling intensity reached minutes after brushing? How does the burning feeling increase and fade away? How long does the numbing effect last and how intense does the numbing get? These three attributes would lead to three separate evaluations, that is, time-intensity analysis of cooling, burning, and numbing for each sample.

Time intensity begins with understanding the attribute of interest, and then providing information that will enable product developers to better understand their product attributes. The time-intensity technique is performed by trained descriptive analysis panelists. Initially, so that panelists can learn the time-intensity task, it is often helpful to experience products while scoring intensity on paper then discuss the scores as a group. In practice, the panel leader calls out times while the panelist rates the intensity of the sensation at that time point. Discussions in the group may focus around how fast the intensity of the attribute seemed to rise then disappear, how long it appeared to last, what was the highest intensity, and so on. Next, panelists should be calibrated on intensity at specific time points, discussing how the attribute changes over time and comparing the different samples characteristics over time. During the initial training the group should come to a decision on how long the chemesthetic sensation lasts for the samples, so that the duration of intensity rating can be standardized for testing. Next, panelists should be oriented

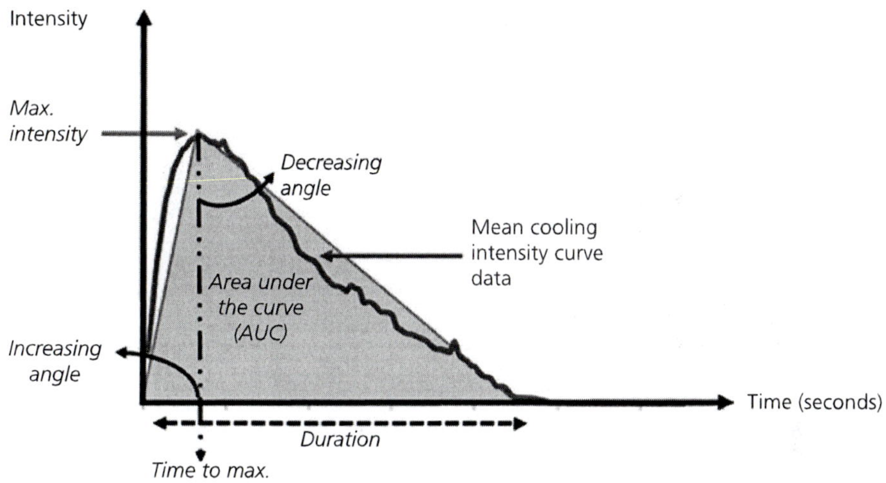

Fig. 10.8 Time–intensity curve labeled with select parameters of the curve.

with respect to the computerized data collection method and idiosyncrasies of the program. Training on the computer with feedback on curves obtained on practice samples is essential to further improve the panelist's rating technique prior to data collection.

Time intensity can be utilized to quantify the evolution of an attribute over time. The data are composed of the attribute intensity over time, which results in a time-intensity curve. Each curve has parameters associated with it that are used to characterize the behavior of an attribute over time. Some of the parameters of the curve include maximum intensity (peak intensity), area under the curve (total amount of the attribute intensity over time), the duration (how long the attribute lasts) in addition to determining how steep the curve is while it is increasing to maximum intensity (increasing angle, higher angles up to 90° are related to faster rates of increase), and how fast it disappears (steepness of the decreasing angle, the smaller angles are related to faster decreases) (see Fig. 10.8). By comparing these types of data across samples, it is possible to determine differences between the temporal behavior of the chemesthetic agents. To better understand the benefits of time intensity, a case study on cooling released from a mint chewing gum is discussed in the following section.

In Fig. 10.9, we can visually inspect the graph to view the overall trends between the samples cooling release. We see that product P1 has a curve that reaches a higher intensity than the other samples, as well as quickly increasing and decreasing in intensity. P5 is slower to reach maximum intensity, slower to loose intensity, and may last longer. However, sample P3 has a slow increase to maximum intensity and is lower in this maximum intensity.

P1 is significantly higher in maximum intensity and similar to P5 in the overall cooling that is experienced, as indicated by area under the curve (Fig. 10.10). Additionally, P5 has a longer duration than the other samples. While P1 has greater intensity, it has a shorter duration but is not as lingering as P5.

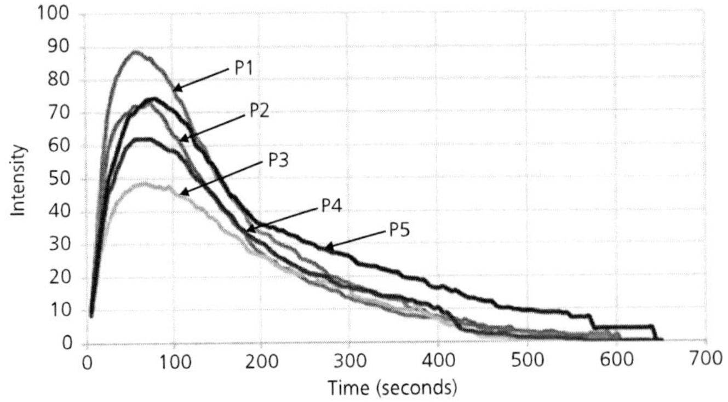

Fig. 10.9 Mean time-intensity curve for cooling released from mint chewing gum samples.

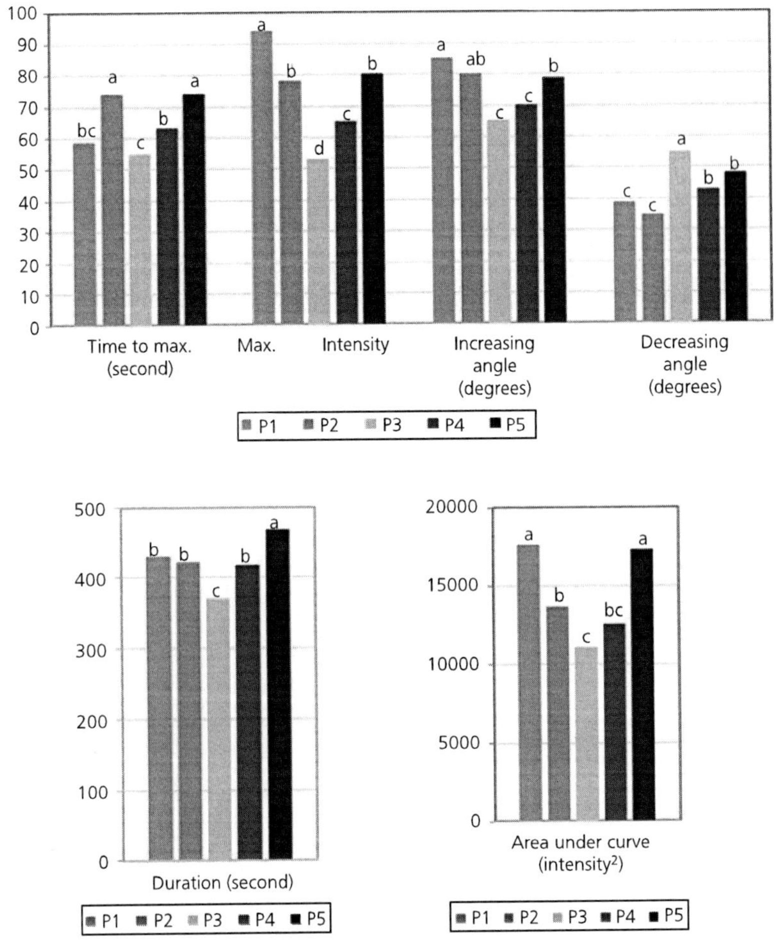

Fig. 10.10 Statistical analysis results of time-intensity attributes for mint chewing gum curves. This bar chart represents the means obtained from the parameters of the curve. The means have been tested by Anova with Fisher's LSD at the 95% confidence level. Within an attribute, bars that do not share a letter are significantly different. Bars, left to right: P1, P2, P3, P4, and P5.

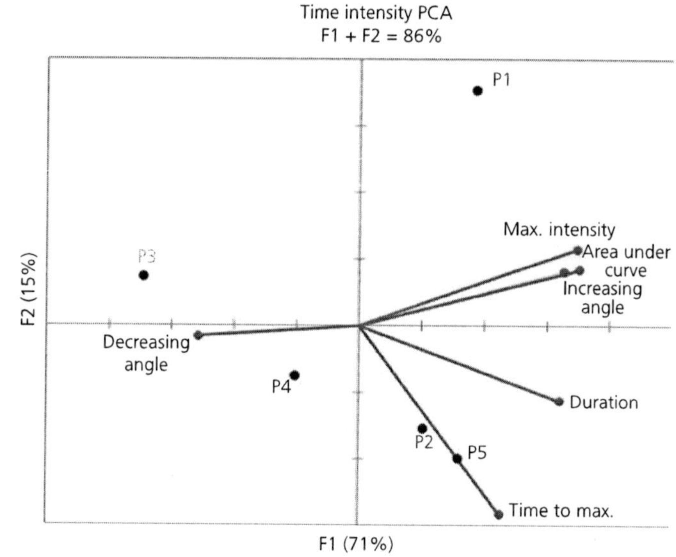

Fig. 10.11 Principal component analysis bi-plot for time intensity parameters for cooling released from mint chewing gum samples.

Another way of looking at these data is by applying a principal component analysis (PCA), as seen in Fig. 10.11. This data reduction approach can be applied to multifaceted temporal data. In this case, PCA is used to clarify the relationship between parameters of the time intensity curves and the products. Attributes that are plotted close together have similar relationships within the samples, those on opposite sides of the plot are related inversely and those at right angles (orthogonal) are unrelated. Similarly, products plotted close together have similar sensory perception, while those far apart are more different.

The PCA plot (Fig. 10.11) shows a variance of 86% for the data set in the first two factors. Factor1 accounts for 71% of the variance while Factor2 explains 15%. Factor1 separates the products from left to right. Products plotted toward the left have a larger Decreasing Angle, while products plotted toward the right are related to a larger Increasing Angle, Total Area Under the Curve, and Maximum Intensity. The vertical axis (Factor2) separates products based on how fast they reach maximum intensity, where products that reach maximum intensity quickly are at the bottom of the plot and those near the bottom are slower. Product P1 is related to a high maximum intensity with a rapid increase in intensity (larger increasing angle) and a high amount of total cooling (area under the curve). However, product P3 is low in intensity, total cooling, and duration; P3 is also relatively low in the decreasing angle due to its broad curve. Products P2 and P5 are similar in time-intensity parameters: these products take longer to reach maximum intensity and they last longer with moderately high intensity and total cooling.

10.11 Consumer testing with chemesthetic agents

While it is outside the scope of this chapter to discuss all the best practices for consumer testing, this section will discuss some aspects for researchers to remember to incorporate into their study. Consumer testing with products containing chemesthetic agents requires close adherence to good sensory practices: restricted numbers of samples (1–4 samples), enforced breaks between samples, and use of palate cleansers. Additionally, since consumers usually experience the products of interest in only one or two sessions, it is both inappropriate and not feasible to spend time becoming oriented to the sensations in the product set. To compensate for this issue, use good consumer testing practices. These include appropriate recruitment/screening, day of test instructions, and product test questionnaire instructions (ASTM, 2007; Meilgaard *et al.*, 1991).

Consumers often fall into several liking groups based on the intensity of the chemesthetic agents: those with an aversion to the sensation, those with a liking for a small amount of the sensation (mild likers), those with a moderate intensity liking, and those who like high-intensity amounts. Since consumers fall into specific groupings, it is critical to make sure the "right" consumer for the product of interest is utilized in any test. The process for finding the "right consumer" depends on recruitment and screening. Consumer recruitment begins with detailed screener development, which includes the targeted demographics and quotas. Along with the normal project requirements, consider detailed screening for product use. It is important to recruit consumers (or a segment of consumers) who use similar products with similar intensities as the samples tested. For example, if salsas are being tested, make sure the consumers in the test consume salsas of similar intensity. A question in the screener could be framed to determine whether the consumer likes mild, medium, or hot salsas. Additionally, an investigation on use of similar intensity products such as hot pepper sauces, specific brand names of products, and liking for specific dishes of Thai and Indian food, and so on, would be informative.

Consumers should be cautioned upon selection into the study, and prior to the day of testing, to remember to refrain from drinking, eating or using the specific products (lists) on the day of or 2 hours before testing. For example, if consumers are participating in a toothpaste evaluation study, it may be necessary that they refrain from brushing their teeth, using mouthwash, drinking alcohol, chewing gum or eating mints for 2 hours prior to testing the products. It may also be necessary for them to refrain from analgesics, or at least to report their analgesic usage as part of the study.

The typical questions utilized in consumer testing can be applied, such as liking, just about right, ideal, and check all that apply questions (Meilgaard *et al.*, 1991). Drivers of liking study designs and experimental designs are used in larger studies of consumer testing (ASTM, 1997). Drivers of liking designs relate liking data to descriptive analysis data through statistical modeling. In experimental design studies, product ingredients or processing conditions are systematically manipulated to determine their effect on the product characteristics, such as liking or descriptive analysis profile. Drivers of liking and experimental design studies may need particular approaches, such as incomplete block designs where the number of samples per consumer is reduced by increasing the number of consumers and having each consumer evaluate fewer products. Other screening designs may be more feasible, such as descriptive analysis of the products, followed by consumer testing of fewer products that cover the design space; then, conducting a follow-up consumer study to test the validity of the consumer liking of experimental models (Ward *et al.*, 1999).

10.12 Conclusions

This chapter discusses the difficulty in testing chemesthetic agents due to fatigue related to the slow recovery of the chemesthetic system caused by the processes involved in sensitization and desensitization. By understanding the unique pitfalls associated with sensory evaluation and consumer testing of chemesthetic agents, the researcher can design the experiments to ensure accurate product guidance data are obtained. It is my opinion that advances in chemesthetic research with a focus on interactions of chemesthetic agents with receptors will continue to enlighten product research so that better product guidance can be designed.

Acknowledgments

Karen A. Sylvester has my sincere gratitude for her gentle encouragement, and editing, as well as imposing some structure on this rather creative sensory scientist. I am thankful for the help of my sweet family especially Kristen Welker-Hood, PhD, Grace A. Ward, and Daniel R. Ward for their help with outlining, reference perusal, graphics making, and editing.

References

Allison, A., Chambers IV, E., Milliken, G., and Chambers, D. 1999. Effects of interstimulus rinsing and time on measurements of capsaicin heat in tomato salsa. *Journal of Sensory Studies*, **14** (4), 401–414.

Allison, A., Chambers, E., and Chambers, D. 2007. Repeated testing and rinsing regimens for toothpastes with various cooling and burn intensities: impact on discrimination and repeatability. *Journal of Sensory Studies*, **22** (6), 695–712.

Almeida, T., Cubero, E., and O'Mahony, M. 1999. Same-different discrimination tests with interstimulus delays up to one day. *Journal of Sensory Studies*, **14** (1), 1–18.

ASTM Standard MNL 26. 1996. *e2, Descriptive analysis*, ASTM International, West Conshohocken, PA, 1996, DOI: 10.1520/MNL10657M, www.astm.org (accessed 6 October 2015).

ASTM Standard MNL 30. 1997. *Relating consumer descriptive and laboratory data to better understand consumer responses*, ASTM International, West Conshohocken, PA, 1997, DOI: 10.1520/MNL30-EB, www.astm.org (accessed 6 October 2015).

ASTM Standard MNL 55, 2007. *International consumer product testing across cultures and countries: General principles*, ASTM International, West Conshohocken, PA, 2007, DOI: 10.1520/MNL11101M, www.astm.org (accessed 6 October 2015).

ASTM E1490-11. 2011a. *Standard guide for two sensory descriptive analysis approaches for skin creams and lotions*, ASTM International, West Conshohocken, PA, 2011, DOI: 10.1520/E1490-11, www astm.org (accessed 6 October 2015).

ASTM Standard E1083. 2011b. *Standard test method for sensory evaluation of red pepper heat*, ASTM International, West Conshohocken, PA, 2011, DOI: 10.1520/E1083-00R11, www.astm.org (accessed 6 October 2015).

ASTM Standard E1909. 2014. *Standard guide for time-intensity evaluation of sensory attributes,* ASTM International, West Conshohocken, PA, 2014, DOI: 10.1520/E1909, www.astm.org (accessed 6 October 2015).

Bharate, S. and Bharate, S. 2012. Modulation of thermoreceptor TRPM8 by cooling compounds. *ACS Chem. Neuroscience,* **3** (4), 248–267.

Cardello, A. and Wise, P. 2008. Taste, smell, and chesthesis in product experience, in *Product Experience.* (eds. H. Schifferstein and P. Hekkert), 1st edn, Amsterdam: Elsevier, pp. 91–131.

Carstens, E., Iodi Carstens, M., Dessirier, J., *et al.* 2002. It hurts so good: oral irritation by spices and carbonated drinks and the underlying neural mechanisms. *Food Quality and Preference,* **13** (7–8), 431–443.

Cliff, M. and Heymann, H. 1992. Descriptive analysis of oral pungency. *Journal of Sensory Studies,* **7** (4), 279–290.

Dessirier, J., O'Mahony, M., Iodi-Cartens, M., and Carstens, E. 2000. Sensory properties of citric acid: psychophysical evidence for sensitization, self-desensitization, cross-desensitization and cross-stimulus-induced recovery following capsaicin. *Chemical Senses,* **25** (6), 769–780.

Eccles, R. 1994. Menthol and related cooling compounds. *Journal of Pharmacy and Pharmacology,* **46** (8), 618–630.

Green, B. 1989. Capsaicin sensitization and desensitization on the tongue produced by brief exposures to a low concentration. *Neuroscience Letters,* **107** (1–3), 173–178.

Green, B. 1991. Temporal characteristics of capsaicin sensitization and desensitization on the tongue. *Physiology & Behavior,* **49** (3), 501–505.

Green, B., Dalton, P., Cowart, B., *et al.* 1996. Evaluating the 'Labeled Magnitude Scale' for measuring sensations of taste and smell. *Chemical Senses,* **21** (3), 323–334.

Hayes, J.E., Allen, A.L. and Bennett, S.M. 2012. Direct comparison of the generalized visual analog scale (gVAS) and general labeled magnitude scale (gLMS). *Food Quality and Preference,* **28** (1), 36–44.

Hutchinson, S., Trantow, L., and Vickers, Z. 1990. The effectiveness of common foods for reduction of capsaicin burn. *Journal of Sensory Studies,* **4** (3), 157–164.

Jacquot, L., Monnin, J., Lucarz, A., and Brand, G. 2005. Trigeminal sensitization and desensitization in the nasal cavity: a study of cross interactions. *Rhinology,* **43**, 93–98.

Klein, A., Carstens, M., and Carstens, E. 2013. Eugenol and carvacrol induce temporally desensitizing patterns of oral irritation and enhance innocuous warmth and noxious heat sensation on the tongue. *Pain,* **154** (10), 2078–2087.

Lawless, H., Hartono, C., and Hernandez, S. 2000. Thresholds and suprathreshold intensity functions for capsaicin in oil and aqueous based carriers. *Journal of Sensory Studies,* **15** (4), 437–477.

Meilgaard, M., Civille, G., and Carr, B. 1991. *Sensory Evaluation Techniques.* Boca Raton: CRC Press.

Nasrawi, C. and Pangborn, R. 1989. The influence of tastants on oral irritation by capsaicin. *Journal of Sensory Studies,* **3** (4), 287–294.

Peryam, D.R. and Pilgrim, F.J. 1957. Hedonic scale method of measuring food preferences. *Food Technology,* (September), 9–14.

Törnwall, O., Silventoinen, K., Hiekkalinna, T., *et al.* 2014. Identifying flavor preference subgroups. Genetic basis and related eating behavior traits. Appetite, **75**, 1–10.

Ward, C., Koeferli, C.S., Schwegler, P.P., Schaeppi, D. and Plemmons, L. 1999. European strawberry yogurt market analysis with a case study on acceptance drivers for children in Spain using principal component analysis and partial least squares regression. *Food Quality and Preference,* **10**, 387–400.

Yang, X. 2008. Aroma constituents and alkylamides of red and green huajiao (Zanthoxylum bungeanum and Zanthoxylum schinifolium). *Journal of Agricultural and Food Chemistry,* **56** (5), 1689–1696.

Yang, X., Rohr, M., and Jordan, J. 2009. Identification of dihydrogalangal acetate in galangal [Alpinia galangal (L.) Swartz] extracts. *Journal of Agricultural and Food Chemistry,* **57** (8), 3286–3290.

CHAPTER 11

Analytical chemistry of chemesthetic compounds

David A. Bolliet
Kalsec©, Inc., Kalamazoo, MI, USA

11.1 Introduction

Chemesthetic compounds can be analyzed using a variety of techniques. The type of method used is typically dictated by a number of factors including, but not limited to, the analyte(s) of interest, concentration levels, sample matrix, desired assay sensitivity, academic field, and production environment. The published peer-reviewed scientific literature contains dozens of unique analytical solutions applied to the detection and quantification of chemesthetic compounds. In addition, since chemesthetic compounds are not only found in food products, as they are also present in many different medicines of the world, both traditional and modern, this chapter covers analytical methods in both food and non-food applications, published between 2005 and 2015.

11.2 Allyl isothiocyanate

Allyl isothiocyanate (AITC) is the chemesthetic compound found in the *Brassicaceae* family that includes mustard, horseradish, radish, and wasabi. Due to its volatile nature, gas chromatography (GC) is the primary method used to detect AITC. In the last 10 years, several research groups have published their work on the detection of AITC, among other compounds, by gas chromatography. One of the most comprehensive bodies of work on AITC lies in the PhD dissertation of Xiao (Xiao, 2005). AITC was analyzed qualitatively and quantitatively by headspace solid phase microextraction (SPME) with GC-MS (mass spectrometry) in three matrices: mustard oil, wasabi, and "Chongcai" (a leaf mustard product). Besides a number of other isothiocyanate peaks, AITC accounted for 73.45, 64.25, and 32.74% of the total volatile compounds found in these matrices, respectively. Separately, static headspace GC-FID (flame ionization detection) was used to analyze AITC in a dissertation on the study of bread molding (Ma, 2012).

Headspace SPME with GC-MS was also used to qualitatively detect AITC in a complex matrix of pickled pothead mustard (Zhao *et al.*, 2007). These workers selected a PDMS (polydimethylsiloxane) SPME fiber for the study, and an OV-1701 capillary column, which has intermediate polarity properties, for the separation of AITC from the other volatile compounds. In further published works, AITC was detected in a variety of matrices, including wasabi (Nakanishi *et al.*, 2014), cabbage leaves (Troncoso *et al.*, 2005), white cabbage (Penas *et al.*, 2012), black mustard (Crespo *et al.*, 2012), and "Chongcai" paste (Li *et al.*, 2012a). In yet another paper, AITC was one the many compounds detected by GC with atomic emission detection (Kanamori-Kataoka *et al.*, 2012).

As an alternative to chromatographic approaches for analyzing for AITC, a Japanese research team has reported their works on the development of several colorimetric detection kits for AITC (Norimine *et al.*, 2008; Ishiwata *et al.*, 2009). These workers claim their detection scheme, a series of chemical reactions taking place in a series of connected glass tubes (and thus applicable to both liquid and gaseous AITC), has a limit of detection (LOD) as low as 1 ppm (vol/vol).

It should be noted that AITC can also be analyzed by reversed-phase high-performance liquid chromatography (RP-HPLC) using a C18 column, with ultraviolet (UV) detection at 242 nm (Pelosi *et al.*, 2014). The limit of quantitation (LOQ) was estimated at 14 μg AITC/g mustard, and the linear range was above 2800 μg AITC/g mustard. The UV absorbance property of AITC has also been applied for the spectrophotometric monitoring of the reaction of AITC with hydroxyl–water and β-cyclodextrin in a number of acidic–alkaline media (Jiang *et al.*, 2006). RP- HPLC analysis with refractive index detection, and (LC) MS of AITC have been presented at a conference in the Philippines (Alea *et al.*, 2014).

11.3 Capsaicinoids

Found in the *Capsicum* genus, and more specifically mostly in the placenta of chilis, capsaicinoids are strong chemesthetic agents. While numerous capsaicinoids have been identified to date (Schweiggert *et al.*, 2006; Kozukue *et al.*, 2005), it is notable that about 90% of the capsaicinoids found in chilis are capsaicin (C) and dihydrocapsaicin (DC) (Yardim and Senturk, 2013). Besides C and DC, a few other capsaicinoids are often mentioned in the published scientific literature: nordihydrocapsaicin (NDC), homocapsaicin (HC), and homodihydrocapsaicin (HDC). Capsaicinoids are best separated by liquid chromatography, and can be detected using many different detection schemes, including UV, mass spectral, electrochemical, fluorescent, spectrophotometric, and colorimetric detections.

By far, RP-HPLC on a C18 column, either under isocratic or gradient conditions, with UV detection between 280 and 282 nm (280 nm is the most common wavelength) is the most popular way to detect capsaicinoids in chilis, for example: C (Li *et al.*, 2006; Gao *et al.*, 2005); C and DC (Bae *et al.*, 2013; Juangsamoot *et al.*, 2012; Luo *et al.*, 2012; Gu *et al.*, 2011; Zhou *et al.*, 2011; Zuo *et al.*, 2010); C, DC, and NDC (Zhang *et al.*, 2011a); and C, DC, NDC, and HC (Han *et al.*, 2012b). HPLC-UV was also performed to quantify the following: C in a nanoemulsion (Jiang *et al.*, 2011); C, and DC in Gochujang, a spicy and savory Korean condiment paste made from red chili, glutinous rice, fermented soybeans and salt (Ha *et al.*, 2010b); C, DC, and NDC in chili-infused oil products using 230 nm for detection (Wang *et al.*, 2007); C and piperine in some Korean food products (Choi *et al.*, 2006); or C and DC in six varieties of *C. annuum* L. with detection at 222 nm (Al Othman *et al.*, 2011). HPLC with UV detection was also used for quantitation of capsaicinoids in *C. annuum* L. from the Italian region of Calabria (Ziino *et al.*, 2009), and from Turkey (Topuz and Ozdemir, 2007; Poyrazoglu *et al.*, 2005). C and DC were quantified in Scotch Bonnet (*C. chinense*) by HPLC with detection at 280 nm (Gahungu *et al.*, 2011). Gonzalez-Zamora and colleagues used a C18 RP-HPLC system with UV detection at 222 and 280 nm to quantify C, DC, NDC, HC, and HDC in eight *Capsicum* varieties grown in Northern Mexico (Gonzalez-Zamora *et al.*, 2013). In his Master of Science thesis, Muchena used a C18 column RP-HPLC with UV detection at 210 nm to study capsaicinoids content in locally-grown and commercial chilis (Muchena, 2009). The effect that cooking techniques of Mexican chilis may have on capsaicinoids was studied by RP-HPLC with UV detection at 236 nm (de Jesus Ornelas-Paz *et al.*, 2010). C, DC, and NDC were monitored in raw and cooked (boiled and grilled) *Capsicum* species (poblano, bell, chilaca, caribe, jalapeño, serrano, habanero, and manzano). These workers observed a moderate decrease of capsaicinoids when chilis were boiled, and a significant increase when grilling.

A couple of stand-alone UV spectrophotometric methods were developed: detection and quantitation of C at 248 and 296 nm in oleoresin *Capsicum* (Wang *et al.*, 2012a); development of a chemometric model based on UV spectral data between 210 and 300 nm for the detection and quantitation of C and DC in habanero (Davis *et al.*, 2007). On average, published analytical methods based on UV detection (with or without HPLC) have LODs around 0.1–1.0 mg l^{-1}. When looking for increased specificity and lower detection limits, one turns to either fluorescent or electrochemical detections.

Fluorescence has been reported for the detection of capsaicinoids in chilis but also in more complex matrices (e.g., foods or plasma), using a number of excitation/emission wavelength combinations (229/320 nm, 280/310 nm, 280/325 nm, etc.), and with reported LODs of between 5 and 50 μg l^{-1}. In another paper, Barbero and colleagues have developed an RP-HPLC method with a monolithic column, with detection by fluorescence, for the quantitation of C, DC, NDC, HC, and HDC in chilis (Barbero *et al.*, 2007).

Electrochemical detection (ECD) is another extremely specific and sensitive detection technique used to detect capsaicinoids as they go through an irreversible oxidation reaction at the electrode site. With LODs as low as 1–2 μg l^{-1}, ECD is ideally suited for the determination of low levels of capsaicinoids. A variety of electrodes have been looked at for the detection of capsaicinoids: a glassy carbon electrode (Lau *et al.*, 2015a; Lau *et al.*, 2015b), a glassy carbon electrode covered with acetylene black nanoparticles (Zhang *et al.*, 2013b), a pencil graphite electrode (Yardim and Senturk, 2013), a screen printed carbon nanotube electrode (Randviir *et al.*, 2013; Surareungchai and Phanthong, 2012; Compton *et al.*, 2009; Kachoosangi *et al.*, 2008), or a boron-doped diamond electrode (Yardim, 2011).

Randviir and coworkers compared their screen printed carbon nanotube electrodes used in both cyclic voltammetry (CV) and electrochemical impedance spectroscopy (EIS), finding that CV is more appropriate for low C content, and EIS is more appropriate for high C concentrations (Randviir *et al.*, 2013). ECD has been used post-HPLC separation for the quantitation of C and DC from different parts of a chili (Supalkova *et al.*, 2007).

Mass spectrometry is another common detection technique used to help identify capsaicinoids in chilis, for example, using Orbitrap MS in *C. frutescens* L. (Bijttebier *et al.*, 2014), combining fluorescence and MS to study the extraction of capsaicinoids from chilis by pressurized liquid extraction (Barbero *et al.*, 2006b), detecting minor and major capsaicinoids in *C. frutescens* L. (Schweiggert *et al.*, 2006) or *C. annuum* L. (Kozukue *et al.*, 2005). A reversed-phase LC-MS/MS method was developed to quantify C, DC, and NDC in microwave-assisted extracted *C. annuum* L., with reported LOD values between 0.01 and 0.11 $\mu g\, l^{-1}$ (Han *et al.*, 2012a). Garces-Claver *et al.* used LC time of flight (TOF) MS to quantify C and DC in 11 different *Capsicum* genotypes, with a reported LOD for C and DC of 20 and 4 pmol (approximately 6 and 1 $ng\, l^{-1}$), respectively (Garces-Claver *et al.*, 2006). Detection of C, the active ingredient in pepper sprays, on pepper-sprayed clothing has been demonstrated by DART (direct analysis in real time) ionization coupled with an AccuTOF (accurate TOF) MS, with GC-MS used for results verification (Pfaff and Steiner, 2011). Localizing C in cross-sections of chili fruits was performed by mass spectrometry imaging using MALDI-TOF-MS (matrix-assisted laser desorption/ionization) (Taira *et al.*, 2012). In this study, it was shown that C was higher in the placenta (higher on the surface than in the internal region) of the chili fruit than in the pericarp. Further, C was not found in the seeds.

Another reported use of MS is for the detection of capsaicinoids at trace levels in used cooking oil. Indeed, it has been shown that C, DC, and NDC are good indicators of used cooking oil in China, as chilis are widely used in Chinese cooking, thus leaving behind some of the capsaicinoids in the used oil (Wang *et al.*, 2012c). For a very similar purpose, the China National Analytical Center, Guangzhou, China developed an LC-MS/MS method to detect C and DC in used cooking oils (Wu *et al.*, 2012a), and at least two other groups are also using LC-MS/MS to detect capsaicinoids in the same matrix (Mao *et al.*, 2013; Zhang *et al.*, 2012d).

In complex biological samples such as urine, C was detected by LC-MS following an enzymatic hydrolysis of capsaicin glucuronide, the form under which C was present in the rat urine samples (Noami *et al.*, 2006). Noami *et al.* reported an LOD of 0.1 $\mu g\, l^{-1}$. At least two other groups have performed LC-MS/MS analyses of C and/or DC in plasma samples (Zhang *et al.*, 2010b; Yu *et al.*, 2008).

Faster and more efficient than HPLC, ultra-performance liquid chromatography (UPLC) coupled with UV, fluorescence, or MS detection has been used for detecting and quantifying capsaicinoids in chilis and food products: C and DC (Usman *et al.*, 2014; Ha *et al.*, 2010a); C, DC, and NDC (Zhao *et al.*, 2011); C, DC, NDC, HC, and HDC (Alothman *et al.*, 2012).

Recent research work on novel HPLC approaches for the analysis of capsaicinoids over the past year has been limited to micellar LC and reversed-phase argentation LC. On a C18 column, with UV detection at 210 nm, C was detected by micellar LC in less than 7 minutes, with an LOD of 0.7 μg l^{-1} (Chin-Chen *et al.*, 2010). Reversed-phase argentation LC, where silver nitrate is added to the mobile phase to enhance selectivity, was used to quantify C, DC, and NDC in chilis, with a C18 column, and UV detection at 280 nm (Jin *et al.*, 2009a; Jin *et al.*, 2009b). Argentation LC was used in a thorough comparison of 12 different column chemistries (C8, C18, C30, phenyl, cation-exchange, etc.), with varying mobile phase conditions, with ESI-MS detection of 11 capsaicinoids (Thompson *et al.*, 2005).

New biosensors are being developed to detect capsaicinoids. These include a surface plasmon resonance biosensor for the detection of capsaicinoids from the use of lachrymators where an LOD of 150 ppb was achieved (Nakamura *et al.*, 2013). A team from Malaysia has developed two biosensors for detection of C, an optical biosensor (Mohammad *et al.*, 2014), and an amperometric biosensor (Mohammad *et al.*, 2013). Even though these research-type biosensors have yet to find a commercial outlet, it should be noted that the amperometric biosensor achieved a lower LOD than its optical counterpart, with approximately 0.7 versus approximately 55 mg l^{-1}, respectively.

Capsaicinoids can easily be detected and quantified by colorimetry. While individual capsaicinoids cannot be separated and quantified, a colorimetric approach for the quantitation of total capsaicinoids can be fast, easy, and cheap to implement. Colorimetric analysis of capsaicinoids, using Gibbs reagent (2,6-dichloroquinone chloroimine) has been used to detect capsaicinoids in crop breeding (Jeong *et al.*, 2012), and eight varieties of chilis in a classroom setting (Thompson *et al.*, 2012). In this reaction, the Gibbs reagent turns blue as a function of the total capsaicinoids content. It should be pointed out that, in both cases, the accuracy of the colorimetric approach compared favorably to that of HPLC results. Another group proposed a different colorimetric scheme for the quantitative analysis of capsaicinoids, through a reaction involving copper(II) and 4-(2-pyridylazo)resorcinol, with a reported LOD of 0.4 μg capsaicinoids/g chili, with HPLC analysis confirming these results with 95% confidence (Tonon de Souza and Rossi, 2014). A different colorimetric assay has been developed to quantify C in capsaicin-based pesticides, using a combination of phosphotungstic acid and phosphomolybdic acid, with detection of a blue color at 660 nm. The method has an LOD of 0.5 mg l^{-1} (Liu and Han, 2008).

Even though GC analysis is not used as much as HPLC as a routine separation technique for capsaicinoids, several groups have reported their GC work on capsaicinoids. As part of a larger paper on the analysis of 17 compounds found in chemical warfare agents, GC with atomic emission detection was used to detect C (Kanamori-Kataoka *et al.*, 2012). GC-MS was chosen by a Nigerian team for the quantitative analysis of C in five different chilis grown in Nigeria (Nwokem *et al.*, 2010), by a Mexican group for the analysis of C and DC in chilis and chili sauces (Pena-Alvarez *et al.*, 2012; Pena-Alvarez *et al.*, 2009), and by a European team in Andean aji (*C. baccatum*) and rocoto (*C. pubescens*) chilis (Kollmannsberger *et al.*, 2011). A Korean group used GC-FID to quantify C in the Korean condiment Gochujang, with a reported LOD of 1.0 mg kg^{-1} (Ha *et al.*, 2008). Ha *et al.* validated their GC-FID method against an HPLC method, AOAC 995.03 (AOAC, 2012). Since capsaicinoids contain nitrogen, one group has decided to combine GC with the selective NPD (nitrogen phosphorus detection), quantifying C, DC, and NDC in chilis of different varieties (*C. chinense, C. frutescens, C. baccatum, C. annuum*, and *C. pubescens*) (Antonious and Jarret, 2006). In their work, Antonious and colleagues also used mass spectral information for C (305 *m/z* molecular ion), DC (307 *m/z*), and NDC (293 *m/z*), focusing on a common benzyl cation fragment (137 *m/z*) used for monitoring capsaicinoids in chilis.

Near-IR (NIR) spectroscopy has been used for the quantitative determination of C (He *et al.*, 2012), and total capsaicinoids in chili powder (Mo *et al.*, 2011). Thin-layer chromatography (TLC) has also been reported for measuring the content of total capsaicinoids in chili extracts (Monforte-Gonzalez *et al.*, 2007). While C and DC could not be separated by TLC, the combined amounts were equal to quantitative results obtained by GC. TLC was also used as a screening technique by the Nigerian team that studied C and DC in Scotch Bonnet (Nwokem *et al.*, 2010). TLC and HPLC were used to show the lack of capsaicinoids (C and DC) in the vegetative parts (leaves and stems) of the habanero (*C. chinense* Jacq.) (Ruiz-Lau *et al.*, 2010). In an undergraduate science protocol, TLC and ^{1}H NMR (nuclear magnetic resonance) were used to detect and quantify C and DC in a chili extract (Wagner *et al.*, 2011). Detection of C and DC in *C. annuum* was demonstrated using capillary electrophoresis (Liu *et al.*, 2010).

11.4 Carbonic acid

In water, dissolved carbon dioxide exists as the weak acid carbonic acid, H_2CO_3. The preferred method for detection of carbonic acid is ion-exclusion chromatography (IEC) with conductivity detection. This approach was used to detect and quantify carbonic acid in water samples (Fan *et al.*, 2005). Using a weakly acidic cation-exchange resin column, carbonic acid was separated from other weak acids (formic acid, acetic acid, and propionic acid) in under 16 minutes. The method was sensitive (LOD of 0.084 mg carbonic acid/l water) and linear over two orders of magnitude (3.0 to 500 mg l^{-1}). The same year, another paper reporting the development of a high-speed IEC method for carbonic acid was published (Mori *et al.*, 2005). In this paper, a combination of columns installed in series was used. A weakly acidic cation-exchange resin column was used for separation, followed by two ion-exchange resin columns: a strongly acidic cation-exchange resin and a strongly basic anion-exchange resin. The retention time of carbonic acid was 60 seconds. The calibration curve was linear from 0.005 to 10 mM, with an LOD of 0.15 μM. More recently, another IEC method with conductivity detection was developed to detect and quantify carbonic acid in oxidized glycerol products (Hou *et al.*, 2012). The method was linear from 2 to 100 mg l^{-1} with an LOD of 0.45 mg l^{-1}. In 2011, a Chinese group patented a new ion chromatography detector, using an electrochemical suppressor and resistance detection system, with applications for carbonic acid detection (Hu *et al.*, 2011b).

Other analytical schemes used to detect carbonic acid include Fourier transform microwave spectroscopic analyses of gas-phase cis–cis and cis–trans H_2CO_3 (Mori *et al.*, 2009; Mori *et al.*, 2011), GC-FID (Yang and Zhang, 2012c), flow analysis (Liao *et al.*, 2015), ultrasonic detection (Sugiyama and Kagatani, 2012b), electrochemical biosensor (Singh *et al.*, 2014), and chemical reactions (Hu *et al.*, 2012). The electrochemical biosensor developed by Singh *et al.* was used to measure low levels of carbonic acid in blood, had a linear range of 1.32–2.32 mM, and stated an LOD of 19.4 μM (Singh *et al.*, 2014). Finally, a couple of Japanese patents described two inventions for the measurement of carbonic acid in an ion-exchange water purification unit (Miyamoto and Horii, 2013), and in bath water (Sugiyama and Kagatani, 2012a).

11.5 Cinnamaldehyde

Cinnamaldehyde is the chemesthetic agent found in a number of species from the *Cinnamomum* genus, including cassia (*C. cassia,* also known as Chinese Cassia or Chinese cinnamon) and Ceylon cinnamon or "true cinnamon" (*C. verum,* syn. *C. zeylanicum*). It should be noted that confusion exists when using the term "cinnamon" around the world. Indeed, in the USA, Canada, the UK, and India, products sold as "cinnamon" are more likely to contain cassia while, in most of Europe, "cinnamon" refers to the more expensive Ceylon Cinnamon. Even though both species do contain cinnamaldehyde, this difference in nomenclature is of importance as *C. cassia,* unlike *C. verum,* also contains significant amounts of coumarin (up to 5%), a regulated compound in the European Union, which has been shown to cause liver and kidney damage in rats and mice (Lungarini *et al.*, 2008). Therefore, in this section, we will try our best to use botanical names for the different species we will cover. It should also be noted that the leaves of *C. verum* contain eugenol, another chemesthetic agent that is covered later in this chapter. Finally, in this section, the term "cinnamaldehyde" will be used to refer to "*t*-cinnamaldehyde".

Even though cinnamaldehyde is a volatile compound, thus making gas chromatography an obvious option for analytical purposes, it is interesting to note liquid chromatography is the second most published technique for the analysis of cinnamaldehyde over the past 10 years. Similarly to other sections in this chapter, 50% of the total published papers on analytical methods for the separation and detection of cinnamaldehyde are written in Chinese.

As mentioned earlier, GC-MS is the most common way to detect cinnamaldehyde in species from the *Cinnamomum* genus: most abundant in *C. cassia* (Dong *et al.*, 2013; Lv *et al.*, 2010), at trace levels in *C. chekiangense* (Yang *et al.*, 2009), at 20–80% in leaves of *C. tamala* (Mohan *et al.*, 2012; Joshi *et al.*, 2008). Wang *et al.* studied the volatile profiles of the essential oils from the leaves of five different cinnamon species: *C. cassia, C. zeylanicum, C. tamala, C. burmannii* (Indonesian cinnamon), and *C. pauciflorum* (Wang *et al.*, 2009a). Cinnamaldehyde was detected in all five species, but was the largest volatile compound in *C. cassia* and *C. burmannii.* GC-MS was also chosen to analyze cinnamaldehyde in the volatile fraction extracted from a beef sauce (Duan *et al.*, 2013), and in a polyherbal traditional Chinese medicine containing *Cortex Cinnamomi* (Yan *et al.*, 2009).

A known skin sensitizer in some people, cinnamaldehyde has been included in various papers describing GC-MS methods to detect a number of allergens, including cinnamaldehyde, in toys (Wei *et al.*, 2013; Zheng *et al.*, 2013), in essential oils destined for the cosmetic industry (Kaloustian *et al.*, 2007), and in fragrances (Cordero *et al.*, 2007). Cordero *et al.* chose comprehensive 2D GC (GC×GC) with mass spectrometry, a wise choice dictated mostly by the complexity of the formulated fragrances and the number of volatile compounds to be separated.

RP-HPLC on a C18 column with photodiode array detection at 290 nm is the most common liquid separation technique for the analysis of cinnamaldehyde (Qiang *et al.*, 2014; Liu *et al.*, 2014a; Gao *et al.*, 2013). However, other wavelengths have been reported, mostly in the analysis of cinnamaldehyde in traditional Chinese medicines: 225 nm (Wang *et al.*, 2010), 254 nm (Zhang *et al.*, 2012c), 276 nm (Chen *et al.*, 2011), 280 nm (Wang *et al.*, 2013e; Wang *et al.*, 2013c), 282 nm (Wang *et al.*, 2011a), and 285 nm (Zhang *et al.*, 2011c; Li *et al.*, 2008a). Separations are evenly divided between isocratic and gradient solvent systems, with mostly acetonitrile–water or methanol–water combinations. In the isocratic mode, solvent ratios of approximately 35 : 65 (v/v) acetonitrile : water or 60 : 40 (v/v) methanol : water seem to be preferred.

Cinnamaldehyde has been analyzed by HPLC in different species of *Cinnamomum*, including *C. cassia* (Xu *et al.*, 2014), *Cinnamomi Ramulus* (another name for *C. cassia*) (Lee *et al.*, 2013b; Cao *et al.*, 2013; Lin *et al.*, 2005a), *C. verum* (Lungarini *et al.*, 2008), *Cinnamomi Ramulus* and *Cortex Cinnamomi* (Yin *et al.*, 2007). In their paper, Lungarini and colleagues used HPLC to detect the adulteration of *C. verum* with cheaper *C. aromaticum* (cassia) in a number (50) of supposedly "true cinnamon" food products collected from an Italian market. They found that 51% of the products they tested were not *C. verum* but instead were cassia, 10% were a blend of both *C. verum* and cassia, and only 39% were "true cinnamon", that is, *C. verum* (Lungarini *et al.*, 2008).

Kumar and coworkers published an interesting sampling technique when analyzing cinnamaldehyde in cassia, where they used SPME, followed by a more standard RP-HPLC on a C18 column, with 70 : 30 (v/v) acetonitrile : water, and UV detection at 254 nm, with a remarkable reported LOD for cinnamaldehyde of 13 pg ml^{-1} (Kumar *et al.*, 2007). Cinnamaldehyde, encapsulated in β-cyclodextrin (Cevallos *et al.*, 2010), can be detected by RP-HPLC on a C18 column with UV detection at 230 nm (Zhang *et al.*, 2013a).

Mass spectrometry, combined with HPLC, is another detection means for cinnamaldehyde. One South Korean group has used it on several traditional Chinese and Korean herbal medicines, either in MS mode (Weon *et al.*, 2012) or in MS/MS mode (Lee *et al.*, 2013a).

Cinnamaldehyde has been isolated from *C. zeylanicum* bark oil, and was confirmed and identified with FTIR (Fourier transform infrared) spectroscopy and HPLC (Al-Bayati and Mohammed, 2009). Another form of spectroscopy, NIR, was used to detect cinnamaldehyde in a traditional Chinese medicine containing cinnamon twig (Xiao *et al.*, 2012). A German research group presented their research on the interesting use of ^{1}H NMR to screen total aldehydes, including cinnamaldehyde, in vegetable oils and cosmetics, concluding that, even though they are useful, ^{1}H NMR results need to be complemented by GC, a more specific assay (Lachenmeier *et al.*, 2010).

A most unique approach, using differential pulse polarography, was chosen to simultaneously detect cinnamaldehyde and benzaldehyde in a number of food and drug samples (Gholivand and Ahmadi, 2008). In their study, Gholivand and Ahmadi found that cinnamaldehyde undergoes a heat-induced degradation at 70 °C, forming benzaldehyde. Capillary electrophoresis with UV detection is yet another interesting analytical way of separating and detecting cinnamaldehyde in *C. cassia* (Li *et al.*, 2007). Detection was at 310 nm, and the LOD for cinnamaldehyde was 2.3 mg l^{-1}. TLC can be used to qualitatively identify cinnamon from a poly-herbal traditional Chinese medicine, but quantitative analysis of cinnamaldehyde still relies on RP-HPLC analysis, this time with UV detection at 277 nm (Chen *et al.*, 2005). TLC has been used to detect cinnamaldehyde in pepper-contaminated cinnamon, that is, cinnamon bark powder adulterated with pepper powder, cinnamon oil, clove powder, clove oil, and a commercial preparation containing cinnamaldehyde and eugenol (Gopu *et al.*, 2008).

11.6 Eugenol

The numbing chemesthetic agent eugenol is found in a number of different herbs and spices, including basil, bay, cinnamon leaf, clove, pimento, and so on. Since it is a volatile compound, eugenol is mostly separated from other compounds using GC. However, eugenol is also separated by LC, as evidenced by the literature published over the last 10 years.

Eugenol, present at 70% or more in clove, is easily detected using GC-MS or GC-FID (Horvath *et al.*, 2013; Sanchez-Cabrera and Pino, 2011). GC-MS is also used to detect eugenol in clove cigarettes (kretek) (Liao *et al.*, 2012; Stanfill *et al.*, 2006). Since clove is also used in a number of medicinal preparations, a handful of groups have developed GC-FID or GC-MS assays to detect eugenol in these formulations (Liu *et al.*, 2011a; Wei and Ni, 2011b; Zhou *et al.*, 2010; Wang and Wei, 2006). Because of its numbing properties, eugenol is used in dental applications as a mild pain reliever aid; GC-FID and GC-MS are used to analyze the eugenol content in these products (Ertas *et al.*, 2007). Eugenol is also a characteristic marker in several alcoholic beverages. A German research team, studying eugenol in blood samples after the ingestion of eugenol-containing alcoholic drinks, found that in blood, eugenol metabolizes rapidly into eugenol glucuronide, requiring an enzymatic cleavage in the sample preparation in order to release the eugenol for analysis by either headspace SPME GC-MS, with an LOD of 3.2 μg l^{-1}(Schulz *et al.*, 2008) or by dynamic headspace GC-MS, with an LOD of 1 mg l^{-1} (Schulz *et al.*, 2007). Eugenol has also been detected in wine samples by headspace SPME GC-MS with an LOD of 0.01 μg l^{-1} (Carrillo and Tena, 2006).

Owing to the potential allergenic effect caused by eugenol in some people, a number of analytical methods have been developed to detect and quantify eugenol in toys and fragrances. A Chinese team has opted for GC-MS to detect eugenol in toys (Wei *et al.*, 2013; Wei *et al.*, 2012; Wei *et al.*, 2011a), while several European groups detected eugenol in fragrances using either GC×GC with MS detection (Cordero *et al.*, 2007), or RP-HPLC with UV detection at 280 nm (Villa *et al.*, 2007).

Reversed-phase liquid chromatography can also be an option for the separation and detection of eugenol. The most common LC setup involves a C18 column and UV detection at 280 nm. LC methods like this have been applied to the development of: a validated method, with an LOD of 0.81 μg l^{-1}, for the quantitation of eugenol in clove (Yun *et al.*, 2010); a validated method for eugenol detection in nanoemulsions with an LOD of 0.44 mg l^{-1} (Pramod *et al.*, 2013); the analysis of eugenol in clove bud with an LOD of 0.2 mg l^{-1} (Ramlingam *et al.*, 2012); a method for the analysis of eugenol in tobacco with an LOD of 0.3 mg kg^{-1} (Zhang *et al.*, 2012b); and a few methods for the quantitation of eugenol in some medicinal and chemical preparations (Wang *et al.*, 2013e; Li *et al.*, 2012e; Lee *et al.*, 2013a). Other wavelengths have been used to detect eugenol after HPLC separation: 215 nm (Chen, 2007), 230 nm (Zhang *et al.*, 2013a), and 293 nm (Yu *et al.*, 2010). Musenga and colleagues separated and quantified eugenol in clove essential oils and clove bud extracts using a C8 RP-HPLC method with UV detection at 210 nm with a reported LOD of 62 μg l^{-1} (Musenga *et al.*, 2006).

Other techniques have been used besides UV for the detection and quantitation of eugenol after HPLC separation. Following RP-HPLC on a C18 column, ECD was used to detect eugenol in clove using amperometric and coulometric detection modes (Cantalapiedra *et al.*, 2014). It was found that a lower LOD was achieved in the coulometric mode (3.0 μg l^{-1}) compared with the amperometric mode (9.7 μg l^{-1}). In comparison, by UV at 280 nm, the LOD was 62 μg l^{-1}. ECD was also used to quantify eugenol in a cherry liqueur (Rodtjer *et al.*, 2006). Fluorescence detection is another sensitive technique applied for the quantitation of eugenol in distilled spirit (Xun *et al.*, 2012). The method uses an excitation wavelength of 272 nm and an emission wavelength of 316 nm.

A Chinese tobacco company has been awarded at least four patents in the last ten years for the analysis of eugenol in tobacco products and byproducts (smoke) using a number of analytical approaches: C18 RP-HPLC with UV detection at 280 nm (Liao *et al.*, 2013b); C18 RP-HPLC with fluorescence detection using 290 nm for excitation and 315 nm for emission (Liao *et al.*, 2013c); and GC-MS (Liao *et al.*, 2012; Liao *et al.*, 2013a).

The electrochemical properties of eugenol have been studied by a handful of research teams, using a variety of electrodes, either commercially available or at the lab prototype stage: a titanium oxide electrode with an LOD of 1.3 μM (Kang *et al.*, 2014), a gold–graphene–MoS_2 modified glassy carbon electrode with an LOD of 0.036 μM (Feng *et al.*, 2014a), a glassy carbon electrode with an LOD of 10 μM (Ziyatdinova *et al.*, 2013), a modified glassy carbon electrode with an LOD of 0.475 μM (Xiong *et al.*, 2011a), a nano-tungsten oxide modified carbon paste electrode with an LOD of 0.01 μM (Xiong *et al.*, 2011b), and gold and platinum electrodes (Wang and Chen, 2011c).

Limited work has been reported on the MS detection of eugenol. One group used eugenol as a model compound to study a newly developed ion source, and claimed detection limits in the picomolar range (Schepler *et al.*, 2013). Another team used reversed-phase LC-MS/MS to detect eugenol in used cooking oils, with a reported LOD of 0.6 μM (Zhang *et al.*, 2012d).

TLC has been used to detect eugenol: in the leaves of a plant used in Indian, Chinese, and Malaysian medicines (Yadav *et al.*, 2012); in a polyherbal formulation containing *Ocimum sanctum* Linn (Holy Basil, also known as Tulasi or Tulsi) (Alam *et al.*, 2012); in clove buds (Ramlingam *et al.*, 2012); in a traditional Siddha medicinal formulation (Patra and Kumar, 2010); in clove, nutmeg, and cinnamon (Rusu *et al.*, 2007); in herbal raw materials (Dhalwal *et al.*, 2007); in *Ocimum sanctum* (Anandjiwala *et al.*, 2006); in a mixture containing cinnamaldehyde, piperine, and eugenol (Gopu *et al.*, 2008); and in the leaves of *C. tamala* Nees and Eberm (Dighe *et al.*, 2005). As part of a review on visualizing indicators used in TLC, eugenol was one of the compounds that was selected (Pyka, 2010).

Analysis of eugenol by capillary electrophoresis has been reported a couple of times within the last 10 years (Rodriguez *et al.*, 2008; Lin *et al.*, 2006). Infrared spectroscopy has been used in a few papers. Mid-IR and near-IR have been studied for the identification of eugenol in a traditional Chinese medicine (Wua *et al.*, 2008), and NIR was used to authenticate cinnamon and clove essential oils against adulteration (Juliani *et al.*, 2006).

11.7 Gingerols and shogaols

Gingerols and shogaols are the chemesthetic compounds found in species from the *Zingiber* genus, ginger (*Zingiber officinale*) being the most common species. The gingerols are [6]-Gingerol, [8]-Gingerol, and [10]-Gingerol. Gingerols, through dehydration, form shogaols, that is, [6]-Shogaol, [8]-Shogaol, and [10]-Shogaol. Zingerone, a compound not found in fresh ginger, is formed by a retro-aldol reaction when gingerol is heated. Shogaols have a stronger chemesthetic effect (about twice as much) than gingerols, while zingerone is about half the strength of gingerols.

Owing to their thermal instability, gingerols are very rarely analyzed by GC, as they end up degrading within the instrument injector port resulting in the formation of zingerone (Li *et al.*, 2012d), and shogaols and zingerone (Zhan *et al.*, 2008; Huang and Chen, 2007). Indeed, there is evidence that [6]-Gingerol begins converting into [6]-Shogaol at 70 °C (Huang *et al.*, 2011a). Other evidence for the thermal lability of gingerols is that shogaols are not found in fresh ginger (Park and Jung, 2012).

HPLC is by far the most popular technique to separate and detect gingerols and shogaols in ginger. Lee and colleagues claim to have published the first validated method for the HPLC separation and detection of [6]-Gingerol, [8]-Gingerol, [10]-Gingerol, and [6]-Shogaol in raw ginger and in a dried aqueous extract of ginger (Lee *et al.*, 2007). It is interesting to notice that they used 200 nm (instead of the more common 280 nm, as seen later) for quantification of the analytes. Some LC-MS/MS was performed to confirm the identities of the analytes. Lee *et al.* did find that the levels of gingerols in the dried aqueous extract were about 5 to 16 times lower than those in the raw herb, and that the level of [6]-Shogaol was higher, confirming the thermal lability of gingerols.

HPLC gradient elution, C18 column, and UV detection at 280 nm are the most common experimental conditions. They have been applied to seeds and fresh ginger (Wang *et al.*, 2013b; Li *et al.*, 2012b; Wu *et al.*, 2012c; Meng, 2011; Ma *et al.*, 2009;

Li *et al.*, 2008b), ginger oleoresin (Liu *et al.*, 2014b), dried ginger (Meng, 2011), ginger juice (Zhang *et al.*, 2008), ginger oil (Yan, 2007), different ginger extraction techniques (Chen *et al.*, 2014; Zhang *et al.*, 2013c; Feng and Yan, 2007), and ginger preparations (Nikam *et al.*, 2013; Qu *et al.*, 2007). A validated HPLC method, combined with chemometrics, was used to classify different ginger types found in China, based on their geographical location (Feng *et al.*, 2014b). Their model identified 16 common peaks within ten different batches of ginger, and managed to classify the ginger samples into two groups. Another group determined [6]-Gingerol and [6]-Shogaol in ginger samples from different districts in China (Zhang *et al.*, 2009a). The content of gingerols and [6]-Shogaol, studied by LC in the oleoresins of Jamaican ginger picked from different locations was reported as a function of maturation (Bailey-Shaw *et al.*, 2008).

[6]-Gingerol has been determined in condiments (Lin *et al.*, 2011) and toothpaste (Wei *et al.*, 2009). However, since ginger is used in many traditional Chinese and Korean medicines, it comes as no surprise there are dozens of published papers covering the analysis of gingerols and/or shogaols in a multitude of formulated products, of which the following papers are only the tip of the iceberg (Wang *et al.*, 2013a; Weon *et al.*, 2013; Liu *et al.*, 2013b; Ye *et al.*, 2012; Li *et al.*, 2012c; Jia *et al.*, 2011b; Jain *et al.*, 2010; Xie *et al.*, 2010; Li *et al.*, 2008b). As a consequence, several groups have developed assays for the analysis of gingerols and shogaols in plasma and tissue materials (Xiang *et al.*, 2013; Yu *et al.*, 2011; Jiang *et al.*, 2010; Wang *et al.*, 2009b; Wang *et al.*, 2009c).

Besides UV detection, mass spectrometry has offered a number of popular detection options for gingerols and shogaols, in particular in complex matrices. In 2005, Jiang *et al.* published one of the earliest MS/MS methods for the analysis of gingerols and shogaols in ginger (Jiang *et al.*, 2005). Since then, several other groups have published their works using MS (Weon *et al.*, 2013; Yang *et al.*, 2012a; Weon *et al.*, 2012; Wang *et al.*, 2009c), MS/MS (Cheng *et al.*, 2013; Wang *et al.*, 2012d; Yu *et al.*, 2011; Tao *et al.*, 2009), TOF-MS (Park and Jung, 2012; Tan *et al.*, 2011; Liu *et al.*, 2009; Wang *et al.*, 2009b), and even TOF-MS/MS (Wang *et al.*, 2011b).

ECD is an underutilized detection technique that offers high selectivity and excellent sensitivity. One requirement of the electrochemical detector is that the analytes of interest have to be able to be either oxidized or reduced. In 2010, ECD, combined with HPLC, was used for the first time to quantify all six gingerols and shogaols in 11 ginger-containing products, with sensitivity ranging between 7.3 and 20.2 pg LOD (Shao *et al.*, 2010). The same year, another team quantified all three gingerols and [6]-Shogaol in human plasma by HPLC with ECD (Zick *et al.*, 2010). This marked the first time ginger chemesthetic compounds were detected in human plasma. The lower LOQ was 0.1 $\mu g\ ml^{-1}$ for all the analytes, except for [10]-Gingerol, which was 0.25 $\mu g\ ml^{-1}$.

UPLC, being a natural progression of HPLC, has been applied a few times for the identification of gingerols and shogaols. Reversed-phase UPLC on a C18 column with detection at 280 nm was first reported in 2012 for the analysis of gingerols in ginger dried using a number of drying technologies (Han *et al.*, 2012c). Drying of ginger in the shade or vacuum drying at 60 °C were the preferred drying methods, as neither negatively impacted the quality of the final dried ginger product. A year later, the same team published additional UPLC work on fresh, dried, and processed ginger from 17 different places in China (Han *et al.*, 2013). A different group combined UPLC with MS/MS to identify gingerols and [6]-Shogaol in a traditional Chinese herbal medicine (Cheng *et al.*, 2013).

TLC or HPTLC (high-performance thin-layer chromatography) have been used a few times in the past decade to detect [6]-Gingerol and [6]-Shogaol in a Japanese medicine (Tsuno, 2006), [6]-Gingerol in herbal extracts of *Z. officinale* and poly-herbal formulations, teas, creams, or dietary supplements (Jain *et al.*, 2011; Alqasoumi, 2009), [8]-Gingerol in *Z. officinale* and some ginger-containing dietary supplements (Alam, 2013), and [10]-Gingerol in herbal extracts of *Z officinale* and teas, creams, dietary supplements (Yusufoglu and Alqasoumi, 2008).

A colorimetric assay was developed to detect [6]-Gingerol (Xu *et al.*, 2012). This involved reacting [6]-Gingerol with potassium ferricyanide–ferric chloride, with a detection at 680 nm. The method was linear between 25.6 and 320.0 $\mu g\ ml^{-1}$. Another group chose capillary LC to detect gingerols and [6]-Shogaol in three ginger varieties from Indonesia (Rafi *et al.*, 2013). Column flow was 5 $\mu l\ min^{-1}$. LOD and LOQ were around 0.036 and 0.120 $\mu g\ ml^{-1}$, respectively.

Some teams have extracted and purified [6]-Gingerol from ginger (Yue *et al.*, 2013; Huang and Zhang, 2005) and ginger oleoresin (Liu *et al.*, 2014b). Others have looked into optimizing the extraction conditions of ginger by supercritical CO_2 fluid extraction, resulting in a product with 23% [6]-Gingerol (Chen *et al.*, 2009a). A recent review presented information on patents granted between *ca.* 2002 and 2012 and related to ginger processing, methods of extraction, and with an emphasis on health claims of ginger-containing products (Kubra and Jaganmohanrao, 2012).

While truly not an analytical method to measure the amount of gingerols, it should be pointed out that electron spin resonance has been used to measure the antioxidant activity of gingerols in ginger oleoresin (Ge *et al.*, 2010).

11.8 Menthol

Menthol is the chemesthetic compound found in the *Mentha* genus. Owing to its lack of a chromophore, menthol is best detected by gas chromatography as it is a volatile compound. Nonetheless, liquid chromatographic detection of menthol is made possible thanks to derivatization schemes.

Menthol, among other compounds, was determined in mint essential oil using three different analytical approaches: GC-FID, GC-MS, and derivatization followed by HPLC-UV (Rodenas-Montano *et al.*, 2013). For HPLC, phthalic anhydride was chosen as the derivatization agent, thus forming phthalate hemi-esters that were then separated by RP-HPLC on a C8 column under gradient conditions. These workers reported the LOD for the HPLC method to be better than that for GC-FID analysis, and similar to GC-MS.

In a different paper, to help with the separation of stereoisomers, derivatization was used prior to chiral GC analysis of a number of chiral alcohols, including menthol (Oromi-Farru *et al.*, 2012). Alcohols were acylated using either acetic acid or trifluoro-acetic acid. NMR (^{1}H and ^{13}C) was used for structural elucidation of the acylated species.

A method based on inverted gated decoupling ^{13}C NMR has been patented by a Chinese team (Zhao *et al.*, 2012a). In their invention, they claim this method can be used to fingerprint the active components contained in a number of medicinal plants, for example, menthol in Chinese mint (*Mentha haplocalyx*).

The increase in separation power offered by GC×GC, when applied to complex mixtures, has been demonstrated in hundreds of different applications and matrices. GC×GC has been applied to the analysis of menthol in a number of *Mentha* species (Cordero *et al.*, 2012), and in complex mixtures containing 362 analytes (van Stee and Brinkman, 2011). Cordero and coworkers used headspace SPME sampling coupled with GC×GC-MS, while van Stee and colleagues used GC×GC coupled with TOF-MS.

In the last 10 years, menthol has been reported, by use of GC-MS, as a minor component in many different herbs and plants (Joshi *et al.*, 2011; Kumar *et al.*, 2011; Kuzmenko, 2009; Zhang and Wang, 2009b; Guedes De Pinho *et al.*, 2009; He *et al.*, 2007). Menthol is also commonly added to a variety of tobacco-based products. Detection of the analyte has been accomplished by GC-MS in bidi cigarettes (Stanfill *et al.*, 2006), and through a patented fluorescence scheme in mentholated tobacco (Cadieux, 2015).

Besides the works by Rodenas-Montano and colleagues (2013), another group has analyzed menthol, more specifically (+)-menthol and (–)-menthol, using RP-HPLC on a C8 column (Lin *et al.*, 2005b). The chromatographic separation was performed after derivatizing the enantiomeric pair with naproxen acyl chloride, a fluorescent reagent. Detection by fluorescence was performed with an excitation wavelength of 235 nm and an emission wavelength of 350 nm. The method was linear from 5.0 to 50 μM, with an LOD of 1 μM.

Finally, it should be noted that near-IR and mid-IR vibrational circular dichroism spectra for nine enantiomeric pairs, including menthol, have been recorded (Guo *et al.*, 2006), and that TLC was reported once for the analysis of menthol in a *Mentha piperita* extract (Jesionek *et al.*, 2013).

11.9 Piperine

The chemesthetic compound piperine is found mostly in black pepper (*Piper nigrum*) and long pepper (*Piper longum*). It is also found in a number of Chinese and Indian medicinal formulations. In fact, over the last 10 years, 44% of the published scientific literature covering analytical detection schemes for piperine was published in Chinese, and almost two-thirds of the total published papers on analytical methods for piperine detection were since 2010. HPLC, coupled with a variety of detectors, was the predominant method of separation, with 55% of the published work. HPTLC came in as a distant second, with close to 10% of the scientific literature. Other analytical approaches used for separation and detection of piperine include spectrophotometry, IR technologies, GC-MS, high-speed countercurrent chromatography (HSCCC), micellar LC, fluorescence, capillary electrochromatography, and others.

RP-HPLC, most typically under isocratic conditions, with UV detection, is the most common means of separating and detecting piperine in a number of matrices and blends, with 343 nm, piperine λ_{max}, being the most widely used detection. Piperine has been detected using a 343 nm wavelength in *Piper longum* (Hu *et al.*, 2014; Bi *et al.*, 2012; Wu *et al.*, 2011b; Yu *et al.*, 2009), in *Piper nigrum* (Yang *et al.*, 2013; Guo *et al.*, 2013; Rao *et al.*, 2011; Hu *et al.*, 2009), in several traditional Mongolian medicinal preparations (Bao *et al.*, 2014; Gao and Zheng, 2011; Guo and Yang, 2011; Bao *et al.*, 2009; Bao *et al.*, 2008; Wang and Xi, 2008b; Fu *et al.*, 2005; Xu *et al.*, 2005b), in Ayurvedic medicinal blends (Mishra *et al.*, 2013; Karpakavalli *et al.*, 2012; Kulkarni and Surana, 2012; Pandey *et al.*, 2010; Rai *et al.*, 2009a), in polyherbal formulation (Rai *et al.*, 2009b), and in a number of other Chinese (Wang *et al.*, 2014b; Zhang and Wang, 2014; Cui *et al.*, 2014a; Pei *et al.*, 2013; Wang *et al.*, 2012b; Chen and Ji, 2012; Zhang *et al.*, 2012a; Ye *et al.*, 2011; Hu *et al.*, 2011a; Zhu *et al.*, 2010; Zhang *et al.*, 2010a; Xin *et al.*, 2009; Yao *et al.*, 2007; Zhang *et al.*, 2006; Zhong *et al.*, 2006) and Indian formulated medicinal products (Kapuriya *et al.*, 2012). It should be noted that the vast majority of these short papers are using C18 columns and an isocratic mobile phase made of 70 : 30 or 77 : 23 (v/v) methanol : water. Some of these groups used 342 nm instead of 343 nm. It is also worth highlighting the more thorough validation work presented by Rai and colleagues, where their method was validated for linearity, precision, accuracy, specificity, system suitability, and robustness (Rai *et al.*, 2009b).

Even though piperine λ_{max} is 343 nm, several other wavelengths were reported for HPLC detection of piperine: 230 nm (Bao *et al.*, 2013a), 240 nm (Kamal *et al.*, 2012), 254 nm (Chauhan *et al.*, 2008), 270 nm (Li *et al.*, 2011), 280 nm (Lin *et al.*, 2011; Kulkarni and Surana, 2011), 290 nm (Ni and Yang, 2012), 328.5 nm (Quan *et al.*, 2005), 335 nm (Sethi *et al.*, 2009), and 340 nm (Kakarala *et al.*, 2010; Nagappan *et al.*, 2009). Unlike other groups, Sethi *et al.* conducted a more thorough validation study on their method.

Owing to the popularity and major economic stability of the *Piper* genus, adulteration is a concern. HPLC has been successfully used to check for adulteration of ground black pepper (*Piper nigrum*) with papaya seeds (Menghani *et al.*, 2010). Ultimately, this team is trying to establish libraries of HPLC chromatograms that could be referenced as typical fingerprints when evaluating and authenticating black pepper samples.

HPLC with mass spectrometry detection of piperine has been reported a few times over the past 10 years. The fragmentation patterns of piperine by positive ESI-MS, MS/MS, and APCI-MS (atmospheric pressure chemical ionization-MS) were reported (Khajuria *et al.*, 2013). Piperine, with four other alkaloids, has been detected by HPLC-ESI-MS/MS in *Piper longum* L. and *Piper nigrum* L. (Liu *et al.*, 2015), and by LC-MS in *Piper nigrum* L. and in wild *Piper* species found in Central America (Scott *et al.*, 2005). Friedman *et al.* used LC-MS to analyze piperine and four other piperamides in ten commercial samples of whole peppercorns and ten samples of ground black, white, green, and red peppers (Friedman *et al.*, 2008). LC-MS, combined with LC-UV at 280 nm, was used to identify and quantify piperine in Korean food products containing both chilis and black pepper (Choi *et al.*, 2006). Curcumin and piperine were simultaneously detected in food products by LC-MS on a C18 column under isocratic conditions (Nagappan *et al.*, 2011). Piperine can undergo light-induced isomerizations, where one of the trans-aliphatic double bonds can convert into the cis-configuration. LC-MS was used to monitor the kinetics of cis–trans isomerization of piperine under a number of different lighting conditions, for example, fluorescent light, sunlight, and UV at 254 nm (Kozukue *et al.*, 2007). Reversed-phase LC-MS/MS of piperine was reported in Longhu Rendan pills (Wang *et al.*, 2014c) and in gutter oils (Wu *et al.*, 2012b). Lastly, piperine has been studied by LC-MS/MS in complex matrices such as human plasma (Wang *et al.*, 2012e) and rat tissues and plasma (Yin *et al.*, 2014; Liu *et al.*, 2013a; Basu *et al.*, 2013; Liu *et al.*, 2011b).

It is worth mentioning that HPLC analysis of piperine combined with fluorescence detection is rarely reported in the recently published scientific literature, even though piperine fluoresces (excitation wavelength of 339 nm and emission wavelength of 450 nm) and this means of detection is both very selective and sensitive (Patel and Vyas, 2012).

Piperine and curcumin (the main curcuminoid found in turmeric) are often found together in formulated products, due to the poor bioavailability of curcumin when used alone. Indeed, curcumin metabolizes rapidly in the liver and intestinal wall. The addition of piperine helps minimize curcumin metabolism, thus resulting in higher bioavailability of the curcuminoid. Besides the previously aforementioned works (Sethi *et al.*, 2009; Nagappan *et al.*, 2009; Zhong *et al.*, 2006), a number of other methods have been developed to separate and detect piperine in the presence of curcumin: HPTLC detection of piperine and curcumin in a transdermal patch used to treat arthritis (Vaykole *et al.*, 2014); RP-HPLC on a C18 column with detection at 262 nm, validated as per ICH (International Conference on Harmonisation) guidelines, with an LOD and LOQ of 0.1 and 0.9 ppm of piperine, respectively (Moorthi *et al.*, 2013); a validated RP-HPLC method on a C8 column, using ESI-MS/MS for detection of piperine and curcumin in human plasma, where the lowest LOQ for piperine was 0.50 ng ml^{-1} (Wang *et al.*, 2012e); an RP-HPLC method on a C18 column using APCI-MS for separation and detection of curcumin and piperine in food products where a mobile phase comprised mostly of 95% (by volume) acetonitrile and 5% of an aqueous solution of formic acid was chosen (Nagappan *et al.*, 2011); a micellar LC method using a C18 column with piperine detection at 210 nm in the presence of curcumin and capsaicin, with a piperine LOD and LOQ of 5 ng ml^{-1} and 15 ng ml^{-1}, respectively (Chin-Chen *et al.*, 2010).

As UPLC instruments are becoming more widespread in laboratories, replacing HPLC systems, it comes as no surprise that a handful of methods for the analysis of piperine has been developed to take advantage of the many benefits offered by UPLC technology: UPLC-MS/MS on a C18 column for detection of piperine in rat plasma (Yin *et al.*, 2014); UPLC on a C18 column with 338 nm detection, for detection of piperine in dietary supplements (Hasegawa *et al.*, 2012; Bao *et al.*, 2013b); UPLC with UV at 343 nm for the detection of piperine in pepper (Yang *et al.*, 2012b).

Owing to their simplicity, thin-layer chromatography techniques (TLC, HPTLC) are popular for the analysis of piperine. It has been performed on cultivars of *Piper nigrum* (De Mey *et al.*, 2014; Shingate *et al.*, 2013; Marutoiu *et al.*, 2006), and *Piper longum* (Rajopadhye *et al.*, 2012; Yu *et al.*, 2009). Similarly to HPLC analysis, TLC and HPTLC have been used to measure piperine, an active ingredient in a number of medicinal treatments or food preparations, for example, in several Ayurvedic formulations (Jain *et al.*, 2014; Shivani *et al.*, 2013a; Shivani *et al.*, 2013b; Vyas *et al.*, 2011; Gupta and Jain, 2011c; Patel *et al.*, 2010; Patra and Kumar, 2010), traditional Chinese medicines (Mirensha *et al.*, 2014; Wang *et al.*, 2012b; Hu *et al.*, 2011a; Jiang *et al.*, 2008), and miscellaneous medicinal products (Vaykole *et al.*, 2014; Samten *et al.*, 2010; Shanmugasundaram *et al.*, 2008). A validated HPTLC method (where linearity, reproducibility, specificity, accuracy, precision, robustness, and ruggedness were determined) has been reported to measure piperine and gallic acid in herbal formulations (De *et al.*, 2013). In yet another paper, HPTLC has been used to detect trace amounts of piperine in pepper-contaminated cinnamon bark powder (Gopu *et al.*, 2008).

While not as popular as HPLC or TLC, GC-MS has been reported for the analysis of piperine in a handful of supercritical CO_2 black pepper extracts. For instance, in a sample from the Chinese province of Hainan, piperine was 39.47% of the peaks detected by GC-MS (Wang *et al.*, 2013d). Besides TLC, Marutoiu *et al.* also used GC-MS in their separation and identification of piperine in black pepper (Marutoiu *et al.*, 2006). In a supercritical CO_2 extract of a species unrelated to the *Piper* genus, piperine was detected at 4.68% by GC-MS (Wu *et al.*, 2005) and, in an ethanolic extract of yet another species not from the *Piper* genus, GC-MS was chosen to analyze the compounds extracted, of which piperine was present at 5.66% (Jenecius and Mohan, 2014). In a supercritical CO_2 extract of a Tibetan medicine called Pazhu, piperine was analyzed by GC-MS (Wei and Du, 2008a). In a different paper, Wei *et al.* analyzed the same capsules by HPLC (Wei *et al.*, 2008b).

Gupta and Jain reported piperine analysis in three Ayurvedic formulations, using three different analytical techniques, that is, HPTLC (Gupta and Jain, 2011c), RP-HPLC (Gupta and Jain, 2011b), and UV spectrophotometry (Gupta and Jain, 2011a).

Besides Gupta and Jain (Gupta and Jain, 2011a), other groups have used spectrophotometry to analyze piperine (Wu *et al.*, 2011a; Singh *et al.*, 2011; Pandey *et al.*, 2009; Maheswari *et al.*, 2008; Shukla *et al.*, 2007; Jain *et al.*, 2007). It should be noted that the same Indian group has used both spectrophotometry (Maheswari *et al.*, 2008) and HPTLC (Shanmugasundaram *et al.*, 2008) to measure piperine in herbal cough syrup. Another group from India has reported their work on developing a spectrophotometric method for the analysis of piperine using a first-order derivative at 238 nm (Khamar and Patel, 2012a) and a second-order derivative at 241 nm (Khamar and Patel, 2012b).

Piperine purification from black pepper extracts has been accomplished by a Chinese team, using HSCCC and preparative HPLC (Jin *et al.*, 2013; Qian *et al.*, 2011). Piperine was structurally identified by ESI-MS, ^{1}H and ^{13}C NMR. Another group purified piperine from an ethanolic extract of black pepper, and used HPLC, ^{13}C NMR, and FTIR to confirm the structure (Zarai *et al.*, 2013). Another group used high-performance centrifugal partition chromatography to isolate and purify piperine from black pepper, and then used ^{1}H and ^{13}C NMR for structural confirmation (Chen *et al.*, 2009b). Piperine was also purified from *Piper trioicum* Roxb (Mallikarjun *et al.*, 2012). A model studying the solubility of piperine in multiple solvents (ethanol, acetic acid, water, and HCl) was developed to optimize the extraction of piperine from black pepper in these solvent systems (Huang *et al.*, 2011b). The parameters for the supercritical CO_2 fluid extraction of piperine from *Piper nigrum* L. were optimized in terms of temperature, pressure, CO_2 volume, and extraction time (Wang *et al.*, 2008a).

Piperine has been analyzed by a few other analytical techniques, including the following: the aforementioned micellar LC (Chin-Chen *et al.*, 2010), capillary electrochromatography (Musenga *et al.*, 2007), pipette-tip ESI-MS (Wang *et al.*, 2014a), NIR (Lv *et al.*, 2013), and FTIR (Zarai *et al.*, 2013). Schulz *et al.* reported results from their thorough spectroscopic investigations of piperine in black and white ground pepper, green whole peppercorn, and pepper oleoresins, where they used NIR-FT-Raman, ATR-IR (attenuated total reflection-IR), and NIR (Schulz *et al.*, 2005).

11.10 Sanshools

Sanshools, a series of alkylamides, are the chemesthetic compounds responsible for the numbing and tingling effect characteristics of Szechuan peppers (*Zanthoxylum* or *Xanthoxylum* genus). The main sanshools are α-, β-, γ-, and δ-sanshools, with their corresponding hydroxy analogues, that is, hydroxy-α-sanshool, and so on. HPLC is the preferred and most popular method for the separation of sanshools and hydroxy sanshools. In the last 10 years, a fair number of research groups have published works on analytical methods for the separation, detection, and identification of sanshools and hydroxy-sanshools. It should be noted that, besides their chemesthetic properties, Szechuan peppers are also prized for their essential oils, described as having fresh, floral, spicy, and green notes. While these volatile compounds are not chemesthetic agents, and therefore fall outside of the scope of this chapter, the interested reader can learn more on the volatile profiles and the aroma constituents in the *Zanthoxylum* genus in recent published works (Jia *et al.*, 2011a; Yang, 2008; Iseli *et al.*, 2007; Yang, 2007).

Most recently, a research team reported their work on developing an RP-HPLC method on a C18 column, with UV detection at 270 nm, for the quantitative analysis of three hydroxy sanshools (α-, β-, and γ-) in *Z. bungeanum* Maxim. extracts (Cui *et al.*, 2014b). About 55–60% hydroxy sanshools were quantified in the sample, with the following breakdown: 41.2–46.9% hydroxy-α-sanshool, 8.2–9.3% hydroxy-β-sanshool, and 4.5–5.1% hydroxy-γ-sanshool. The UV spectra of sanshools and hydroxy sanshools have three absorption maxima around 270 nm, a feature characteristic of the aliphatic double bond structure found in these compounds (Iseli *et al.*, 2007). Iseli and colleagues also reported semi-quantitatively the detection, by HPLC-UV-MS with peak assignment confirmation by HPLC-NMR, of hydroxy-α-sanshool and hydroxy-β-sanshool in a CO_2 extract of *Z. schinifolium*. These were the predominant alkylamides found in the extract (Iseli *et al.*, 2007). Sanshools have also been detected by RP-HPLC with UV at 254 nm in *Z. bungeanum* Maxim and *Z. schinifolium* Sieb. et Zucc (Yu *et al.*, 2012). Yu and coworkers inferred that hydroxy-γ-sanshool could be used to differentiate the two *Zanthoxylum* species from each other.

Besides UV detection, a number of other detection techniques have been reported in the identification of sanshools. For instance, among other peaks, hydroxy-α-sanshool, hydroxy-β-sanshool, and the minor hydroxy-ε-sanshool, hydroxy-ζ-sanshool, and hydroxy-γ-isosanshool were analyzed in *Z. piperitum* by LC-MS, LC-TOF-MS, and 1D/2D NMR (Bader *et al.*, 2014). This paper marks the first time hydroxy-ζ-sanshool has been reported in the literature. The same year, a team from the Natural Plant Products Division of CSIR-Institute of Himalayan Bioresource Technology, Palampur, India published a paper where they developed a new UPLC method for the quantitative analysis of amides in *Z. armatum* (Kumar *et al.*, 2014). Photodiode array and electrospray MS/MS were used for the detection and identification of 16 compounds, including hydroxy-α-sanshool. The use of HPLC-MS/MS was also reported by a team from Sichuan University, Chengdu, China when identifying the major alkylamides in *Z. bungeanum* and *Z. schinifolium* (Zhao *et al.*, 2013). Hydroxy sanshools identified in the work were hydroxy-α-sanshool, hydroxy-β-sanshool, hydroxy-γ-sanshool, hydroxy-ε-sanshool, hydroxy-γ-isosanshool, with hydroxy-α-sanshool being the most abundant alkylamide. Interestingly, Zhao *et al.* used principal component analysis (PCA) on HPLC data to help classify *Z. bungeanum* and *Z. schinifolium*. From the PCA, they identified hydroxy-γ-sanshool and bungeanool (another compound typically found in the *Zanthoxylum* genus, but with no known chemesthetic properties) as potential fingerprinting markers to help differentiate *Z. bungeanum* and *Z. schinifolium*. In yet another paper, MS, IR, and 1D/2D NMR were used to confirm the structures of new alkylamides from the pericarp of *Z. bungeanum* (Huang *et al.*, 2012). Hydroxy-α-sanshool was included in this work. Yang analyzed a number of sanshool compounds by reversed-phase LC-UV-QTOF/MS (quadrupole TOF) on a C18 column (Yang, 2008). A number of NMR experiments (^{1}H, ^{13}C, HMQC, HMBC, and NOESY) were performed by Yang to help with structure elucidation. Further, as we already mentioned in the previous paragraph, Iseli *et al.* have also explored HPLC-MS for the detection of pungent principles in *Z. schinifolium* (Iseli *et al.*, 2007). Lastly, an LC-MS method has been developed for the analysis of (among other analytes) hydroxy-α-isosanshool, hydroxy-β-isosanshool, and γ-sanshool in human plasma and urine, following the ingestion of Daikenchuto, a traditional Japanese herbal (Kampo) medicine based on ancient Chinese medicine, made of *Zanthoxylum* fruit, processed dried ginger, ginseng, and malt sugar (Iwabu *et al.*, 2010).

Sugai and coworkers have contributed significantly to the knowledge base on alkylamides, by not only isolating pure sanshool compounds, determining their detection thresholds, and testing them as TRPV1 (transient receptor potential cation channel subfamily V member 1) agonists (Sugai *et al.*, 2005a; Sugai and Kubota, 2006), but also by looking at the quantitative distribution of sanshools in the leaves, flowers, and fruits of *Z. piperitum* DC., and the effect of fruit maturation stages on sanshool content (Sugai *et al.*, 2005b). Sanshool standards were isolated through a well described semi-preparative HPLC with a silica gel column and detection at 275 nm. Structures were confirmed by spectral analyses (NMR, MS). Quantitative analytical work was done by HPLC with a silica gel column (100 Å pore size) with UV detection set at 275 nm. While hydroxy-α-sanshool was only found at trace levels in young leaves and flowers, it was the main alkylamide in fruits. In young leaves and flowers, α-sanshool was the dominant chemesthetic agent. Further, the *Z. piperitum* DC. fruits had more chemesthetic effect than the young leaves and flowers. Finally, as the fruit of *Z. piperitum* DC. matured, Sugai and colleagues observed a decrease in the amount of α-sanshool and an increase in the levels of γ-sanshool, hydroxy-α-sanshool, and hydroxy-β-sanshool, while β-sanshool and δ-sanshool remained virtually unchanged at low levels (Sugai *et al.*, 2005b).

In addition to the aforementioned work by Sugai *et al.*, a few other groups have worked on small or preparative scale extractions, purification, and separation of sanshools. High-speed counter-current chromatography was successfully applied to isolate and purify hydroxy-α-sanshool, hydroxy-β-sanshool, and hydroxy-ε-sanshool (a minor hydroxy sanshool) in *Z. bungeanum* Maxim (Wang *et al.*, 2011d). Compound confirmation was achieved by MS, IR, and ^{1}H NMR analyses. These workers found 21.4% hydroxy sanshools in their *Z. bungeanum* Maxim extract, with hydroxy-α-sanshool being the highest (15.7%). A Chinese team patented a process for the extraction of hydroxy sanshool from *Z. bungeanum* oil using preparative liquid chromatography with UV detection at 270 nm (Zhao *et al.*, 2012b). The Japanese company Tsumura & Co., a company specializing in Kampo medicine, patented a method to produce sanshools (α-sanshool, β-sanshool, hydroxy-α-sanshool, hydroxy-β-sanshool, etc.) with structural confirmations through ^{1}H and ^{13}C NMR (Aoki, 2013). Lastly, a South Korean team from Seoul National University isolated, from the seeds of *Z. piperitum*, hydroxy-α-sanshool, hydroxy-β-sanshool, hydroxy-ε-sanshool, and two new compounds identified as dihydroxy-α-sanshool and dihydroxy-β-sanshool (Jang *et al.*, 2008). These workers performed the lab scale separation using reversed-phase vacuum flash chromatography with a C18 stationary phase. Peak elucidation was confirmed by IR, UV profiles, high resolution fast-atom bombardment mass spectrometry (HR FAB-MS), and ^{1}H and ^{13}C NMR experiments.

11.11 Spilanthol

The alkylamide spilanthol is another chemesthetic agent imparting tingling and numbing effects, similarly to sanshools (see Section 11.10). Spilanthol, a volatile compound, is found in species from the *Acmella* genus in general and in particular in paracress, also known as jambu, or toothache plant (*Acmella oleracea*, also commonly known as *Spilanthes oleracea* L., or *Spilanthes acmella*). Spilanthol has been analyzed through an extensive number of NMR experiments (^{1}H, ^{13}C, DEPT, COSY, HSQC, and HMBC) and mass spectrometric characterization with electron ionization (EI) and FAB-MS in a paper from a Mexican research team (Lopez-Martinez *et al.*, 2011). Spilanthol can be detected using either liquid or gas chromatography.

Eleven *N*-alkylamides, including spilanthol, have been profiled by HPLC/ESI-MS in *Spilanthes acmella* (Boonen *et al.*, 2010). These workers, operating the mass spectrometer in MS/MS mode with the use of collision induced dissociation, were able to perform structural elucidations of these analytes, and included molecular weight and characteristics fragmentation patterns of *N*-alkylamides. Compound separation was achieved using a gradient method, in RP-HPLC on a C18 stationary phase. In addition to MS detection, they used UV detection at 237 nm, the absorption maximum of spilanthol. A group from The University of North Carolina

Greensboro, NC, USA, has been working on a validated isocratic RP-HPLC/ESI-MS method to separate, identify, and quantify alkylamides, including spilanthol, in *Spilanthes acmella* (Bae *et al.*, 2010; Stinson *et al.*, 2006). UV absorbance at 229 nm was used to detect spilanthol, which was isolated by semi-preparative HPLC from hydroethanolic extracts of the plant, and used as a standard for quantitation. The calibration range was 0.45–450 μM, with an LOD of 0.27 μM, and an LOQ of 0.45 μM. Structural elucidations were done based on MS-MS fragmentation patterns, and NMR experiments. In a different study, spilanthol was quantified by HPLC-MS in *in vitro* cultures of *Spilanthes acmella* Murr. (Singh and Chaturvedi, 2012). From the data, higher levels of spilanthol were reported in the leaves of *in vitro* plants (3294 μg spilanthol/g of dry weight leaves) than in *in vivo* plants (2704 μg spilanthol/g of dry weight leaves). Spilanthol was chosen as a model compound in the evaluation of the effects that injection volume has on quantitative analyses performed in fused-core chromatography (Boonen *et al.*, 2013).

As we alluded earlier in this section, spilanthol, due to its volatile nature, is also easily detected and quantified by GC. Spilanthol has been measured by GC-MS in the essential oils of the leaves and inflorescences of *Spilanthes oleracea* cv. Jambuarana (Borges *et al.*, 2012), in young *Acmella radicans* plants (Cortez-Espinosa *et al.*, 2011), in the plant, flower heads, an *in vitro* plantlets of *Spilanthes acmella* (Leng *et al.*, 2011), and in the essential oil of *Spilanthes acmella* Murr. (Jirovetz *et al.*, 2005). The last group reported 2.5% spilanthol in *Spilanthes acmella* Murr.

In a couple of instances, liquid and gas chromatography were both applied to identify spilanthol. Recently, a Chinese group published their work where headspace SPME coupled with GC-MS, and HPLC with APCI-IT-TOF-MS^n (atmospheric pressure chemical ionization–ion trap–time of flight–multistage mass spectrometry) were both used to analyze a number of compounds in two complex matrices, that is, rabbit plasma and cerebrospinal fluid. Spilanthol was one of the 47 analytes to be monitored (Li *et al.*, 2014). In a different effort, spilanthol was fractionated by preparative HPLC from commercial Jambu oleoresin and was analyzed by LC-MS under APCI conditions, and GC-MS in the EI mode, with additional structural work performed by ^1H and ^{13}C NMR, and high-resolution mass spectrometry (Ley *et al.*, 2006).

11.12 Conclusions

Chemesthetic compounds have been analyzed by a number of techniques. In the last 10 years, due to its wide adoption in analytical laboratories, chromatography (mostly GC, LC, and TLC) is the most popular technique reported in the published peer-reviewed literature for the separation and quantitation of chemesthetic compounds in a variety of matrices, including raw spices, extracts, food products, medicine, biological fluids, and so on. The variety of available detectors (UV, ECD, fluorescence, MS, MS/MS) adds to the strength and versatility of the chromatographic approach by providing ease of use, specificity, and sensitivity.

Often complementing chromatography, several spectroscopic techniques have also been used to analyze and characterize chemesthetic compounds, for example, IR, NIR, FTIR, ^1H, and ^{13}C NMR.

With the current and future commercialization of newer and more advanced analytical instruments and detectors, laboratories will have faster and more sensitive tools at their disposal to analyze chemesthetic compounds.

Abbreviations

AccuTOF	Accurate time of flight
AITC	Allyl isothiocyanate
APCI	Atmospheric pressure chemical ionization
ATR	Attenuated total reflection
C	Capsaicin
CV	Cyclic voltammetry
DART	Direct analysis in real time
DC	Dihydrocapsaicin
ECD	Electrochemical detection
EI	Electron ionization
EIS	Electrochemical impedance spectroscopy
FAB	Fast-atom bombardment
FID	Flame ionization detection
FTIR	Fourier transform infrared
GC	Gas chromatography
GC×GC	Comprehensive 2D gas chromatography
HC	Homocapsaicin
HDC	Homodihydrocapsaicin
HPLC	High-performance liquid chromatography
HPTLC	High-performance thin-layer chromatography
HSCCC	High-speed countercurrent chromatography
ICH	International Conference on Harmonisation
IEC	Ion-exclusion chromatography
IR	Infrared
LC	Liquid chromatography
LOD	Limit of detection
LOQ	Limit of quantitation
MALDI	Matrix-assisted laser desorption/ionization
MS	Mass spectrometry
NDC	Nordihydrocapsaicin
NIR	Near-infrared
NMR	Nuclear magnetic resonance
PCA	Principal component analysis
PDMS	Polydimethylsiloxane
RP-HPLC	Reversed-phase high-performance liquid chromatography
SPME	Solid phase microextraction

TLC	Thin-layer chromatography
TOF	Time of flight
TRPV1	Transient receptor potential cation channel subfamily V member 1
UPLC	Ultra-performance liquid chromatography
UV	Ultraviolet

References

Al Othman, Z., Ahmed, Y., Habila, M., and Abdel Ghafar, A., 2011. Determination of capsaicin and dihydrocapsaicin in Capsicum fruit samples using high performance liquid chromatography. *Molecules*, **16**, 8919–8929.

Alam, P., 2013. Densitometric HPTLC analysis of 8-gingerol in Zingiber officinale extract and ginger-containing dietary supplements, teas and commercial creams. *Asian Pacific Journal of Tropical Biomedicine*, **3** (8), 634–638.

Alam, P. *et al.*, 2012. HPTLC method for qualitative and quantitative estimation of eugenol from Ocimum sanctum Linn in polyherbal formulation. *Pharmacie Globale (International Journal of Comprehensive Pharmacy)*, **3** (10), 1–3.

Al-Bayati, F.A., and Mohammed, M.J., 2009. Isolation, identification, and purification of cinnamaldehyde from Cinnamomum zeylanicum bark oil. An antibacterial study. *Pharmaceutical Biology*, **47** (1), 61–66.

Alea, A.C.H., Co, D.E.T., and Noel, M.G., 2014. *An Investigative Study of Reactions Involving Glucosinolates and Isothiocyanates*. Proceedings of the De La Salle University, 2014 Research Congress. Manila, The Philippines, http://www.dlsu.edu.ph/conferences/dlsu_research_congress/2014/food-nutrition-proceedings.asp http://www.dlsu.edu.ph/conferences/dlsu_research_congress/2014/food-nutrition-proceedings.asp (accessed 3 October 2015).

Alothman, Z. *et al.*, 2012. Determination of capsaicinoids in Capsicum species using ultra performance liquid chromatography-mass spectrometry. *Journal of Separation Science*, **35** (21), 2892–2896.

Alqasoumi, S.I., 2009. Quantification of 6-gingerol in Zingiber officinale extract, ginger-containing dietary supplements, teas and commercial creams by validated HPTLC densitometry. *FABAD Journal of Pharmaceutical Sciences*, **34**, 33–42.

Anandjiwala, S., Kalola, J., and Rajani, M., 2006. Quantification of eugenol, Luteolin, ursolic acid, and oleanolic acid in black (Krishna Tulasi) and green (Sri Tulasi) varieties of Ocimum sanctum Linn. using high-performance thin-layer chromatography. *Journal of AOAC International*, **89** (6), 1467–1474.

Antonious, G.F. and Jarret, R.L., 2006. Screening Capsicum accessions for capsaicinoids content. *Journal of Environmental Science and Health, Part B: Pesticides, Food Contaminants, and Agricultural Wastes*, **41** (5), 717–729.

Aoki, K., 2013. Method for producing sanshool. Japan, Patent No. EP2644612 A1 (similar patents in China and USA), European Patent 2,644,612, filed Mar. 23, 2013 and issued Oct. 2, 2013.

Association of Official Analytical Chemists. 2012. *Official Methods of Analysis of AOAC International*, 19th edn, Gaithersburg, MD: AOAC International, Official Method 995.03.

Bader, M. *et al.*, 2014. All-trans-configuration in Zanthoxylum alkylamides swaps the tingling with a numbing sensation and diminishes salivation. *Journal of Agricultural and Food Chemistry*, **62** (12), 2479–2488.

Bae, H. *et al.*, 2013. Simultaneous quantification of capsaicinoids and ascorbic acid from pungent peppers. *Journal of Chromatographic Science*, **51** (5), 412–418.

Bae, S.S., Ehrmann, B.M., Ettefagh, K.A., and Cech, N.B., 2010. A validated liquid chromatography-electrospray ionization-mass spectrometry method for quantification of spilanthol in Spilanthes acmella (L.) Murr. *Phytochemical Analysis*, **21** (5), 438–443.

Bailey-Shaw, Y.A. *et al.*, 2008. Changes in the contents of oleoresin and pungent bioactive principles of Jamaican ginger (Zingiber officinale Roscoe.) during maturation. *Journal of Agricultural and Food Chemistry*, **56** (14), 5564–5571.

Bao, B., Qi, G., and Yang, H., 2008. Determination of piperine in Qiqirigan-8 by HPLC. *Zhongguo Yaoshi (Beijing, China)*, **22** (8), 699–700.

Bao, G.-H., Kang, T.-G., Bagana, and Bai, M.-G., 2009. Determination of piperine in Ganlu Yigan capsule by HPLC. *Zhonghua Zhongyiyao Xuekan*, **27** (9), 2001–2002.

Bao, G.-R., Ling, X., Meng, H., and Gereltu, B., 2013b. UPLC simultaneous determination of piperine and piperlonguminine in Mongolian medicine Zhuangxi-6. *Yaowu Fenxi Zazhi*, **33** (7), 1223–1226.

Bao, G.-R., Meng, H., Hu, C.-F., and Bo, G., 2013a. Simultaneous determination of piperine and piperlonguminine in Mongolian medicine tonglaga-5 by HPLC. *Guoji Yaoxue Yanjiu Zazhi*, **40** (1), 108–110.

Bao, L., Wu, L., Na, S., and Fan, Y., 2014. Content determination of piperine in Monglia medicine harigaburi-10 tablets. *Zhongguo Shiyan Fangjixue Zazhi*, **20** (1), 81–83.

Barbero, G.F., Liazid, A., Palma, M., and Barroso, C.G., 2007. Fast determination of capsaicinoids from peppers by high-performance liquid chromatography using a reversed phase monolithic column. *Food Chemistry*, **107** (3), 1276–1282.

Barbero, G.F., Palma, M., and Barroso, C.G., 2006a. Determination of capsaicinoids in peppers by microwave-assisted extraction-high-performance liquid chromatography with fluorescence detection. *Analytica Chimica Acta*, **578** (2), 227–233.

Barbero, G.F., Palma, M., and Barroso, C.G., 2006b. Pressurized liquid extraction of capsaicinoids from peppers. *Journal of Agricultural and Food Chemistry*, **54** (9), 3231–3236.

Basu, S., Patel, V.B., Jana, S., and Patel, H., 2013. Liquid chromatography tandem mass spectrometry method (LC-MS/MS) for simultaneous determination of piperine, cinnamic acid and gallic acid in rat plasma using a polarity switch technique. *Analytical Methods*, **5** (4), 967–976.

Bi, Y. *et al.*, 2012. Simultaneous determination of piperine and piperlongumine in Piper longum by HPLC. *Zhongguo Shiyan Fangjixue Zazhi*, **18** (1), 47–50.

Bijttebier, S. *et al.*, 2014. Generic characterization of apolar metabolites in red chili peppers (*Capsicum frutescens* L.) by Orbitrap mass spectrometry. *Journal of Agricultural and Food Chemistry*, **62** (20), 4812–4831.

Boonen, J. *et al.*, 2010. LC-MS profiling of N-alkylamides in Spilanthes acmella extract and the transmucosal behaviour of its main bio-active spilanthol. *Journal of Pharmaceutical and Biomedical Analysis*, **53** (3), 243–249.

Boonen, J. *et al.*, 2013. A critical quality parameter of injection volume in quantitative fused-core chromatography. *Journal of Pharmaceutical Analysis*, **3** (5), 330–334.

Borges, L.d. *et al.*, 2012. Influence of organic and mineral soil fertilization on essential oil of Spilanthes oleracea cv. Jambuarana. *American Journal of Plant Physiology*, **7** (3), 135–142.

Cadieux, E.J.J., 2015. Menthol detection on tobacco. US Patent 20,150,008,162, filed Sept. 24, 2014 and issued Jan. 8, 2015.

Cantalapiedra, A., Gismera, M.J., Sevilla, M.T., and Procopio, J.R., 2014. Sensitive and selective determination of phenolic compounds from aromatic plants using an electrochemical detection coupled with hplc method. *Phytochemical Analysis*, **25** (3), 247–254.

Cao, J. *et al.*, 2013. Determination of cinnamaldehyde in *Cinnamomi ramulus* by high pressure liquid chromatography. *Shipin Anquan Zhiliang Jiance Xuebao*, **4** (2), 463–466.

Carrillo, J.D. and Tena, M.T., 2006. Determination of volatile oak compounds in aged wines by multiple headspace solid-phase microextraction and gas chromatography-mass spectrometry (MHS-SPME-GC-MS). *Analytical and Bioanalytical Chemistry*, **385** (5), 937–943.

Cevallos, P.A.P., Buera, M.P., and Elizalde, B.E., 2010. Encapsulation of cinnamon and thyme essential oils components (cinnamaldehyde and thymol) in β-cyclodextrin: effect of interactions with water on complex stability. *Journal of Food Engineering*, **99** (1), 70–75.

Chauhan, R., Dwivedi, J., and Siddiqui, A.A., 2008. Chemical standardization and quantification of piperin from methanolic extract of Piper nigrum by HPLC method on the basis of isolated markers. *International Journal of Chemical Sciences*, **6** (4), 1726–1733.

Chen, D. *et al.*, 2005. Quality analysis of Shenrongguben pills. *Zhongguo Yiyuan Yaoxue Zazhi*, **25** (5), 468–470.

Chen, H., 2007. Detection of eugenol in zhenzhutongluojiaonang by HPLC. *Qinghai Yiyao Zazhi*, **37** (1), 51–53.

Chen, H.-J. and Ji, L.-J., 2012. Content determination of piperine in Shiwei Shoushen powder by RP-HPLC. *Zhongguo Shiyan Fangjixue Zazhi*, **18** (13), 122–124.

Chen, X. *et al.*, 2011. Simultaneous determination of four components in Guizhi-Gancao decoction by HPLC with double-wavelength switch. *Zhongguo Xiandai Yingyong Yaoxue*, **28** (3), 265–267.

Chen, Y., He, R., Zeng, M., and Yan, H., 2009a. Optimizing the extraction condition of ginger oil by supercritical CO_2 fluid by orthogonal test. *Zhongguo Yaoshi*, **12** (9), 1196–1197.

Chen, Y., Zhou, D., and Zhang, M., 2014. Extraction of effective ginger compositions by three different methods and determination of 6-gingerol. *Zhongguo Yaoshi (Wuhan, China)*, **17** (7), 1099–1102.

Chen, Z. *et al.*, 2009b. One step purification of piperine directly from *Piper nigrum* L. by high performance centrifugal partition chromatography. *Separation Science and Technology*, **44** (8), 1884–1893.

Cheng, X.-L. *et al.*, 2013. Highly efficient sample preparation and quantification of constituents from traditional Chinese herbal medicines using matrix solid-phase dispersion extraction and UPLC-MS/MS. *Analyst*, **138** (8), 2279–2288.

Chin-Chen, M.-L., Carda-Broch, S., Bose, D., and Esteve-Romero, J., 2010. Direct injection and determination of the active principles of spices using micellar liquid chromatography. *Food Chemistry*, **120** (3), 915–920.

Choi, S.-H. *et al.*, 2006. Analysis of the contents of pungent compounds in fresh Korean red peppers and in pepper-containing foods. *Journal of Agricultural and Food Chemistry*, **54** (24), 9224–9231.

Compton, R.G., Wildgoose, G.G., and Kachoosangi, R.T., 2009. Method of electrochemically detecting a capsaicinoid in a sample. World Patent 2,009,115,840, filed March 17, 2009 and issued Sept. 24, 2009.

Cordero, C., Bicchi, C., Joulain, D., and Rubiolo, P., 2007. Identification, quantitation and method validation for the analysis of suspected allergens in fragrances by comprehensive two-dimensional gas chromatography coupled with quadrupole mass spectrometry and with flame ionization detection. *Journal of Chromatography A*, **1150** (1–2), 37–49.

Cordero, C. *et al.*, 2012. HS-SPME-GC×GC-qMS volatile metabolite profiling of Chrysolina herbacea frass and Mentha spp. leaves. *Analytical and Bioanalytical Chemistry*, **402** (5), 1941–1952.

Cortez-Espinosa, N. *et al.*, 2011. Valine and phenylalanine as precursors in the biosynthesis of alkamides in Acmella radicans. *Natural Product Communications*, **6** (6), 857–861.

Crespo, E. *et al.*, 2012. On-line detection of root-induced volatiles in Brassica nigra plants infested with *Delia radicum* L. root fly larvae. *Phytochemistry*, **84**, 68–77.

Cui, C. *et al.*, 2014a. Determination of piperine capsules by HPLC. *Zhongguo Yaoshi (Wuhan, China)*, **17** (1), 77–79.

Cui, M.-Y. *et al.*, 2014b. Simultaneous determination of three sanshools in Zanthoxylum bungeanum Maxim. extract by RP-HPLC. *Shenyang Yaoke Daxue Xuebao*, **31** (4), 272–275.

Davis, C.B., Markey, C.E., Busch, M.A., and Busch, K.W., 2007. Determination of capsaicinoids in habanero peppers by chemometric analysis of UV spectral data. *Journal of Agricultural and Food Chemistry*, **55** (15), 5925–5933.

de Jesus Ornelas-Paz, J. *et al.*, 2010. Effect of cooking on the capsaicinoids and phenolics contents of Mexican peppers. *Food Chemistry*, **119** (4), 1619–1625.

De, S., Nariya, P., and Jirankalgikar, N., 2013. A rapid validated high-performance thin-layer chromatographic-densitometric method for the simultaneous estimation of different chemcial-nature compounds piperine and gallic acid in pharmaceutical dosage forms. *Journal of Planar Chromatography - Modern TLC*, **26** (4), 325–340.

De Mey, E. *et al.*, 2014. Application of accelerated solvent extraction (ASE) and thin layer chromatography (TLC) to determination of piperine in commercial samples of pepper (*Piper nigrum* l.). *Journal of Liquid Chromatography & Related Technologies*, **37** (20), 2980–2988.

Dhalwal, K., Shinde, V.M., Mahadik, K.R., and Namdeo, A.G., 2007. Rapid densitometric method for simultaneous analysis of umbelliferone, psoralen, and eugenol in herbal raw materials using HPTLC. *Journal of Separation Science*, **30** (13), 2053–2058.

Dighe, V.V. *et al.*, 2005. Quantification of eugenol in Cinnamomum tamala nees and eberm. Leaf powder by high-performance thin-layer chromatography. *Journal of Planar Chromatography – Modern TLC*, **18** (104), 305–307.

Dong, Y., Lu, N., and Cole, R.B., 2013. Analysis of the volatile organic compounds in Cinnamomum cassia bark by direct sample introduction thermal desorption gas chromatography-mass spectrometry. *Journal of Essential Oil Research*, **25** (6), 458–463.

Duan, Y. *et al.*, 2013. Analysis of volatiles in sauced beef by microwave-assisted extraction coupled with solvent-assisted flavor evaporation and gas chromatography-mass spectrometry. *Shipin Kexue (Beijing, China)*, **34** (14), 250–254.

Ertas, E. *et al.*, 2007. Determination of eugenol contents of some zinc oxide-eugenol based dental cements. *Asian Journal of Chemistry*, **19** (4), 3105–3112.

Fan, H., Ding, M., Tanaka, K., and Hu, W., 2005. Analysis of carbonic acid in water samples by ion-exclusion chromatography with pure water as eluent. *Analytical Sciences*, **21** (2), 121–123.

Feng, K. and Yan, D., 2007. Determination of 6-gingerol in the supercritical fluid extract of Rhizoma Zingiberis Recens by HPLC. *Zhongchengyao*, **29** (7), 1041–1043.

Feng, Q. *et al.*, 2014a. A novel way for detection of eugenol via poly(diallyldimethylammonium chloride) functionalized graphene-MoS_2 nano-flower fabricated electrochemical sensor. *Sensors and Actuators, B: Chemical*, **192**, 1–8.

Feng, X. *et al.*, 2014b. HPLC fingerprint analysis combined with chemometrics for pattern recognition of ginger. *Pharmaceutical Biology* , **52** (3), 362–367.

Friedman, M. *et al.*, 2008. Analysis by HPLC and LC/MS of pungent piperamides in commercial black, white, green, and red whole and ground peppercorns. *Journal of Agricultural and Food Chemistry*, **56** (9), 3028–3036.

Fu, Y., Zhao, C., Wu, D., and Wen, A., 2005. Determination of piperine in Mongolian medicine Huangbaibaweisan by HPLC. *Zhongguo Yiyuan Yaoxue Zazhi*, **25** (11), 1095–1096.

Gahungu, A. *et al.*, 2011. Volatile compounds and capsaicinoid content of fresh hot peppers (Capsicum chinense) scotch bonnet variety at red stage. *Advance Journal of Food Science and Technology*, **3** (3), 211–218.

Gao, L. and Zheng, J., 2011. Determination of piperine in Namujige-14 by HPLC. *Zhongguo Yaoshi (Beijing, China)*, **25** (8), 804–805, 815.

Gao, Y. *et al.*, 2013. Quality control method for analyzing Chinese medicinal capsule comprising Angelica and Cnidium and others for treating male premature ejaculation diseases. Chinese Patent 103,018,395, filed Jan. 1, 2013 and issued April 3, 2013.

Gao, Y., Zhong, S., Zhou, C., and Chen, L., 2005. Determination of capsaicin in chilli (*Capsicum annuum* L.) by HPLC. *Tianran Chanwu Yanjiu Yu Kaifa*, **17** (6), 797–799.

Garces-Claver, A. *et al.*, 2006. Determination of capsaicin and dihydrocapsaicin in capsicum fruits by liquid chromatography-electrospray/time-of-flight mass spectrometry. *Journal of Agricultural and Food Chemistry*, **54** (25), 9303–9311.

Ge, Q., Song, M., and Gu, L., 2010. Using ESR for measuring DPPH free radical scavenging activity of gingerol oleoresin. *Shipin Kexue*, **31** (15), 121–125.

Gholivand, M.B. and Ahmadi, F., 2008. Simultaneous determination of trans-cinnamaldehyde and benzaldehyde in different real samples by differential pulse polarography and study of heat stability of trans-cinnamaldehyde. *Analytical Letters*, **41** (18), 3324–3341.

Gonzalez-Zamora, A. *et al.*, 2013. Characterization of different capsicum varieties by evaluation of their capsaicinoids content by high performance liquid chromatography, determination of pungency and effect of high temperature. *Molecules*, **18** (11), 13471–13486.

Gopu, C.L. *et al.*, 2008. Simultaneous determination of cinnamaldehyde, eugenol, and piperine by HPTLC densitometric method. *Phytochemical Analysis*, **19** (2), 116–121.

Gu, Y. *et al.*, 2011. Determination of capsaicin and dihydrocapsaicin in pepper by high performance liquid chromatography. *Huaxue Fenxi Jiliang*, **20** (6), 74–76.

Guedes De Pinho, P. *et al.*, 2009. Volatile composition of *Catharanthus roseus* (L.) G. Don using solid-phase microextraction and gas chromatography/mass spectrometry. *Journal of Pharmaceutical and Biomedical Analysis*, **49** (3), 674–685.

Guo, C. *et al.*, 2006. Fourier transform vibrational circular dichroism from 800 to 10,000cm^{-1}: Near-IR-VCD spectral standards for terpenes and related molecules. *Vibrational Spectroscopy*, **42** (2), 254–272.

Guo, F.-Y. *et al.*, 2013. The determination of piperine in Hainan Piper nigrum by HPLC. *Shipin Yanjiu Yu Kaifa*, **34** (1), 74–76.

Guo, H. and Yang, H., 2011. HPLC determination of piperine in Mongolian medicine Tasiyin-10 pills. *Baotou Yixueyuan Xuebao*, **27** (2), 20–21.

Gupta, V. and Jain, U.K., 2011a. Quantitative analysis of piperine in ayurvedic formulation by UV spectrophotometry. *International Journal of Pharma Sciences and Research*, **2** (2), 58–61.

Gupta, V. and Jain, U.K., 2011b. Quantitative estimation of piperine in Pancasama churna by RP-HPLC. *Pharmacia Lettre*, **3** (1), 400–406.

Gupta, V. and Jain, U.K., 2011c. Status of piperine content in Ayurvedic formulation: method standardization by HPTLC. *Research Journal of Pharmaceutical, Biological and Chemical Sciences*, **2** (2), 524–532.

Ha, J., Han, K.-J., Kim, K.-J., and Jeong, S.-W., 2008. Gas chromatographic analysis of capsaicin in Gochujang. *Journal of AOAC International*, **91** (2), 387–391.

Ha, J. *et al.*, 2010b. Rapid method for the determination of capsaicin and dihydrocapsaicin in Gochujang using ultra-high-performance liquid chromatography. *Journal of AOAC International*, **93** (6), 1905–1911.

Ha, J. *et al.*, 2010a. Determination of capsaicinoids in foods using ultra high performance liquid chromatography. *Food Science and Biotechnology*, **19** (4), 1005–1009.

Han, C. *et al.*, 2012a. Microwave-assisted extraction and determination of three capsaicinoids in Capsicum annuum by HPLC-MS/MS. *Yaowu Fenxi Zazhi*, **32** (8), 1365–1369.

Han, Y. *et al.*, 2013. UPLC fingerprint comparison among fresh, dried and processed ginger from different places of origin. *Zhongchengyao*, **35** (2), 356–359.

Han, Y. *et al.*, 2012c. UPLC characteristic fingerprint of rhizoma zingiberis by different drying technology. *Zhongchengyao*, **34** (6), 987–990.

Han, Y., Shi, L., Jin, S., and Zhang, J., 2012b. RP-HPLC determination of 4 capsaicinoids in chili peppers. *Shipin Kexue (Beijing, China)*, **33** (16), 257–260.

Hasegawa, T.*et al.*, 2012. Determination of piperine in dietary supplements containing black pepper using UPLC. *Chiba-ken Eisei Kenkyusho Nenpo*, **59**, 70–73.

He, H. *et al.*, 2012. Quantitative analysis method of capsaicin by near-IR spectroscopy. Chinese Patent 102,313,708, filed June 29, 2010 and issued Jan. 11, 2012.

He, L.-J., Liang, Y.-Z., and Zhao, C.-X., 2007. GC/MS study on chemical constituents of essential oil of lemiaceae plants. *Huaxue Xuebao*, **65** (3), 227–232.

Horvath, G. *et al.*, 2013. Detection of antibacterial activity of essential oil components by TLC-bioautography using luminescent bacteria. *Journal of Planar Chromatography–Modern TLC*, **26** (2), 114–118.

Hou, S.-J., Liu, S.-S., Yang, C.-D., and Ding, M.-Y., 2012. Determination of carbonic acid and formic acid in the glycerol oxidation products by ion-exclusion chromatography. *Fenxi Shiyanshi*, **31** (8), 90–92.

Hu, J., Hu, L., Tang, C., and Fang, W., 2012. Detection method for carbonic acid content in sevelamer carbonate. China, Patent No. CN 102680309.

Hu, J. *et al.*, 2009. Determination of piperine in root of Piper nigrum L. by HPLC. *Zhongyao Xinyao Yu Linchuang Yaoli*, **20** (3), 260–262.

Hu, J., Yi, F., and Liu, S., 2011a. Identification of five active ingredients and content determination of piperine in Shiwuwei Heyaowan pills by TLC and HPLC. *Di-San Junyi Daxue Xuebao*, **33** (6), 621–624.

Hu, R., Huang, W., and Chen, H., 2011b. Integrated device of electrochemical suppressor and resistance detection cell of ion chromatography. Chinese Patent 102,141,559, filed Jan. 5, 2011 and issued May 9, 2012.

Hu, Y. *et al.*, 2014. Content determination of piperine in the roots of Piper longum by HPLC. *Zhongguo Yaofang*, **25** (7), 633–635.

Huang, S. *et al.*, 2012. New alkylamides from pericarps of Zanthoxylum bungeanum. *Chinese Chemical Letters*, **23** (11), 1247–1250.

Huang, T.-C. *et al.*, 2011a. Formation of 6-Shogaol of ginger oil under different drying conditions. *Drying Technology*, **29** (16), 1884–1889.

Huang, X. and Chen, Y., 2007. Comparison of flavor material between fresh and dried ginger by GC-MS. *Zhongguo Shipin Xuebao*, **7** (5), 133–138.

Huang, X., Lin, X., Guo, M., and Zou, Y., 2011b. Characteristics of piperine solubility in multiple solvent. *Advanced Materials Research (Durnten-Zurich, Switzerland)*, 236–238.

Huang, X. and Zhang, N., 2005. Crystal gingerol produced and identified from ginger. *Shipin Yu Fajiao Gongye*, **31** (10), 48–50.

Iseli, V. *et al.*, 2007. Characterization of the pungent principles and the essential oil of Zanthoxylum schinifolium pericarp. *Pharmazie*, **62** (5), 396–400.

Ishiwata, T., Norimine, E., and Ishizawa, F., 2009. Development of allyl isothiocyanate detection kit. *Japanese Journal of Clinical Toxicology*, **22** (3), 236–237.

Itoh, K., Yoshioka, K., and Yamazaki, K., 2006. The establishment of a method of the rapid and sensitive analysis capsaicinoid of Capsicum annuum. *Kenkyu Hokoku - Tochigi-ken Sangyo Gijutsu Senta*, **3**, 39–42.

Iwabu, J. *et al.*, 2010. Profiling of the compounds absorbed in human plasma and urine after oral administration of a traditional Japanese (Kampo) medicine, Daikenchuto. *Drug Metabolism and Disposition*, **38** (11), 2040–2048.

Jain, K.L., Patel, R.K., and Patel, H.P., 2014. HPTLC method for simultaneous determination of piperine, embeline, and carvone in the Ayurvedic formulation Catpusphadhya churna. *Journal of AOAC International*, **97** (3), 773–777.

Jain, P.S., Tatiya, A.U., Bagul, S.A., and Surana, S.J., 2011. Development and validation of a method for densitometric analysis of 6-gingerol in herbal extracts and polyherbal formulation. *Journal of Analytical & Bioanalytical Techniques*, **2** (4), 1000124.

Jain, V., Saraf, S., and Saraf, S., 2007. Spectrophotometric determination of piperine in Trikatu churna: an ayurvedic formulation. *Asian Journal of Chemistry*, **19** (7), 5331–5335.

Jain, V., Thakur, A., Soman, G., and Laddha, K.S., 2010. Validated HPLC method development for simultaneous analysis of withaferin-A and 6-gingerol. *Acta Chromatographica*, **22** (1), 153–159.

Jang, K.H. *et al.*, 2008. New polyunsaturated fatty acid amides isolated from the seeds of Zanthoxylum piperitum. *Archives of Pharmacal Research*, **31** (5), 569–572.

Jenecius, A. and Mohan, V.R., 2014. GC-MS analysis of bioactive components on the stem extract of Bacolepsis nervosa Decne. ex. Moq. (periplocaceae). *World Journal of Pharmacy and Pharmaceutical Sciences*, **3**, 1004–1059.

Jeong, H.-J. *et al.*, 2012. Development of a simple method for detecting capsaicinoids using Gibb's reagent in pepper. *Weon'ye Gwahag Gi'sulji*, **30** (3), 294–300.

Jesionek, W., Grzelak, E.M., Majer-Dziedzic, B., and Choma, I.M., 2013. Thin-layer chromatography-direct bioautography for the screening of antimicrobial properties of plant extracts. *Journal of Planar Chromatography–Modern TLC*, **26** (2), 109–113.

Jia, C.-X., Wang, R.-L., Mao, D.-B., and Yue, Z.-Y., 2011a. Analysis of aroma components from Zanthoxylum bungeanum Maxin by microwave-assisted extraction coupled to headspace solid-phase microextraction and gas chromatography-mass spectrometry. *China Condiment*, **36** (2), 109–114.

Jia, X. *et al.*, 2011b. RP-HPLC determination of cinnamic acid, cinnamaldehyde, capsaicin, dihydrocapsaicin and 6-gingerol in Dai-wen-jiu-paste. *Yaowu Fenxi Zazhi*, **31** (8), 1467–1470.

Jiang, H., Solyom, A.M., Timmermann, B.N., and Gang, D.R., 2005. Characterization of gingerol-related compounds in ginger rhizome (Zingiber officinale Rosc.) by high-performance liquid chromatography/electrospray ionization mass spectrometry. *Rapid Communications in Mass Spectrometry*, **19** (20), 2957–2964.

Jiang, J. *et al.*, 2008. Study on quality standards for Qiwei powder. *Zhongguo Yaoshi (Wuhan, China)*, **11** (3), 297–299.

Jiang, S., Mi, S., and Wang, N., 2010. Determination of [6]-gingerol in rat plasma by HPLC-UV. *Zhongyao Xinyao Yu Linchuang Yaoli*, **21** (2), 171–174.

Jiang, Z., Liu, X., Zhang, G., and Guo, H., 2011. Determination of encapsulation efficiency of capsaicin in compound capsaicin nanoemulsion. *Chongqing Yike Daxue Xuebao*, **36** (2), 195–197.

Jiang, Z., Zhang, Q.-F., Tian, H.-L., and Li, R., 2006. The reaction of allyl isothiocyanate with hydroxyl/water and b-cyclodextrin using ultraviolet spectrometry. *Food Technology and Biotechnology*, **44** (3), 423–427.

Jin, R. *et al.*, 2009a. Separation and quantitative analysis of capsaicinoids in chili peppers by reversed-phase argentation LC. *Chromatographia*, **70** (5/6), 1011–1013.

Jin, R. *et al.*, 2009b. Separation and quantitative analysis of capsaicinoids in chili peppers by reversed-phase argentation LC - ERRATUM. *Chromatographia*, **70** (5/6), 1015.

Jin, Y., Qian, D., and Du, Q., 2013. Preparation of bioactive amide compounds from black pepper by countercurrent chromatography and preparative HPLC. *Industrial Crops and Products*, **44**, 258–262.

Jirovetz, L. *et al.*, 2005. Essential oil analysis of Spilanthes acmella Murr. fresh plants from Southern India. *Journal of Essential Oil Research*, **17** (4), 429–431.

Joshi, R. *et al.*, 2011. Characterization of volatile components of tea flowers (Camellia sinensis) growing in Kangra by GC/MS. *Natural Product Communications*, **6** (8), 1155–1158.

Joshi, S.C. *et al.*, 2008. Quality assessment of flavor constituents of natural and commercial Cinnamomum tamala leaves. *Journal of Essential Oil-Bearing Plants*, **11** (3), 278–283.

Juangsamoot, J., Ruangviriyachai, C., Techawongstien, S., and Chanthai, S., 2012. Determination of capsaicin and dihydrocapsaicin in some hot chilli varities by RP-HPLC-PDA after magnetic stirring extraction and clean up with C18 cartridge. *International Food Research Journal*, **19** (3), 1217–1226.

Juliani, H.R. *et al.*, 2006. Application of near-infrared spectroscopy in quality control and determination of adulteration of African essential oils. *Phytochemical Analysis*, **17** (2), 121–128.

Kachoosangi, R.T., Wildgoose, G.G., and Compton, R.G., 2008. Carbon nanotube-based electrochemical sensors for quantifying the heat of chilli peppers: the adsorptive stripping voltammetric determination of capsaicin. *Analyst*, **133** (7), 888–895.

Kakarala, M. *et al.*, 2010. Ultra-low flow liquid chromatography assay with ultraviolet (UV) detection for piperine quantitation in human plasma. *Journal of Agricultural and Food Chemistry*, **58** (11), 6594–6599.

Kaloustian, J. *et al.*, 2007. GC-MS analysis of allergens in plant oils meant for cosmetics. *Oleagineux, Corps Gras, Lipides*, **14** (2), 110–115.

Kamal, Y.T. *et al.*, 2012. Development and validation of HPLC method for simultaneous estimation of piperine and guggul sterones in compound Unani formulation (tablets) and a nanoreservoir system. *Biomedical Chromatography*, **26** (10), 1183–1190.

Kanamori-Kataoka, M. *et al.*, 2012. Analysis of chemical warfare agents and their related compounds using gas chromatograph with atomic emission detector. *Bunseki Kagaku*, **61** (11), 969–977.

Kang, S.-Z. *et al.*, 2014. Electrochemical behavior of eugenol on TiO_2 nanotubes improved with Cu_2O clusters. *RSC Advances*, **4** (2), 538–543.

Kapuriya, K.G., Parmar, P.M., Topiya, H.R., and Faldu, S.D., 2012. Method development and validation of Rifampicin and Piperine in their combined dosage form. *International Bulletin of Drug Research*, **1** (2), 71–80.

Karpakavalli, M., Sini, K.R., and Arthi, I., 2012. Microwave assisted extraction and estimation of piperine, andrographolide using HPLC techniques. *Pharmacie Globale*, **3** (5), 3.

Khajuria, R.K., Sharma, N., Koul, J.L., and Verma, M.K., 2013. Concentration dependent electrospray ionisation mass spectrometry and tandem mass spectrometry (MS/MS) studies on (E,E)-1-[5-(1,3-benzodioxol-5yl)-1-oxo-2,4-pentadienyl]-piperidine (piperine) and its analogues. *SpringerPlus*, **2** (1), 427/1–427/8.

Khamar, J.C. and Patel, S.A., 2012a. First derivative spectrophotometric method for the simultaneous estimation of Rifampicin and Piperine in their combined capsule dosage form. *Asian Journal of Pharmacy & Life Science*, **2** (1), 49–55.

Khamar, J.C. and Patel, S.A., 2012b. Second derivative spectrophotometric method for the estimation of Rifampicin and Piperine in their combined dosage form. *International Research Journal of Pharmacy*, **3** (4), 305–309.

Kollmannsberger, H., Rodríguez-Burruezo, A., Nitz, S., and Nuez, F., 2011. Volatile and capsaicinoid composition of ají (Capsicum baccatum) and rocoto (Capsicum pubescens), two Andean species of chile peppers. *Journal of the Science of Food and Agriculture*, **91** (9), 1598–1611.

Kozukue, N. *et al.*, 2005. Analysis of Eight capsaicinoids in peppers and pepper-containing foods by high-performance liquid chromatography and liquid chromatography-mass spectrometry. *Journal of Agricultural and Food Chemistry*, **53** (23), 9172–9181.

Kozukue, N. *et al.*, 2007. Kinetics of light-induced cis-trans isomerization of four piperines and their levels in ground black peppers as determined by HPLC and LC/MS. *Journal of Agricultural and Food Chemistry*, **55** (17), 7131–7139.

Kubra, I.R. and Jaganmohanrao, L., 2012. An overview on inventions related to ginger processing and products for food and pharmaceutical applications. *Recent Patents on Food, Nutrition & Agriculture*, **4** (1), 31–49.
Kulkarni, V.S. and Surana, S.J., 2011. HPLC analysis of an ayurvedic formulation. *Asian Journal of Chemistry*, **23** (9), 3915–3918.
Kulkarni, V.S. and Surana, S.J., 2012. Simultaneous estimation of piperne and 6-gingerol in an ayurvedic formulation by*HPLC*. **24** (5), 1983–1986.
Kumar, A., Singh, B., Malik, A.K., and Tiwary, D.K., 2007. Determination of some aldehydes by using solid-phase microextraction and high-performance liquid chromatography with UV detection. *Journal of AOAC International*, **90** (6), 1689–1694.
Kumar, K.A. *et al.*, 2011. Determination of chemical composition of essential oil portion of reputed marketed unani formulation Zinda Tilismath. *International Journal of Pharmacy and Pharmaceutical Sciences*, **3** (3), 67–68.
Kumar, V., Kumar, S., Singh, B., and Kumar, N., 2014. Quantitative and structural analysis of amides and lignans in Zanthoxylum armatum by UPLC-DAD-ESI-QTOF-MS/MS. *Journal of Pharmaceutical and Biomedical Analysis*, **4**, 23–29.
Kuzmenko, A.N., 2009. Investigation of the medicinal herbal mixture composition. *Vestnik Moskovskogo Universiteta, Seriya 4: Geologiya*, (3), 212–216.
Lachenmeier, D.W. *et al.*, 2010. Rapid NMR screening of total aldehydes to detect oxidative rancidity in vegetable oils and decorative cosmetics. *Spectroscopy Europe*, **22** (6), 11–14.
Lau, B.B.Y., Panchompoo, J., and Aldous, L., 2015a. Extraction and electrochemical detection of capsaicin and ascorbic acid from fresh chilli using ionic liquids. *New Journal of Chemistry*, **39** (2), 860–867.
Lau, B.B.Y., Panchompoo, J., and Aldous, L., 2015b. Extraction and electrochemical detection of capsaicin and ascorbic acid from fresh chilli using ionic liquids [Erratum to document cited in CA162:216935]. *New Journal of Chemistry*, **39** (2), 1526.
Lee, J.-S. *et al.*, 2013b. Analysis of cinnamic acid, cinnamaldehyde and 2-methoxycinnamaldehyde in Cinnamomi Ramulus on the market in Seoul by HPLC. *Yakhak Hoechi*, **57** (4), 235–240.
Lee, J. *et al.*, 2013a. Simultaneous determination of six components in the traditional herbal medicine Oryeongsan by HPLC-DAD and LC-MS/MS. *Natural Product Sciences*, **19** (1), 20–27.
Lee, S. *et al.*, 2007. Liquid chromatographic determination of 6-, 8-, 10-gingerol, and 6-shogaol in ginger (Zingiber officinale) as the raw herb and dried aqueous extract. *Journal of AOAC International*, **90** (5), 1219–1226.
Leng, T.C., Ping, N.S., Lim, B.P., and Keng, C.L., 2011. Detection of bioactive compounds from Spilanthes acmella (L.) plants and its various in vitro culture products. *Journal of Medicinal Plants Research*, **5** (3), 371–378.
Ley, J.P. *et al.*, 2006. Isolation and synthesis of acmellonate, a new unsaturated long chain 2-ketol ester from Spilanthes acmella. *Natural Product Research*, **20** (9), 798–804.
Li, C. *et al.*, 2011. Determination of four components in Tibetan medicine Dangzuo from different tibetan regions by HPLC. *Zhongguo Zhongyao Zazhi*, **36** (8), 1046–1049.
Li, C. *et al.*, 2012a. Influence of different gel complexes on flavour and colour change in Chongcai paste during storage. *Food Chemistry*, **130** (3), 632–637.
Li, C. *et al.*, 2014. Identification of absorbed constituents in the rabbit plasma and cerebrospinal fluid after intranasal administration of Asari Radix et Rhizoma by HS-SPME-GC-MS and HPLC-APCI-IT-TOF-MSn. *Molecules*, **19** (4), 4857–4879.
Li, J. *et al.*, 2006. Determination of capsaicin in different kinds of Capsicum frutescens by reversed phase HPLC. *Zhongyao Xinyao Yu Linchuang Yaoli*, **17** (3), 202–205.
Li, L. *et al.*, 2007. Determination of cinnamaldehyde and cinnamic acid in Cinnamomum cassia by nonaqueous capillary electrophoresis with UV detection. *Fenxi Ceshi Xuebao*, **26** (2), 274–276.
Li, L., Zhu, X., and Li, S., 2012b. Determination of 6-gingerol in ginger by high performance liquid chromatography. *Tianjin Zhongyiyao Daxue Xuebao*, **31** (1), 36–38.
Li, Q. *et al.*, 2012c. Preparation of ginger element sustained-release dropping pills, and quantitative determination of 6-gingerol. *Zhongchengyao*, **34** (6), 1171–1174.

Li, X., Du, A., Wang, Y., and Du, A., 2012d. A conditional study of analyzing the main active compositions in ginger oleoresin with gas chromatography-mass spectrometry. *Zhongguo Tiaoweipin*, **37** (10), 88–92.

Li, X., Ge, X., and Li, Q., 2008a. HPLC determination of cinnamaldehyde in Zushimafengshi plasters. *Yaowu Fenxi Zazhi*, **28** (9), 1559–1561.

Li, X. *et al.*, 2008b. Rapid and accurate analytical method for the determination of gingerols in three medicinal gingers (Zingiber officinale Roscoe) by high performance liquid chromatography. *Analytical Letters*, **41** (10), 1732–1741.

Li, Y. *et al.*, 2012e. Discussion about assay method of eugenol in Flos Caryophylli in Chinese Pharmacopoeia (edition 2010). *Zhongyao Xinyao Yu Linchuang Yaoli*, **23** (5), 579–582.

Liao, H., Dasgupta, P.K., Srinivasan, K., and Liu, Y., 2015. Mixing characteristics of mixers in flow analysis. Application to two-dimensional detection in ion chromatography. *Analytical Chemistry*, **87** (1), 793–800.

Liao, H. *et al.*, 2012. Determination of eugenol content in cigarette mainstream smoke by GC/MS-SIM. Chinese Patent 102,809,621, filed Aug. 30, 2012 and issued Sept. 25, 2013.

Liao, H. *et al.*, 2013a. Method for detecting eugenol in essence spice by gas chromatography-mass spectrometry. Chinese Patent 103,344,734, filed July 10, 2013 and issued Jan. 21, 2015.

Liao, H. *et al.*, 2013b. Method for measuring eugenol in edible flavor using HPLC. Chinese Patent 103,399,089, filed July 10, 2013 and issued Feb. 25, 2015.

Liao, H. *et al.*, 2013c. Method for measuring eugenol in the cigarette mainstream smoke based on high performance liquid chromatography (HPLC). Chinese Patent 103,336,085, filed July 10, 2013 and issued Oct. 22, 2014.

Lin, J., Xu, L., Liu, J., and Zou, Z., 2005a. Study on contents of cinnamaldehyde and cinnamic acid and distribution of Ramulus Cinnamomi. *Zhongguo Yaoxue Zazhi (Beijing, China)*, **40** (23), 1784–1787.

Lin, W. *et al.*, 2006. Determination of eugenol in several Chinese herbal medicines by capillary electrophoresis with high frequency conductivity method. *Gaodeng Xuexiao Huaxue Xuebao*, **27** (11), 2070–2072.

Lin, X. *et al.*, 2011. Simultaneous determination of piperine and 6-gingerol in condiment by HPLC internal standard method. *Shipin Gongye Keji*, **32** (12), 452–454.

Lin, Y.-T. *et al.*, 2005b. Enantiomeric analysis of (+)-menthol and (−)-menthol by fluorogenic derivatization and liquid chromatography. *Journal of Chromatography A*, **1087** (1–2), 223–228.

Liu, F. *et al.*, 2011a. Determination eugenol, borneol and isoborneol of Guanxin Shutong capsule by gas chromatography. *Zhongguo Shiyan Fangjixue Zazhi*, **17** (18), 56–59.

Liu, H. and Han, B., 2008. Determining general capsaicin in capsaicin-based pesticides by spectrophotometry. *Xiandai Nongyao*, **7** (1), 24–26.

Liu, H. *et al.*, 2015. Identification and simultaneous quantification of five alkaloids in Piper longum L. by HPLC-ESI-MSn and UFLC-ESI-MS/MS and their application to Piper nigrum L. *Food Chemistry*, **177**, 191–196.

Liu, H. *et al.*, 2013a. Tissue distribution profiles of three antiparkinsonian alkaloids from Piper longum L. in rats determined by liquid chromatography-tandem mass spectrometry. *Journal of Chromatography B: Analytical Technologies in the Biomedical and Life Sciences*, **928**, 78–82.

Liu, H., Zhang, Y., and Nie, L., 2014a. Simultaneous determination of puerarin, daidzin, baicalin, daidzein, cinnamaldehyde and baicalein in Xiaoer Chaigui Tuire Keli by HPLC. *Yaowu Fenxi Zazhi*, **34** (6), 1038–1042.

Liu, J., Bi, Y., Luo, R., and Wu, X., 2011b. Simultaneous UFLC-ESI-MS/MS determination of piperine and piperlonguminine in rat plasma after oral administration of alkaloids from Piper longum L.: Application to pharmacokinetic studies in rats. *Journal of Chromatography B: Analytical Technologies in the Biomedical and Life Sciences*, **879** (27), 2885–2890.

Liu, L. *et al.*, 2010. Determination of capsaicin and dihydrocapsaicin in Capsicum annuum and related products by capillary electrophoresis with a mixed surfactant system. *Food Chemistry*, **119** (3), 1228–1232.

Liu, M. *et al.*, 2013b. Simultaneous determination of six marker components in Sini decoction by high performance liquid chromatography. *Dier Junyi Daxue Xuebao*, **34** (6), 682–686.

Liu, W. *et al.*, 2014b. Purification of 6-gingerol from oleoresin with macroporous resin. *Nongye Jixie Xuebao*, **45** (6), 237–241.

Liu, X. *et al.*, 2009. HPLC-TOF/MS in rapid identification of chemical compositions in Xiaochaihu decoction. *Dier Junyi Daxue Xuebao*, **30** (8), 941–946.

Lopez-Martinez, S., Aguilar-Guadarrama, A.B., and Rios, M.Y., 2011. Minor alkamides from Heliopsis longipes S.F. Blake (Asteraceae) fresh roots. *Phytochemistry Letters*, **4** (3), 275–279.

Lungarini, S., Aureli, F., and Coni, E., 2008. Coumarin and cinnamaldehyde in cinnamon marketed in Italy: A natural chemical hazard?. *Food Additives & Contaminants, Part A: Chemistry, Analysis, Control, Exposure & Risk Assessment*, **25** (11), 1297–1305.

Luo, J. *et al.*, 2012. Study on determination of capsaicinoids in capsicum chilli and its storage stability. *Shipin Gongye Keji*, **33** (11), 333–337.

Lv, D. *et al.*, 2013. Method for detecting content of piperine and moisture and evaluating quality of white pepper. Chinese Patent 103,278,473, filed May 14, 2013 and issued Sep. 4, 2013.

Lv, G.-P. *et al.*, 2010. Pressurized liquid extraction and GC-MS analysis for simultaneous determination of seven components in Cinnamomum cassia and the effect of sample preparation. *Journal of Separation Science*, **33** (15), 2341–2348.

Ma, J., 2012. Allyl isothiocyanate derived from oriental mustard meal as a natural antimicrobial to inhibit the growth of moulds on bread, Master of Science Thesis, Food Science Department at the University of Guelph, Canada.

Ma, W. *et al.*, 2009. Determination of content of 6-gingerol in ginger by HPLC. *Weiliang Yuansu Yu Jiankang Yanjiu*, **26** (6), 27–28.

Maheswari, R., Abtheen, K.S.S.A., Shanmugasundaram, P., and Vijeyaanandhi, M., 2008. Spectrophotometric method for the quantitative estimation of piperine in herbal cough syrup. *Acta Ciencia Indica, Chemistry*, **34** (4), 619–621.

Mallikarjun, M., Jayaveera, K.N., Sailaja, S., and Yogananda, R.K., 2012. Isolation and structure confirmation of piperine from Piper trioicum Roxb. *Journal of Pharmacy and Chemistry*, **6** (2), 33–36.

Mao, L. *et al.*, 2013. Research on detection technology of capsaicinoids in hogwash oil. *Zhongguo Weisheng Jianyan Zazhi*, **23** (3), 544–546, 550.

Marutoiu, C. *et al.*, 2006. Separation and identification of piperine and chavicine in black pepper by TLC and GC-MS. *Journal of Planar Chromatography–Modern TLC*, **19** (109), 250–252.

Meng, X., 2011. Content determination of 3 gingerols in dried ginger, baked ginger and fresh ginger by RP-HPLC. *Zhongguo Yaofang*, **22** (23), 2182–2184.

Menghani, E., Jain, S.C., and Jain, R., 2010. Scientific validation of Piper nigrum by HPLC and anti-oxidative assay markers. *Asian Journal of Biotechnology*, **2** (2), 133–138.

Mirensha, Y. *et al.*, 2014. Quality standards analysis for Xiaoaifei preparation. *Xinjiang Yike Daxue Xuebao*, **37** (4), 436–439.

Mishra, A., Mishra, A.K., Tiwari, O.P., and Jha, S., 2013. HPLC analysis and standardization of Brahmi vati – An Ayurvedic poly-herbal formulation. *Journal of Young Pharmacists*, **5** (3), 77–82.

Miyamoto, S. and Horii, S., 2013. Method for breakthrough detection and operation of ion exchange unit in water purification. Japanese Patent 2,013,208,566, filed March 30, 2012 and issued Oct. 10, 2013.

Mo, C.Y. *et al.*, 2011. Apparatus and method for measuring pungency of red pepper powder. US Patent 20,110,218,739, filed July 1, 2010 and issued Sep. 8, 2011.

Mohammad, R., Ahmad, M., and Heng, L.Y., 2013. An amperometric biosensor utilizing a ferrocene-mediated horseradish peroxidase reaction for the determination of capsaicin (chili hotness). *Sensors*, **13**, 10014–10026.

Mohammad, R., Ahmad, M., and Heng, L.Y., 2014. Chilli hotness determination based on optical capsaicin biosensor using stacked immobilisation technique. *Sensors and Actuators, B: Chemical*, **190**, 593–600.

Mohan, M. *et al.*, 2012. Antimicrobial activity and composition of the volatiles of Cinnamomum tamala Nees. and Murraya koenigii (L.) Spreng. from Uttarakhand (India). *Asian Pacific Journal of Tropical Disease*, **2** (Suppl. 1), 5324–5327.

Monforte-Gonzalez, M., Medina-Lara, F., Gutierrez-Carbajal, G., and Vazquez-Flota, F., 2007. Capsaicinoid quantitation by in situ densitometry of thin layer chromatography plates. *Journal of Liquid Chromatography & Related Technologies*, **30** (11), 1697–1704.

Moorthi, C. *et al.*, 2013. Application of validated RP-HPLC-PDA method for the simultaneous estimation of curcumin and piperine in Eudragit E 100 nanoparticles. *Journal of Pharmacy Research (Gurgaon, India)*, **7** (3), 224–229.

Mori, M. *et al.*, 2005. High-speed ion-exclusion chromatography of dissolved carbon dioxide on a small weakly acidic cation-exchange resin column with ion-exchange enhancement columns of conductivity detection. *Journal of chromatography, A*, **1092** (2), 250–253.

Mori, T., Suma, K., Sumiyoshi, Y., and Endo, Y., 2009. Spectroscopic detection of isolated carbonic acid. *Journal of Chemical Physics*, **130** (20), 204308/1–204308/7.

Mori, T., Suma, K., Sumiyoshi, Y., and Endo, Y., 2011. Spectroscopic detection of the most stable carbonic acid, cis-cis H_2CO_3. *Journal of Chemical Physics*, **134** (4), 044319/1-044319/6.

Muchena, J.K., 2009. Studies of capsaicinoids contents of locally grown and commercial chilies using reversed-phase high performance liquid chromatography, Master of Science thesis, Chemistry Department, East Tennessee State University, TN.

Musenga, A. *et al.*, 2006. Simultaneous determination of aromatic and terpenic constituents of cloves by means of HPLC with diode array detection. *Journal of Separation Science*, **29** (9), 1251–1258.

Musenga, A. *et al.*, 2007. Analysis of aromatic and terpenic constituents of pepper extracts by capillary electrochromatography. *Journal of Separation Science*, **30** (4), 612–619.

Nagappan, K.V., Meyyanathan, S.N., Rajan, S., and Elango, K., 2011. A liquid chromatography-atmospheric Pressure Chemical Ionisation-mass spectrometry method for the simultaneous determination of curcumin and piperine in food products. *Recent Progress in Medicinal Plants*, **31**, 151–159.

Nagappan, K.V., Meyyanathan, S.N., Raja, R.B., and Kannan, E., 2009. A liquid chromatography method for the simultaneous determination of curcumin and piperine in food products using diode array detection. *Asian Journal of Research in Chemistry*, **2** (2), 115–118.

Nakamura, S., Yatabe, R., Onodera, T., and Toko, K., 2013. Sensitive detection of capsaicinoids using a surface plasmon resonance sensor with anti-homovanillic acid polyclonal antibodies. *Biosensors*, **3** (4), 374–384.

Nakanishi, A. *et al.*, 2014. Determination of the absolute configuration of a novel odour-active lactone, cis-3-methyl-4-decanolide, in wasabi. *Flavour and Fragrance Journal*, **29** (4), 220–227.

Ni, L. and Yang, X., 2012. Determination of cinnamaldehyde and piperine in Shiliu Jianwei pill by HPLC. *Zhongguo Xiandai Yingyong Yaoxue*, **29**(1), 67–69.

Nikam, A.R., Sathiyanarayanan, L., and Mahadik, K.R., 2013. Validation of reversed-phase high-performance liquid chromatography method for simultaneous determination of 6-, 8-, and 10-shogaol from ginger preparations. *International Journal of Pharmacy and Pharmaceutical Sciences*, **5** (Suppl. 1), 432–437.

Noami, M. *et al.*, 2006. Study on the analysis of capsaicin glucuronide in rat urine by liquid chromatography-mass spectrometry after enzymatic hydrolysis. *Journal of Health Science*, **52** (6), 660–665.

Norimine, E., Ishiwata, T., and Ishizawa, F., 2008. Devising a detector tube for allyl isothiocyanate. *Japanese Journal of Toxicology*, **21** (3), 319–320.

Nwokem, C.O., Agbaji, E.B., Kagbu, J.A., and Ekanem, E.J., 2010. Determination of capsaicin content and pungency level of five different peppers grown in Nigeria. *New York Science Journal*, **3** (9), 17–21.

Oromi-Farrus, M., Torres, M., and Canela, R., 2012. Acylation of chiral alcohols: a simple procedure for chiral GC analysis. *Journal of Analytical Methods in Chemistry*, DOI:10.1155/2012/452949.

Pandey, R.K., Saraf, S., and Saraf, S., 2010. Development of fingerprint for single component analysis of an ayurvedic formulation (Sitopaladi Churna) by high performance liquid chromatography. *Pharmacia Lettre*, **2** (2), 464–470.

Pandey, R., Saraf, S., and Saraf, S., 2009. Development of UV spectrophotometric fingerprint method for evaluation of Sitopaladi churna. *Biosciences, Biotechnology Research Asia*, **6** (1), 309–312.

Park, J.S. and Jung, M.Y., 2012. Development of high-performance liquid chromatography-time-of-flight mass spectrometry for the simultaneous characterization and quantitative analysis of gingerol-related compounds in ginger products. *Journal of Agricultural and Food Chemistry*, **60** (40), 10015–10026.

Patel, R.K., Kanani, R.J., Patel, V.R., and Patel, M.G., 2010. Development and validation of HPTLC method for simultaneous quantification of vasicine and piperine in Vasavaleha. *International Journal of Pharmaceutical Research*, **2** (3), 14–17.

Patel, S. and Vyas, N., 2012. Validated spectrofluorimetric method for estimation of piperine in an Ayurvedic formulation. *Asian Journal of Pharmaceutical and Clinical Research*, **5** (Suppl. 4), 231–233.

Patra, K.C. and Kumar, K.J., 2010. A validated HPTLC method for simultaneous analysis of eugenol and piperine in a Siddha formulation. *Journal of Planar Chromatography–Modern TLC*, **23** (4), 293–297.

Pei, G.-Z. *et al.*, 2013. Determination of apigenin, piperine, costunolide and dehydrocostuslactone in Victoria drug Aimisi pills by HPLC. *Zhongguo Yiyuan Yaoxue Zazhi*, **33** (20), 1731–1732.

Pelosi, C. *et al.*, 2014. A new method to measure allyl isothiocyanate (AITC) concentrations in mustard: comparison of AITC and commercial mustard solutions as earthworm extractants. *Applied Soil Ecology*, **80**, 1–5.

Pena-Alvarez, A., Alvarado, L.A., and Vera-Avila, L.E., 2012. Analysis of capsaicin and dihydrocapsaicin in hot peppers by ultrasound assisted extraction followed by gas chromtaography–mass spectrometry. *Instrumentation Science & Technology*, **40** (5), 429–440.

Pena-Alvarez, A., Ramirez-Maya, E., and Alvarado-Suarez, L.A., 2009. Analysis of capsaicin and dihydrocapsaicin in peppers and pepper sauces by solid phase microextraction–gas chromatography–mass spectrometry. *Journal of Chromatography A*, **1216** (14), 2843–2847.

Penas, E., Pihlava, J.M., Vidal-Valverde, C., and Frias, J., 2012. Influence of fermentation conditions of Brassica oleracea L. var. capitata on the volatile glucosinolate hydrolysis compounds of sauerkrauts. *LWT–Food Science and Technology*, **48** (1), 16–23.

Pfaff, A.M. and Steiner, R.R., 2011. Development and validation of AccuTOF-DART as a screening method for analysis of bank security device and pepper spray components. *Forensic Science International*, **206** (1–3), 62–70.

Poyrazoglu, E.S., Yemis, O., Kadakal, C., and Artık, N., 2005. Determination of capsaicinoid profile of different chilli peppers grown in Turkey. *Journal of the Science of Food and Agriculture*, **85** (9), 1435–1438.

Pramod, K. *et al.*, 2013. Development and validation of RP-HPLC-PDA method for the quantification of eugenol in developed nanoemulsion gel and nanoparticles. *Journal of Analytical Science and Technology*, **4** (9–23), 16/1–16/6.

Pyka, A., 2010. Phenolic drugs: TLC detection. *Encyclopedia of Chromatography* (ed. J. Cazes), 3rd edn, Taylor & Francis, pp. 1777–1789.

Qian, D. *et al.*, 2011. Separation of piperine from pepper by high-speed countercurrent chromatography. *Shipin Kexue (Beijing, China)*, **32** (24), 274–276.

Qiang, W., Wang, C., Pu, W., and Liu, X., 2014. Establishment and evaluation of determination of cinnamaldehyde in baoyuanqingxue granules by HPLC. *Advanced Materials Research (Durnten-Zurich, Switzerland)*, 998–999.

Qu, X., Lu, X., and Huang, X., 2007. Determination of 6-gingerol in ginger and its produces by using 6-gingerol oxime as internal standard. *Shipin Yu Fajiao Gongye*, **33** (6), 123–125.

Quan, L., Zhao, S., and Xu, L., 2005. RP-HPLC method for determination of piperine content in Zhitong capsule. *Zhongguo Zhongyao Zazhi*, **30** (3), 235.

Rafi, M., Lim, L.W., Takeuchi, T., and Darusman, L.K., 2013. Simultaneous determination of gingerols and shogaol using capillary liquid chromatography and its application in discrimination of three ginger varieties from Indonesia. *Talanta*, **103**, 28–32.

Rai, P.D., Pathak, A., and Rajput, S.J., 2009a. RP-LC Simultaneous determination of aconitine, solanine and piperine in an Ayurvedic medicine. *Chromatographia*, **69** (11/12), 1275–1282.

Rai, P., Pathak, A., and Rajput, S.J., 2009b. Stability-indicating reversed-phase liquid chromatographic methods for the determination of aconitine and piperine in a polyherbal formulation. *Journal of AOAC International*, **92** (4), 1044–1054.

Rajopadhye, A.A., Namjoshi, T.P., and Upadhye, A.S., 2012. Rapid validated HPTLC method for estimation of piperine and piperlongumine in root of Piper longum extract and its commercial formulation. *Revista Brasileira de Farmacognosia*, **22** (6), 1355–1361.

Ramlingam, P. *et al.*, 2012. Quality control of marketed clove buds - Reference to their quality and purityas per WHO guidelines. *International Journal of Bioassays*, **1** (5), 1–7.

Randviir, E.P., Metters, J.P., Stainton, J., and Banks, C.E., 2013. Electrochemical impedance spectroscopy versus cyclic voltammetry for the electroanalytical sensing of capsaicin utilising screen printed carbon nanotube electrodes. *Analyst*, **138** (10), 2970–2981.

Rao, V.R.S. *et al.*, 2011. Simultaneous determination of bioactive compounds in Piper nigrum L. and a species comparison study using HPLC-PDA. *Natural Product Research*, **25** (13), 1288–1294.

Rodenas-Montano, J. *et al.*, 2013. Determination of alcohols in essential oils by liquid chromatography with ultraviolet detection after chromogenic derivatization. *Journal of Chromatography A*, **1296**, 157–163.

Rodriguez, A., Pedersen-Bjergaard, S., Rasmussen, K.E., and Nerin, C., 2008. Selective three-phase liquid phase microextraction of acidic compounds from foodstuff simulants. *Journal of Chromatography A*, **1198–1199**, 38–44.

Rodtjer, A., Skibsted, L.H., and Andersen, M.L., 2006. Identification and quantification of phenolics in aromatic bitter and cherry liqueur by HPLC with electrochemical detection. *European Food Research and Technology*, **223** (5), 663–668.

Ruiz-Lau, N. *et al.*, 2010. Capsaicinoids are absent in habanero pepper vegetative organs (Capsicum chinense Jacq.). *HortScience*, **45** (2), 323–326.

Rusu, M.L. *et al.*, 2007. HPTLC and GC-MS for separation and identification of eugenol in plants. *Journal of Planar Chromatography – Modern TLC*, **20** (2), 139–140.

Samten, Wetwitayaklung, P., Kitcharoen, N., and Sotanaphun, U., 2010. TLC image analysis for determination of the piperine content of the traditional medicinal preparations of Bhutan. *Acta Chromatographica*, **22** (2), 227–236.

Sanchez-Cabrera, Y. and Pino, J.A., 2011. Headspace solid-phase microextraction analysis of volatile compounds from spice essential oils in dry flavourings. *International Journal of Food Science and Technology*, **46** (10), 2118–2123.

Schepler, C. *et al.*, 2013. Flow injection of liquid samples to a mass spectrometer with ionization under vacuum conditions: a combined ion source for single-photon and electron impact ionization. *Analytical and Bioanalytical Chemistry*, **405** (22), 6953–6957.

Schulz, H. *et al.*, 2005. Characterization of peppercorn, pepper oil, and pepper oleoresin by vibrational spectroscopy methods. *Journal of Agricultural and Food Chemistry*, **53** (9), 3358–3363.

Schulz, K. *et al.*, 2007. Determination of aroma compounds from alcoholic beverages in spiked blood samples by means of dynamic headspace GC-MS. *Chromatographia*, **66** (11/12), 879–886.

Schulz, K. *et al.*, 2008. Headspace solid-phase microextraction-gas chromatography-mass spectrometry for the quantitative determination of the characteristic flavoring agent eugenol in serum samples after enzymatic cleavage to validate post-offence alcohol drinking claims. *Journal of Chromatography A*, **1211** (1-2), 113–119.

Schweiggert, U., Carle, R., and Schieber, A., 2006. Characterization of major and minor capsaicinoids and related compounds in chili pods (Capsicum frutescens L.) by high-performance liquid chromatography/atmospheric pressure chemical ionization mass spectrometry. *Analytica Chimica Acta*, **557** (1–2), 236–244.

Scott, I.M. *et al.*, 2005. Analysis of Piperaceae germplasm by HPLC and LC-MS: A method for isolating and identifying unsaturated amides from Piper spp extracts. *Journal of Agricultural and Food Chemistry*, **53** (6), 1907–1913.

Sethi, P. *et al.*, 2009. Development and validation of a reversed phase HPLC method for simultaneous determination of curcumin and piperine in human plasma for application in clinical pharmacological studies. *Journal of Liquid Chromatography & Related Technologies*, **32** (20), 2961–2974.

Shanmugasundaram, P., Maheswari, R., and Vijayaanandhi, M., 2008. Quantitative determination of piperine in herbal cough syrup by HPTLC method. *Rasayan Journal of Chemistry*, **1** (2), 212–217.

Shao, X. *et al.*, 2010. Quantitative analysis of ginger components in commercial products using liquid chromatography with electrochemical array detection. *Journal of Agricultural and Food Chemistry*, **58** (24), 12608–12614.

Shingate, P.N., Dongre, P.P., and Kannur, D.M., 2013. New method development for extraction and isolation of piperine from black pepper. *International Journal of Pharmaceutical Sciences and Research*, **4** (8), 3165–3170.

Shivani, R.I. *et al.*, 2013a. Development of HPTLC method for the determination of piperine in Lavangadi vati – an Ayurvedic formulation. *World Journal of Pharmaceutical Research*, **2** (5), 1720–1730.

Shivani, R.I. *et al.*, 2013b. Development of HPTLC method for the determination of piperine in sitopladi Churna – an Ayurvedic formulation. *International Journal of Pharmaceutical and Phytopharmacological Research*, **3** (1), 46–51.

Shukla, K., Saraf, S., and Saraf, S., 2007. Development of finger printing methods of an Ayurvedic formulation bhaskar lavan churna: a spectrophotometric approach. *Natural Products: An Indian Journal*, **3** (3), 178–181.

Singh, M. and Chaturvedi, R., 2012. Screening and quantification of an antiseptic alkylamide, spilanthol from in vitro cell and tissue cultures of Spilanthes acmella Murr. *Industrial Crops and Products*, **36** (1), 321–328.

Singh, M. *et al.*, 2014. Electrochemical biosensor with ceria-polyaniline core shell nano-interface for the detection of carbonic acid in blood. *Journal of Colloid and Interface Science*, **425**, 52–58.

Singh, N.K. *et al.*, 2011. UV-spectrophotometric method development for estimation of piperine in Chitrakadi Vati. *Pharmacia Lettre*, **3** (3), 178–182.

Stanfill, S.B. *et al.*, 2006. Quantification of flavor-related compounds in the unburned contents of bidi and clove cigarettes. *Journal of Agricultural and Food Chemistry*, **54** (22), 8580–8588.

Stinson, S.L., Craven, B.M., and Cech, N.B., 2006. HPLC/ESI-MS method validated for the analysis of alkylamides in *Spilanthes acmella*. Proceedings of the 58th SE Regional Meeting (SERMACS '06), Augusta, GA, USA.

Sugai, E. and Kubota, K., 2006. The pungency of Japanese pepper and characteristics of sanshool-related compounds. *Koryo*, **229**, 129–137.

Sugai, E. *et al.*, 2005a. Pungent qualities of sanshool-related compounds evaluated by a sensory test and activation of rat TRPV1. *Bioscience, Biotechnology, and Biochemistry*, **69** (10), 1951–1957.

Sugai, E., Morimitsu, Y., and Kubota, K., 2005b. Quantitative analysis of sanshool compounds in Japanese pepper (Xanthoxylum piperitum DC.) and their pungent characteristics. *Bioscience, Biotechnology, and Biochemistry*, **69** (10), 1958–1962.

Sugiyama, T. and Kagatani, I., 2012a. Apparatus for automated detection and display of dissolved carbonic acid gas in bath water. Japanese Patent 2,012,237,689, filed May 13, 2011 and issued Dec. 6, 2012.

Sugiyama, T. and Kagatani, I., 2012b. Ultrasonic-type detectors for determining carbonic acid dissolved in water specimens. Japanese Patent 2,012,112,860, filed Nov. 26, 2010 and issued June 14, 2012.

Supalkova, V. *et al.*, 2007. Study of capsaicin content in various parts of pepper fruit by liquid chromatography with electrochemical detection. *Acta Chimica Slovenica*, **54** (1), 55–59.

Surareungchai, W. and Phanthong, C., 2012. Electrochemical detection of capsaicinoid compounds in a sample. World Patent 2,012,134,408, filed March 30, 2012 and issued Oct. 4, 2012.

Taira, S. *et al.*, 2012. Mass spectrometry imaging of the capsaicin localization in the Capsicum fruits. *International Journal of Biotechnology for Wellness Industries*, **1** (1), 61–65.

Tan, G. *et al.*, 2011. Characterization of constituents in Sini decoction and rat plasma by high-performance liquid chromatography with diode array detection coupled to time-of-flight mass spectrometry. *Biomedical Chromatography*, **25** (8), 913–924.

Tao, Y., Li, W., Liang, W., and Van Breemen, R.B., 2009. Identification and quantification of gingerols and related compounds in ginger dietary supplements using high-performance liquid chromatography-tandem mass spectrometry. *Journal of Agricultural and Food Chemistry*, **57** (21), 10014–10021.

Thompson, R.Q. *et al.*, 2012. Visualizing Capsaicinoids: Colorimetric Analysis of Chili Peppers. *Journal of Chemical Education*, **89** (5), 610–612.

Thompson, R.Q., Phinney, K.W., Sander, L.C., and Welch, M.J., 2005. Reversed-phase liquid chromatography and argentation chromatography of the minor capsaicinoids. *Analytical & Bioanalytical Chemistry*, **381** (7), 1432–1440.

Tonon de Souza, P., and Rossi, A.V., 2014. Indirect spectrophotometric determination of capsaicinoids in Capsicum peppers by reaction with Co(II) complex with 4-(2-pyridylazo)resorcinol. *Quimica Nova*, **37** (4), 631–637.

Topuz, A. and Ozdemir, F., 2007. Assessment of carotenoids, capsaicinoids and ascorbic acid composition of some selected pepper cultivars (Capsicum annuum L.) grown in Turkey. *Journal of Food Composition and Analysis*, **20** (7), 596–602.

Troncoso, R. *et al.*, 2005. Analysis of the isothiocyanates present in cabbage leaves extract and their potential application to control*Alternaria rot in bell peppers*. **38** (6), 701–708.

Tsuno, T., 2006. Identification of ginger. *Toyama-ken Yakuji Kenkyusho Nenpo*, **33**, 53–54.

Usman, M.G. *et al.*, 2014. Capsaicin and dihydrocapsaicin determination in chili pepper genotypes using ultra-fast liquid chromatography. *Molecules*, **19** (5), 6474–6488.

van Stee, L.L.P. and Brinkman, U.A.T., 2011. Peak clustering in two-dimensional gas chromatography with mass spectrometric detection based on theoretical calculation of two-dimensional peak shapes: The 2DAid approach. *Journal of Chromatography A*, **1218** (43), 7878–7885.

Vaykole, A.M., Nirmal, S.A., Jadhav, R.S., and Pattan, S.R., 2014. Development and validation of HPTLC method to detect curcumin, piperine, and boswellic acid in polyherbal transdermal patch. *Journal of Liquid Chromatography & Related Technologies*, **37** (3), 367–378.

Villa, C., Gambaro, R., Mariani, E., and Dorato, S., 2007. High-performance liquid chromatographic method for the simultaneous determination of 24 fragrance allergens to study scented products. *Journal of Pharmaceutical and Biomedical Analysis*, **44** (3), 755–762.

Vyas, A. *et al.*, 2011. TLC densitometric method for the estimation of piperine in ayurvedic formulation Trikatu churna. *Oriental Journal of Chemistry*, **27** (1), 301–304.

Wagner, C.E., Cahill, T.M., and Marshall, P.A., 2011. Extraction, purification, and spectroscopic characterization of a mixture of capsaicinoids. *Journal of Chemical Education*, **88** (11), 1574–1579.

Wang, D. *et al.*, 2012a. Method for detecting capsaicin content in water-soluble oleoresin capsicum by ultraviolet spectrophotometry. Chinese Patent 102,798,603, filed Aug. 17, 2012 and issued Nov. 28, 2012.

Wang, H. *et al.*, 2011a. Simultaneous determination of four effective compounds in Juejin Decoction by high performance liquid chromatography. *Asian Journal of Traditional Medicines*, **6** (4), 168–174.

Wang, H., So, P.-K., and Yao, Z.-P., 2014a. Direct analysis of herbal powders by pipette-tip electrospray ionization mass spectrometry. *Analytica Chimica Acta*, **809**, 109–116.

Wang, H. *et al.*, 2014b. Simultaneous determination of eight effective components in Uighur medicine Dawayijin pills by HPLC. *Zhongguo Xinyao Zazhi*, **23** (11), 1316–1320.

Wang, H. *et al.*, 2013a. HPLC determination of benzoyl aconitine and 6-ginger in Sini decoction prepared by three extraction methods. *Xiandai Yiqi Yu Yiliao*, **19** (5), 55–59.

Wang, J. *et al.*, 2012b. Quality inspection method of Ershiwuweiguijiu pills. Chinese Patent 103,175,907, filed Dec. 5, 2012 and issued Dec. 3, 2014.

Wang, L.-H. and Chen, J.-C., 2011c. Studies on the electrochemical behavior of eugenol and its application using a flow – through chronoamperometric sensor. *Current Pharmaceutical Analysis*, **7** (2), 88–94.

Wang, L. *et al.*, 2011b. Analysis of gingerol-related compounds in fresh ginger by HPLC-ESI-Q-TOF-MS/MS. *Zhongguo Zhongyao Zazhi*, **36** (24), 3467–3471.

Wang, L. *et al.*, 2012c. A novel method for the identification of illegal cooking oil (1): detection of three capsaicinoids with liquid chromatography-mass spectrometry. *Sepu*, **30** (11), 1094–1099.

Wang, L. *et al.*, 2013b. High performance liquid chromatography based determination of 6-gingerol, 8-gingerol and 10-gingerol in seed and tender ginger. *Shandong Kexue*, **26** (1), 12–15.

Wang, L., Zhou, L., and Li, T., 2008a. Supercritical CO_2 extraction process of piperine in Piper nigrum Linn by orthogonal design. *Shizhen Guoyi Guoyao*, **19** (7), C3–C4.

Wang, R., Wang, R., and Yang, B., 2009a. Extraction of essential oils from five cinnamon leaves and identification of their volatile compound compositions. *Innovative Food Science & Emerging Technologies*, **10** (2), 289–292.

Wang, S. *et al.*, 2012d. Structural characterization and identification of major constituents in Jitai tablets by high-performance liquid chromatography/diode-array detection coupled with electrospray ionization tandem mass spectrometry. *Molecules*, **17**, 10470–10493.

Wang, S. *et al.*, 2007. Determination of capsaicinoids in highly oil-containing pepper products by high performance liquid chromatography. *Fenxi Shiyanshi*, **26** (12), 39–43.

Wang, S., Xie, J., Yang, W., and Sun, B., 2011d. Preparative separation and purification of alkylamides from Zanthoxylum bungeanum maxim by high-speed counter-current chromatography. *Journal of Liquid Chromatography & Related Technologies*, **34** (20), 2640–2652.

Wang, S. and Xi, H., 2008b. Determination of piperine in Mongolian recipe of Haodaodun-10 by HPLC. *Zhongyao Xinyao Yu Linchuang Yaoli*, **19** (4), 302–303.

Wang, T.-M. *et al.*, 2014c. Simultaneous determination of 14 major components in Longhu Rendan pills by ultra-high-performance liquid chromatography coupled with electrospray ionisation tandem mass spectrometry. *Analytical Methods*, **6** (9), 2923–2929.

Wang, W. *et al.*, 2009b. Plasma pharmacokinetics, tissue distribution and excretion study of 6-gingerol in rat by liquid chromatography-electrospray ionization time-of-flight mass spectrometry. *Journal of Pharmaceutical and Biomedical Analysis*, **49** (4), 1070–1074.

Wang, W. *et al.*, 2009c. Simultaneous determination of 6-gingerol, 8-gingerol, 10-gingerol and 6-shogaol in rat plasma by liquid chromatography-mass spectrometry: Application to pharmacokinetics. *Journal of Chromatography B: Analytical Technologies in the Biomedical and Life Sciences*, **877** (8–9), 671–679.

Wang, X., Jia, X., Chen, Y., and Jiang, J., 2010. Determination of ferulic acid, cinnamic acid, cinnamaldehyde, costunolide and dehydrocostuslactone contents in Tongjingbao granules by HPLC. *Zhongcaoyao*, **41** (10), 1643–1645.

Wang, X., Liu, Y., Li, X., and Wang, C., 2013c. Simultaneous determination of four effective components in Jishengshenqi tablets by HPLC. *Zhongguo Yaoxue Zazhi (Beijing, China)*, **48** (18), 1589–1593.

Wang, X. *et al.*, 2008c. Determination of capsaicin and dihydrocapsaicin in food by high-performance liquid chromatography. *Shipin Kexue (Beijing, China)*, **29** (7), 378–381.

Wang, X.-M. *et al.*, 2012e. Validated HPLC-MS/MS method for simultaneous determination of curcumin and piperine in human plasma. *Tropical Journal of Pharmaceutical Research*, **11** (4), 621–629.

Wang, Y., Wei, N., Li, H.-F., and Li, Y.-H., 2013d. Analysis of chemical constituents in the supercritical CO_2 extract of black pepper in Hainan by GC-MS. *Zhongguo Shiyan Fangjixue Zazhi*, **19** (12), 121–123.

Wang, Y. *et al.*, 2013e. Content determination of cinnamaldehyde and eugenol in Ruikeweirui disinfectant by HPLC. *Xibei Yaoxue Zazhi*, **28** (3), 258–260.

Wang, Z. and Wei, X., 2006. Detection of eugenol in zangyaojingxinjiaonang by GC. *Zhongchengyao*, **28** (10), 1536–1538.

Wei, B. *et al.*, 2013. Detection of sensibilized odorant in toys material with GC-MS. *Huanjing Huaxue*, **32** (12), 2406–2410.

Wei, B., Huang, G., Long, Z., and Fang, K., 2009. Determination of content of 6-gingerol in toothpaste by HPLC. *Yagao Gongye*, **2**, 40–42.

Wei, B. *et al.*, 2012. Method for simultaneously detecting contents of multiple perfuming agents capable of causing allergic reaction in toy via solid phase extraction. Chinese Patent 102,628,837, filed June 23, 2011 and issued June 26, 2013.

Wei, B. *et al.*, 2011a. Detection of allergenic fragrants in toys by gas chromatography-mass spectrometry or GC-MS and thermal desorption. Chinese Patent 102,175,791, filed Jan. 12, 2011 and issued June 26, 2013.

Wei, L. and Du, Y., 2008a. Chemical component analysis by GC-MS of supercritical-CO_2 extracts from Tibetan medicine Pazhu capsule. *Zhongchengyao*, **30** (2), 236–238.

Wei, L., Du, Y., Zou, X., and Xiao, Y., 2008b. Determination of piperine in Tibet medicine Pazhu capsules by HPLC. *Zhongchengyao*, **30** (3), 454–455.

Wei, P. and Ni, M., 2011b. Method for determining content of eugenol in plaster with stomach warming and analgesic effects by gas chromatography. Chinese Patent 102,253,146, filed July 14, 2011 and issued Nov. 23, 2011.

Weon, J.B. *et al.*, 2013. Qualitative and quantitative analysis of nine major compounds in the Bozhougyiqi-Tang using a high-performance liquid chromatography coupled with a diode array detector and electrospray ionization mass spectrometer. *Pharmacognosy Magazine,* **9** (35), 271–282.

Weon, J.B. *et al.*, 2012. Simultaneous quantification of eleven chemical components in traditional herbal medicinal formula Socheongryongtang by HPLC-DAD and LC-MS. *Journal of Liquid Chromatography & Related Technologies,* **35** (16), 2243–2254.

Wu, G. *et al.*, 2005. GC-MS analysis of extracts from bark of Albizia julibrissin by supercritical fluid CO_2. *Zhongcaoyao,* **36** (6), 832–833.

Wu, H. *et al.*, 2012a. Method for identifying used cooking oil via liquid chromatography-tandem mass spectrometry (LC-MS/MS). Chinese Patent 102,680,608, filed June 13, 2012 and issued Oct. 2, 2013.

Wu, H.-Q. *et al.*, 2012b. A novel method for the identification of gutter oils using LC-MS/MS. *Fenxi Ceshi Xuebao,* **31** (9), 1031–1036.

Wu, Q., Tan, J.-J., Liu, Y.-M., and Shi, Z.-F., 2011a. Study on purity of piperine by ultraviolet-visible spectrophotometry. *Anhui Nongye Kexue,* **39** (7), 3973–3974.

Wu, W. *et al.*, 2012c. Component analysis on ethanol extract of RHIZOMA ZINGIBERIS RECENS from different producing areas and assay of 6-gingerol. *Medicinal Plant,* **3** (1), 62–65.

Wu, X. *et al.*, 2011b. Method for simultaneously measuring contents of piperine and piperlonguminine in Piper longum by HPLC. Chinese Patent 102,288,712, filed May 6, 2011 and issued June 19, 2013.

Wua, Y., Sun, S., Zhou, Q., and Leung, H., 2008. Fourier transform mid-infrared (MIR) and near-infrared (NIR) spectroscopy for rapid quality assessment of Chinese medicine preparation Honghua Oil. *Journal of Pharmaceutical and Biomedical Analysis,* **46** (3), 498–504.

Xiang, Y., Jiang, S., and Huang, Z., 2013. Determination of 6-gingerol in rat liver microsomes by HPLC. *Zhongyao Xinyao Yu Linchuang Yaoli,* **24** (1), 74–76.

Xiao, H., 2005. Study on allyl isothiocyanate (AITC), a pungent flavor substance in food, PhD thesis, the China Agricultural University, Beijing, China.

Xiao, W. *et al.*, 2012. Method for detecting Guizhifuling capsule by near-infrared spectroscopy combined with correction models established by use of HPLC. Chinese Patent 102,759,509, filed April 25, 2011 and issued Oct. 31, 2012.

Xie, Z., Guo, L., Long, R., and Jiang, X., 2010. Determination of 6-gingerol in gingerol soft capsules by HPLC. *Zhongnan Yaoxue,* **8** (2), 133–135.

Xin, Y., Wang, S., Liu, Z., and Sheng, W., 2009. Determination of piperine in Meng medicine Narusanwei pills. *Shizhen Guoyi Guoyao,* **20** (8), 1881–1882.

Xiong, J. *et al.*, 2011a. Electrochemical behavior and determination of eugenol on modified montmorillonite modified glassy carbon electrode. *Fenxi Shiyanshi,* **30** (11), 117–120.

Xiong, X., Hu, W., and Qu, W., 2011b. Determination of eugenol at a nano-WO3 modified carbon paste electrode. *Hubei Minzu Xueyuan Xuebao, Ziran Kexueban,* **29** (4), 419–422.

Xu, A. *et al.*, 2005a. Determination of capsaicin and dihydrocapsaicin by high performance liquid chromatography (HPLC). *Huazhong Nongye Daxue Xuebao,* **24** (6), 592–595.

Xu, L., Li, Q., Ma, J., and Wang, Y., 2012. Preparation and quality control of ginger's extract dropping pills. *Zhongguo Yiyao Daobao,* **9** (9), 45–47.

Xu, L., Xi, H., Zhang, X., and Bi, K., 2005b. Determination of piperine in Mongolian medicine Yizhiwenshenshiwei pills by HPLC. *Huaxue Fenxi Jiliang,* **14** (3), 23–24.

Xu, Y. *et al.*, 2014. Determination of cinnamaldehyde in 5 different standards of Cinnamomum cassia and two kinds of adulterants by HPLC-DAD. *Zhongguo Shiyan Fangjixue Zazhi,* **20** (1), 90–93.

Xun, S. *et al.*, 2012. Method for simultaneously detecting ten volatile phenols in distilled spirit via dispersive liquid-liquid microextraction-fluorescence detection-high performance liquid chromatography (dllme-fld-hplc). Chinese Patent 102,590,373, filed Jan. 20, 2012 and issued June 19, 2013.

Yadav, H. *et al.*, 2012. TLC-densitometric quantification of negundoside, ursolic acid, eugenol, lupeol, and β-sitosterol using HPTLC from Vitex negundo leaves. *Journal of Liquid Chromatography & Related Technologies,* **35** (11), 1565–1584.

Yan, D., 2007. HPLC determination of 6-gingerol in ginger oil. *Yaowu Fenxi Zazhi,* **27** (5), 733–734.

Yan, S. *et al.*, 2009. Chemical fingerprinting and quantitative analysis of volatiles in Shexiang Baoxin Pill by gas chromatography with flame ionization and mass spectrometric detection. *Journal of Analytical Chemistry*, **64** (2), 149–155.

Yang, H.J. *et al.*, 2012a. Qualitative and quantitative simultaneous determination of six marker compounds in soshiho-tang by HPLC-DAD-ESI-MS. *Archives of Pharmacal Research*, **35** (10), 1785–1791.

Yang, Q., Chen, S., Zou, L., and Lian, P., 2009. GC-MS analysis of chemical components of essential oil from leaves of Cinnamomum chekiangense. *Shengwuzhi Huaxue Gongcheng*, **43** (3), 25–27.

Yang, X., 2007. Flavor of the unique spice huajiao (Zanthoxylum bungeanum Maxim). Abstracts, 39th Central Regional Meeting of the American Chemical Society, Covington, KY, United States, May 20–23, 2007, pp. CRM-167.

Yang, X., 2008. Aroma constituents and alkylamides of red and green huajiao (Zanthoxylum bungeanum and Zanthoxylum schinifolium). *Journal of Agricultural and Food Chemistry*, **56** (5), 1689–1696.

Yang, X.-L., Ding, H.-M., Hong, R.-Y., and Shi, Y.-Y., 2013. Comparison of the two methods for measurement of piperine in pepper. *Fujian Fenxi Ceshi*, **22** (4), 28–30.

Yang, X., Yin, G., and Liu, C., 2012b. Rapid detection of piperine in pepper by ultra performance liquid chromatography. *Zhongguo Tiaoweipin*, **37** (1), 74–76.

Yang, Z. and Zhang, Q., 2012c. Gas chromatography (GC) method for detecting contents of free carbon dioxide, carbonate and bicarbonate in water. Chinese Patent 102,507,762, filed Oct. 18, 2011 and issued Aug. 28, 2013.

Yao, R., Hou, K., and Xu, H., 2007. Determination of piperine in Shilioujianwen jiaonang by HPLC. *Zhonghua Yiyao Zazhi*, **7** (6), 545–547.

Yardim, Y., 2011. Sensitive detection of capsaicin by adsorptive stripping voltammetry at a boron-doped diamond electrode in the presence of sodium dodecylsulfate. *Electroanalysis*, **23** (10), 2491–2497.

Yardim, Y. and Senturk, Z., 2013. Electrochemical evaluation and adsorptive stripping voltammetric determination of capsaicin or dihydrocapsaicin on a disposable pencil graphite electrode. *Talanta*, **112**, 11–19.

Ye, J., Xiao, M., and Zhang, X., 2011. RP-HPLC determination of piperine in Honghua Ruyi capsules. *Yaowu Fenxi Zazhi*, **31** (9), 1812–1814.

Ye, J. *et al.*, 2012. Determination of eight drug potential components in the Xiaobanxia plus Fuling Decoction by HPLC. *Liaoning Zhongyi Zazhi*, **39** (8), 1595–1598.

Yin, L. *et al.*, 2007. Analysis of effective constituents in Ramulus Cinnamomi and Cortex Cinnamomi from different habitats. *Zhongcaoyao*, **38** (7), 1094–1096.

Yin, Y. *et al.*, 2014. Simultaneous determination of benzoylmesaconine and piperine in rat plasma after oral administration of Naru-3 by an ultra fast liquid chromatography-tandem mass spectrometry method and its application in a comparative pharmacokinetic study. *Analytical Methods*, **6** (10), 3420–3428.

Yu, H., Yu, X., Jiao, Y., and Liu, X., 2009. Study on quality analysis of 24 batches commercial Fructus Piperis Longi. *Zhongguo Xiandai Zhongyao*, **11** (5), 35–38.

Yu, N.H. *et al.*, 2008. Comprehensive screening of acidic and neutral drugs in equine plasma by liquid chromatography-tandem mass spectrometry. *Journal of Chromatography A*, **1189** (1–2), 426–434.

Yu, X., Shu, J., Huang, H., and Zheng, M., 2010. Determination of metronidazole and eugenolum in Fengjiao Yatong Ding by HPLC. *Zhonghua Zhongyiyao Xuekan*, **28** (2), 429–431.

Yu, X.-Q., Zhang, L.-P., and Kan, J.-Q., 2012. Comparative analysis on HPLC fingerprints of Zanthoxylum bungeanum Maxim and Zanthoxylum Schinifolium Sieb. et Zucc. *Shipin Yu Fajiao Gongye*, **38** (4), 171–174.

Yu, Y. *et al.*, 2011. Examination of the pharmacokinetics of active ingredients of ginger in humans. *AAPS Journal*, **13** (3), 417–426.

Yue, H., Shao, Y., Mei, L., and Tao, Y., 2013. *Method for extracting and purifying 6-gingerol from rhizoma Zingiberis*. Chinese Patent 102,976,909, filed Dec. 19, 2012 and issued Dec. 3, 2014.

Yun, S.-M. *et al.*, 2010. Quantitative analysis of eugenol in clove extract by a validated HPLC method. *Journal of AOAC International*, **93**(6), 1806–1810.

Yusufoglu, H. and Alqasoumi, S.I., 2008. High performance thin layer chromatographic analysis of 10-gingerol in Zingiber officinale extract and ginger-containing dietary supplements, teas and commercial creams. *FABAD Journal of Pharmaceutical Sciences*, **33**, 199–204.

Zarai, Z. *et al.*, 2013. Antioxidant and antimicrobial activities of various solvent extracts, piperine and piperic acid from Piper nigrum. *LWT–Food Science and Technology*, **50** (2), 634–641.

Zhan, K., Wang, C., Xu, K., and Yin, H., 2008. Analysis of volatile and non-volatile compositions in ginger oleoresin by gas chromatography-mass spectrometry. *Sepu*, **26** (6), 692–696.

Zhang, C. *et al.*, 2012a. Determination or piperine in Longluo Qutong ointment by HPLC. *Zhongguo Yaoye*, **21** (13), 33–34.

Zhang, H. *et al.*, 2012b. Method for detecting content of eugenol in tobacco shred. Chinese Patent 102,495,162, filed Nov. 16, 2011 and issued Nov. 20, 2013.

Zhang, H., Su, S., and Lin, Y., 2006. Determination of piperine content in Hufan buccal pellicle by HPLC. *Zhongguo Yiyuan Yaoxue Zazhi*, **26** (5), 630–631.

Zhang, J. *et al.*, 2013b. Sensitive and rapid determination of capsaicin using acetylene black nanoparticles modified electrode. *Nanoscience and Nanotechnology Letters*, **5** (6), 707–711.

Zhang, J. *et al.*, 2011a. RP-HPLC simultaneous determination of three capsaicinoids constituents in fruits of Capsicum annuum. *Yaowu Fenxi Zazhi*, **31** (2), 244–246.

Zhang, J. *et al.*, 2013a. HPLC determination of cinnamaldehyde, eugenol, atractylodin in inclusion complex of compound oil-β-cyclodextrin. *Zhongguo Shiyan Fangjixue Zazhi*, **19** (3), 86–89.

Zhang, K. *et al.*, 2010a. Method for measuring content of piperine, and application thereof. Chinese Patent 101,852,790, filed May 26, 2010 and issued Oct. 6, 2010.

Zhang, K., Ma, C., and Miao, L., 2013c. Changes of active constituents in zingiberis recens rhizoma-citri reticulatae pericarpium decoction. *Zhongchengyao*, **35** (8), 1697–1700.

Zhang, K., Song, S., Cui, X., and Xu, Z., 2009a. Determination of 6-gingerol and 6-shogaol in ginger of different districts in China. *Zhongguo Yaoxue Zazhi* , **44** (22), 1692–1694.

Zhang, L., Wang, Z., Wang, W., and Gao, H., 2008. RP-HPLC fingerprint evaluating different ginger juice as processing material. *Zhongguo Zhongyao Zazhi*, **33** (9), 1010–1013.

Zhang, P. *et al.*, 2011b. Determination of plasma concentration of capsaicin nanoemulsions in rabbits by HPLC-FLU and study of its pharmacokinetics. *Zhongguo Yaofang*, **22** (21), 1938–1940.

Zhang, Q., Hu, J., Sheng, L., and Li, Y., 2010b. Simultaneous quantification of capsaicin and dihydrocapsaicin in rat plasma using HPLC coupled with tandem mass spectrometry. *Journal of Chromatography B: Analytical Technologies in the Biomedical and Life Sciences*, **878** (24), 2292–2297.

Zhang, S. *et al.*, 2012c. Simultaneous determination of 6 chemical constituents in Yigan Fuzheng capsules by HPLC. *Zhongguo Yaofang*, **23** (27), 2546–2549.

Zhang, W.-W. and Wang, Y.-H., 2014. HPLC determination of piperine in Aoyou Bawei powder. *Zhongnan Yaoxue*, **12** (4), 373–375.

Zhang, Y. and Wang, Z., 2009b. Study on chemical constituents of essential oil from semen cannabis. *Tianran Chanwu Yanjiu Yu Kaifa*, **21** (2), 259–262.

Zhang, Y. *et al.*, 2011c. Determination of cinnamaldehyde content in Huanfuyiha microemulsion by HPLC. *Zhongguo Yiyuan Yaoxue Zazhi*, **31** (19), 1657–1659.

Zhang, Z., Ren, F., and Zhang, P., 2012d. Determination of capsaicinoids and eugenol in waste-edible-oil by liquid-liquid extraction and liquid chromatography-tandem mass spectrometry. *Sepu*, **30** (11), 1108–1112.

Zhao, C., Hou, Q., and Yu, J., 2011. Rapid determination of capsaicinoids by ultra performance liquid chromatography. *Huaxue Fenxi Jiliang*, **20** (5), 38–40.

Zhao, D., Tang, J., and Ding, X., 2007. Analysis of volatile components during potherb mustard (Brassica juncea, Coss.) pickle fermentation using SPME–GC-MS. *LWT*, **40**, 439–447.

Zhao, T. *et al.*, 2012a. Method for identifying active component group in medicinal plant by inverted gated decoupling (IGD) 13C nuclear magnetic resonance (NMR) spectra finger print detection. Chinese Patent 102,749,348, filed June 18, 2012 and issued June 11, 2014.

Zhao, Z. *et al.*, 2012b. Method for extracting hydroxysanshool from Zanthoxylum bungeanum oil. Chinese Patent 102,690,208, filed May 31, 2012 and issued April 16, 2014

Zhao, Z.-F. *et al.*, 2013. Characterization and comparison of the pungent components in commercial Zanthoxylum bungeanum oil and Zanthoxylum schinifolium oil. *Journal of Food Science*, **78** (10), C1516–C1522.

Zheng, L. *et al.*, 2013. Method for detecting prohibited and restricted compounds in toys by accelerated solvent extraction and GC-MS. Chinese Patent 103,235,054, filed April 17, 2013 and issued Sept. 10, 2014.

Zhong, M.-Y., Quan, S.-C., and Hu, J.-H., 2006. Simultaneous determination of curcumin and piperine in turmeric capsule by HPLC. *Yaoxue Fuwu Yu Yanjiu*, **6** (1), 54–56.

Zhou, L., Lian, Y., Cheng, Y., and Zhang, Q., 2011. Study of capsaicin and dihydrocapsaicin content in different parts of peppers by HPLC. *Zhongguo Shipin Tianjiaji*, **2**, 231–234.

Zhou, M., Yang, C., and Fu, S., 2010. Determination of limonene and eugenol in soft capsules of health food by GC/MS. *Yufang Yixue Qingbao Zazhi*, **26** (9), 763–765.

Zhu, J., He, W., Li, Y., and Ma, J., 2010. Determination of piperine in Tongluo Zhitong patch by RP-HPLC. *Guangdong Yaoxueyuan Xuebao*, **26** (4), 355–357.

Zick, S.M. *et al.*, 2010. Quantitation of 6-, 8- and 10-gingerols and 6-shogaol in human plasma by high-performance liquid chromatography with electrochemical detection. *International Journal of Biomedical Science* , **6** (3), 233–240.

Ziino, M. *et al.*, 2009. Volatile compounds and capsaicinoid content of fresh hot peppers (Capsicum annuum L.) of different Calabrian varieties. *Journal of the Science of Food and Agriculture*, **89** (5), 774–780.

Ziyatdinova, G., Ziganshina, E., and Budnikov, H., 2013. Voltammetric sensing and quantification of eugenol using nonionic surfactant self-organized media. *Analytical Methods*, **5** (18), 4750–4756.

Zuo, H., Bao, J., and Mao, X., 2010. Quality specification for Fructus Capsici Annui. *Yaowu Fenxi Zazhi*, **30** (12), 2293–2298.

CHAPTER 12

Chemesthesis and health

Richard D. Mattes[1] and Mary-Jon Ludy[2]

[1]*Department of Nutrition Science, Purdue University, West Lafayette, IN, USA*

[2]*Department of Public and Allied Health, Bowling Green State University, Bowling Green, OH, USA*

12.1 Introduction

Spices and herbs – non-leafy and leafy parts of plants with selected sensory properties, respectively – have a long history of use for the prevention and treatment of a wide variety of health conditions. Their use predates recorded history and they are the mainstay of Ayurvedic and Chinese medicine. These spices and herbs are often used in preparations that include much of the original plant, and in combinations with other plant types, as the balance between elements in the preparations are believed to hold diagnostic and therapeutic value. This practice of combining multiple plant types confounds efforts by Western methodologies to ascribe efficacy to a specific chemical or spice that can be isolated and encapsulated into a dietary supplement or pharmaceutical remedy. Consequently, application of historical knowledge and beliefs about plant-based health remedies has been limited in Western nations. However, with increased awareness of, and interest in, alternative health management techniques has come a growing openness to their potential efficacy for numerous conditions. In the present chapter, we limit consideration to effects of single pungent spices (i.e., black pepper, chili peppers, cinnamon, ginger, mustard, and saffron) on dimensions of energy balance (i.e., appetite, food intake, energy expenditure, and body weight).

The use of pungent spices is growing exponentially in the United States. Compared with the 1950s when households cooked with ten or fewer spices, the contemporary American spice rack contains a minimum of 40 different seasonings (Reinagel, 2012). Parallel to this expansion of the American spice palette, 2013 marked the first year that the majority (52%) of consumers reported finding hot and spicy flavors appealing. Fruity flavors paired with pungent spices – chipotle lime, mango habanero, and orange chili – are among the flavor combinations topping consumer preferences (Technomic, Inc., 2013, cited in Sloan, 2014). While the preferred level of spice varies dramatically worldwide – with high consumers in Mexico ingesting the equivalent of 9–25 jalapeño peppers per day (Lopez-Carrillo *et al.*, 2003), and high consumers in the United States eating less than half a pepper daily (Ludy and Mattes, 2011) – demand is undoubtedly increasing. On a global scale, it is estimated that one in four individuals consume hot red peppers on a daily basis (Szallasi and Blumberg, 1999). The health implications of spices in the food supply are multifold in that consumers, across ethnic groups (Glanz and Basil, 1998), rank flavor (90%) as the most influential factor in food choice (Tepper and Trail, 1998; Dressler and Smith, 2013; International Food Information Council Foundation, 2014). In addition, the bioactive compounds present in these spices purportedly reduce the risk for a wide range of health conditions (e.g., heart disease, diabetes, cancer, and obesity).

12.2 Cultural patterns of intake

Cultural and environmental conditions play key roles in individual motivation to consume pungent spices. Use – characterized by the proportion of recipes calling for spices and number of spices per recipe – is positively associated with average annual regional temperature (Billing and Sherman, 1998). Tropical countries such as India, Indonesia, Malaysia, and Thailand (average temperature = 27 °C) enjoy the greatest intake of pungent spices (e.g., black pepper, garlic, ginger, and onion). Scandinavian countries, such as Finland and Norway (average temperature = 3 °C), have the lowest consumption of pungent spices. Similarly, within countries, such as China and the United States, where there is a wide variance in latitude, use of pungent spices increases when approaching the equator (Sherman and Billing, 1999). There are multiple potential health-related explanations for this variance, with the most plausible hypotheses relating to body temperature regulation and microbial protection.

From the perspective of body temperature regulation, individuals residing in warmer climates believe that the distinctive "gustatory sweating" – facial flushing extending to the chest and shoulders, and perspiration on the chin, ears, face, forehead, lips, neck, nose, and scalp – evoked by pungent spices produces a cooling effect (Nabhan, 2004). This theory is highly prevalent in the historical literature, with demonstrations of marked gustatory sweating when chili peppers were chewed and chili paste was painted on various regions of the oral cavity (Lee, 1954; Haxton, 1948). For example, pungent foods, among them a native ginger soup, are consumed continuously for 30 days by indigenous postpartum women in Indonesia and Malaysia for the purpose of cooling the body and "sweating away impurities" (Henry and Piggott, 1987). Alternatively, residents of cooler climates perceive that pungent spices promote a warming sensation. Among villagers of the Mexican highlands, 69% agree that spicy food "makes me feel warm inside" and 15% believe that "it cools me off" (Rozin and Schiller, 1980). This aligns with recent survey data collected in North America at a Northern latitude with a cooler, non-tropical climate where 57% of university students agreed that spicy food provided an internal warming sensation while only 4% indicated that it evoked cooling (Ludy, 2013).

Although seemingly contradictory, both warming and cooling effects have been demonstrated in the experimental literature with hot red peppers, mustard, and ginger. Core (tympanic) temperature increased in a Japanese study where healthy young adults (n = 5M (male), 7F (female)) ingested whole hot red peppers containing 0.1 mg kg^{-1} capsaicin (the compound responsible for their pungency)

(Hachiya *et al.*, 2007). Similarly, core (rectal) temperature increased in an Australian study where male endurance athletes ($n = 6$) consumed a traditional meal with a spicy sauce containing 3 g each of Tabasco sauce and hot English mustard (Edwards *et al.*, 1992). In an American study with healthy young adults (n = 14M, 11F), 1 g cayenne red pepper (2 mg capsaicin) mixed into a meal increased core (ingested sensor) temperature, while promoting evaporative cooling by reducing skin (neck) temperature (Ludy and Mattes, 2011). Animal data support this dual role of capsaicin in body temperature perception. When a single dose of capsaicin (5 mg kg^{-1}) was injected into Wistar rats, it independently facilitated heat production (i.e., oxygen consumption) via the brainstem and induced heat loss (i.e., decreased tail temperature suggesting vasodilation) via the forebrain (Osaka *et al.*, 2000). Another study showed no variation in body temperature when male Wistar rats were chronically fed a 2% ginger-containing chow, the highest dose that rats will reliably eat. In contrast, rapid, dose-dependent decreases in body temperature and metabolic rate were demonstrated following single intraperitoneal injections of ginger (2.5 or 25 mg kg^{-1}) (Ueki *et al.*, 2008).

From a microbial perspective, use of pungent spices may serve as a cultural response to protect populations from foodborne microorganisms. Analyses of "traditional" recipes across every continent and most major linguistic groups – those that predate widespread access to refrigeration – show great variability in the individual types of spices used. However, a few overall trends can be detected. Firstly, annual regional temperature and spice use in general are positively correlated. For example, Indian cuisine incorporates 25 different spices with an average recipe calling for 9.3 spices, while Norwegian cuisine includes 10 different spices with an average of only 1.6 spices per recipe. This is noteworthy because annual temperature is a relative indicator of food spoilage and there is no correlation between the number of spices grown in a country and its temperature (Sherman and Billing, 1999). Secondly, meat-based recipes contain more spices than plant-based recipes regardless of geographical region (3.9 versus 2.4, respectively) (Sherman and Hash, 2001). This is notable given that unrefrigerated meats spoil more quickly and are more likely to result in foodborne illness outbreaks than plant-based foods (Sockett, 1995). Thirdly, there is a positive relationship between use of highly antimicrobial spices and average temperature, within and between countries. A review compiling the data from laboratory testing of 30 common spices (e.g., black pepper and ginger) – both powdered or as purified active ingredients – demonstrated that all inhibited (i.e., killed or slowed the growth of) at least 25% of foodborne bacterial species, such as *Clostridium botulinum, Listeria monocytogenes, Escherichia coli, Salmonella pullorum,* and *Staphylococcus aureus,* which are common in foodborne illness outbreaks and widely distributed globally. Fifteen highly antimicrobial spices, including chili peppers, cinnamon, and mustard, killed at least 75% of foodborne bacteria (Billing and Sherman, 1998). This suggests that adding pungent spices to the food prepared in hot climates, where foodborne illness is more likely to occur, has greater potential to provide antibacterial protection. Finally, the concentrations of spices used in traditional cooking, generally 0.25–3 g kg^{-1}, are sufficient to yield bacterial inhibition (Sherman and Hash, 2001). This is important, because it implies that health benefits can be achieved by eating well-accepted spicy foods that are part of traditional cuisines, not swallowing encapsulated powders or employing pharmacologic intervention, probably increasing consumer acceptance and sustainability of effects.

12.3 Appetite

It is intuitive that appetitive sensations are important drivers of ingestive behavior and, as a consequence, health. This has spawned interest in the possible role of pungent spices on appetite and intake. There are numerous reviews summarizing this perspective (Ludy *et al.*, 2012; Whiting *et al.*, 2012; Mattes, 2012). Of course, spices are often added to foods to enhance their appeal and, presumably, stimulate intake. Such use has not attracted the same level of scientific scrutiny. Both roles are reviewed here following a brief consideration of the indices of appetite.

Briefly, appetite is commonly divided into more focused sensations in the literature. "Hunger" refers to the sensations that prompt the initiation of an eating or drinking event. It is often regarded as a reflection of energy-based need. "Desire to eat" sensations may also lead to meal initiation, but are related more strongly to cognitive and sensory cues. So, it is possible to not be hungry and have a strong desire to eat, as may occur when presented with palatable desserts after a large meal. These two sensations largely determine eating frequency. "Prospective consumption", or a rating of how much an individual could consume at a point in time, is a purported predictor of portion size. The term "fullness" is primarily used to describe the sensations that terminate an eating event. Thus, it determines portion size. It is not simply the opposite of hunger, as different mechanisms underlie the two sensations. For example, hunger has been linked to the concentrations of ghrelin, whereas fullness is more tied to cholecystokinin (CCK), glucagon-like peptide-1 (GLP-1), and peptide YY (PYY), among many other gut peptides (Hameed *et al.*, 2009). Total energy intake is a function of eating frequency and portion size, so the balance between hunger, desire to eat, and fullness is critical. The literature also contains references to satiation and satiety. "Satiation" is the term for sensations that determine the amount, in grams or kilocalories, consumed within a single ingestive event. It is a determinant of meal or portion size. It is commonly measured as "fullness". "Satiety" is the sensation influencing the inter-meal interval or eating frequency. It is commonly measured in units of time. However, in some instances, the amount consumed in an *ad libitum* challenge or test meal provided after a given intervention (e.g., high protein snack) is used as a metric for satiety.

12.3.1 Suppression of appetitive sensations

12.3.1.1 Hot red pepper

There is substantial literature on the appetitive effects of capsaicin, a lipophilic alkaloid responsible for the irritancy of hot red peppers, though much of this work fails to acknowledge the importance of customary exposure. As described previously, high concentrations of hot red peppers are common in some cuisines and not in others. Because familiarity is an important determinant of acceptability, discrepant responses may be obtained across cultures. In an early Japanese trial, healthy females

($n = 13$ habitual spicy food consumers) ingested high-fat and high-carbohydrate breakfast meals with and without 10 g dried red pepper (providing 30 mg capsaicin). Following the high-fat meal, ratings of hunger, desire to eat, and prospective consumption all rose in the 3 hours before an *ad libitum* lunch when no red pepper was consumed, but were suppressed when red pepper was included (Yoshioka *et al.*, 1999). Fullness was not altered by red pepper ingestion, nor was any rating after the high-carbohydrate meals. Thus, capsaicin ingestion dampened a number of the appetitive responses that would presumably promote energy intake after the high-fat meal. This occurred at a higher level of spiciness than customary meals, but without reports that the red pepper dose rendered the food unpleasant. In contrast to the favorable hedonic ratings provided by Japanese and Korean participants to the high capsaicin doses consumed in early studies (Yoshioka *et al.*, 1998; Lim *et al.*, 1997), the same doses were deemed highly unacceptable by less familiar Western populations in the United States and the Netherlands (Ludy *et al.*, 2012; Reinbach *et al.*, 2010) and likely led to lower appetitive ratings and intake based on the discomfort they produced.

Several acute and longer-term trials have addressed whether moderate concentrations of pungent spices in foods suppress appetitive sensations. In one acute Danish study, participants ($n = 17$M, 23F who liked hot spices) received lower capsaicin doses (0.375 mg, divided equally between dried chili powder and freshly ground chili) mixed into a starter meal and followed by an *ad libitum* buffet (Reinbach *et al.*, 2010). The mean palatability rating of the food with chili pepper was about 6 on a 9-point scale (9 = like very much). While there was a small increase in the desire to eat items with selected qualities (e.g., sweetness), overall effects on appetite were weak. Similar findings were obtained with other spicy foods including horseradish, ginger, mustard, and wasabi. Another acute study conducted in the United States evaluated the effects of capsaicin provided through doses indicated by individual participants as palatable (mean dose ≈1 g dried cayenne red pepper, providing 2 mg capsaicin and 53 800 Scoville Heat Units (SHU) per gram) in a serving of tomato soup (Ludy and Mattes, 2011). No effect of the red pepper was observed for hunger, fullness, prospective consumption, or desire to eat any food.

It has been proposed that the addition of capsaicin to the diet of energy-restricted individuals would augment satiety sensations and aid dietary adherence (Smeets *et al.*, 2013; Janssens *et al.*, 2014). In a 36-hour Dutch study where individuals were administered a diet providing either 100 or 75% of estimated energy needs, participants ($n = 8$M, 7F) were administered 1.03 g red pepper (2.5 mg capsaicin and 39 050 SHU) prior to eating a meal (Janssens *et al.*, 2014). Palatability ratings were comparable across the conditions. Under energy balance, the addition of red pepper led to heightened area under the curve (AUC) fullness ratings, but no difference in hunger or desire to eat. When energy intake was restricted, there was no difference in any appetitive index between periods with and without added capsaicin. In another similar 36-hour Dutch trial ($n = 12$M, 12F) with a 20% energy restriction and including manipulation of protein concentrations, 24-hour AUC values (36-hour values were not reported) were higher when capsaicin was consumed relative to control conditions under energy balance or 20% energy restriction (Smeets *et al.*, 2013). However, the addition of capsaicin did not offset the reduction of fullness due to energy restriction and the addition of protein led to a greater appetitive effect than the capsaicin. Hunger and desire to eat ratings were not reported. In a longer-term trial where Dutch participants were administered 135 mg day^{-1} encapsulated capsaicin (45 mg before each of 3 meals daily for 3 months; n = 12M, 30F) or placebo (n = 11M, 38F), no significant effects were noted for hunger or satiety (Lejeune *et al.*, 2003).

The effects of capsaicin on appetite are heavily dependent on the perceived irritation it provides. In Dutch work (n = 12M, 12F) where the addition of capsaicin to a meal resulted in augmented fullness, this was not observed when an equivalent dose was ingested in capsule form (Westerterp-Plantenga *et al.*, 2005). In addition, American participants (n = 4M, 8F) who do not customarily consume spicy foods reportedly reduce their desire to eat foods with selected properties (e.g., fatty, salty) relative to the same foods consumed without red pepper (Ludy and Mattes, 2011). However, these differences are not observed in spicy food consumers (n = 10M, 3F).

Because of the confounding effects of dose, palatability, customary exposure level, and current state of energy balance, it is not possible to determine the magnitude of an independent appetite-suppressing effect of capsaicin. In one Dutch trial (n = 12M, 12F; Smeets *et al.*, 2013), capsaicin exerted a weaker effect than manipulation of protein, but the latter's effects are complex (Masic and Yeomans, 2014) hampering a direct comparison. Based on current evidence, it can only be concluded that larger effects are apparent with higher doses, especially if they are perceived as irritating and exposure occurs when individuals are in a state of relative energy balance as opposed to energy deficit. It is not possible to discern differential effects on the various dimensions of appetitive sensations. Nevertheless, several mechanisms have been proposed for an appetite suppressive effect of capsaicin. One holds that it prompts the release of gut peptides associated with satiety, such as GLP-1 (Reinbach *et al.*, 2009). Evidence that an ingested dose in which the pungency is not sensed is ineffective suggests that appetitive regulation is dependent on neural activation in the oral cavity (i.e., a cephalic phase response) rather than post-ingestive signaling or metabolism. Other work documents that oral exposure activates the sympathetic nervous system, leading to the hypothesis that augmented fullness could be attributable to elevated catecholamine concentrations (Yoshioka *et al.*, 1999). Again, the oral irritation is possibly responsible and this is modulated by customary levels of capsaicin use and learned flavor preferences. Aversive sensations are more effective at promoting fullness than palatable sensations.

12.3.1.2 Cinnamon

Among individuals with type 2 diabetes, cinnamon reportedly moderates serum glucose, triglyceride, low-density lipoprotein cholesterol, and total cholesterol concentrations when ingested at doses of 3 (Solomon and Blannin, 2009), 5 (Solomon and Blannin, 2007), and 6 g day^{-1} (Khan *et al.*, 2003). Less consistent findings have been reported with 1.5 (Khan *et al.*, 2010; Vanschoonbeek *et al.*, 2006) and 3 g day^{-1} (Mang *et al.*, 2006), or an extract providing the equivalent of 20 g day^{-1} (Ziegenfuss *et al.*, 2006). Based on the hypothesis that changes of blood glucose and/or insulin drive hunger (Flint *et al.*, 2007), cinnamon has been hypothesized to reduce hunger. However, in a Swedish crossover study where healthy participants without diabetes

(n = 8M, 6F) consumed rice pudding with or without 6 g cinnamon, there was no difference in reported hunger/satiety rated on a category scale ranging from extreme hunger (−10) to extreme satiety (+10) (Hlebowicz *et al.*, 2007). In healthy overweight American men (n = 6), a 14 g high antioxidant spice blend containing cinnamon, as well as black pepper, cloves, garlic powder, ginger, oregano, paprika, rosemary, and turmeric, lowered postprandial insulin and triglycerides by 21 and 31%, respectively (Skulas-Ray *et al.*, 2011).

12.3.1.3 Ginger

Various gastrointestinal tract disorders – including diarrhea, dyspepsia, flatulence, nausea, and vomiting – are reportedly moderated by ginger as a traditional remedy. This is attributed in part to ginger's enhancement of gastric emptying (Wu *et al.*, 2008; Hu *et al.*, 2011). More rapid emptying would move the ingesta into the small intestine more quickly. This would thereby stimulate the release of satiety peptides such as CCK, GLP-1, and PYY, which would theoretically increase fullness ratings. However, the appetitive response to this property is unclear. Ginger could also reduce fullness ratings by diminishing signaling from gastric stretch receptors. In a small American study, where overweight males (n = 10) ingested 2 g ginger in hot water, participants reported significantly lower hunger and prospective consumption, as well as a trend for higher fullness relative to a placebo treatment (Mansour *et al.*, 2012). In contrast, studies in male Wistar rats indicate that ginger increases energy intake (Ueki *et al.*, 2008; Wadikar and Premavalli, 2011). Studies on the relationship between ginger (1.2 g loads administered in capsules), gastric emptying, and appetite document the effect of the spice on gastric emptying, but fail to support an effect on appetitive sensations (Wu *et al.*, 2008; Hu *et al.*, 2011) – even among those with symptoms of dyspepsia (Hu *et al.*, 2011). Possibly, the lower concentration and mode of ginger administration (as capsules that bypass oral sensory stimulation), as well as the small experimental energy loads provided, resulted in the failure of these latter trials to identify treatment effects on appetite.

12.3.1.4 Saffron

Anxiolytic properties have led to the hypothesis that saffron may ameliorate a stress-related desire to eat and snack (Gout *et al.*, 2010). This was tested in a French trial where overweight female participants were administered capsules with 176.5 mg of a saffron extract named "Satiereal" (n = 31) or a matched placebo (n = 29). A small, but statistically significant larger proportion of treated participants indicated the extract reduced their hunger and need to snack. No effect on satiation, or the desire to eat sweet or fatty snacks, was observed. Whether the posited mode of action was responsible for the outcome will require additional testing.

The limited evidence available on the effects of pungent spices other than capsaicin on appetitive sensations precludes conclusions about their efficacy. At present, there is only suggestive evidence of modest activity for cinnamon, ginger, and saffron. However, this should not be interpreted as grounds to dismiss their potential influence, as spice use may be more sustainable than many other proposed dietary modifications aimed at modifying appetite. Small, but sustained effects may ultimately exert a stronger impact than a large, but transient effect.

12.3.2 Enhancement of appetitive sensations

Rather than attributing an orexigenic property to any particular spice, work in this area has focused on how enhancing the flavor of foods can promote appetitive sensations and energy intake. A number of researchers posited that flavor fortification or enhancement could boost appetite and intake for populations suffering from chemosensory deficits, such as the elderly. These studies have yielded mixed effects (Schiffman and Warwick, 1988; Schiffman, 1993; de Graaf *et al.*, 1994; Griep *et al.*, 2000; Koskinen *et al.*, 2003; Essed *et al.*, 2007; Mathey *et al.*, 2001; Best and Appleton, 2011). This is probably attributable to the fact that deficits of chemosensory function in older adults are often due primarily to the effects of chronic diseases and medication use, leading to idiosyncratic changes and, presumably, remedial approaches. Thus, there is no clear single regimen to enhance the sensory experience appropriately. Other work has attempted to use spices to enhance the appeal of less palatable foods (Savage *et al.*, 2013; Henry *et al.*, 2003). This work has focused on intake rather than appetite as an outcome.

12.3.3 Decreased energy intake

Although it is often assumed that appetitive sensations are closely correlated with energy intake, this is not necessarily the case. Energy intake is modulated by numerous non-appetitive factors such as social custom, food availability, cost, and health beliefs (International Food Information Council Foundation, 2014). Given the impact of appetitive sensations on the quality of life, independent of energy balance, it is important that the effects of pungent spices on appetite and energy intake are explored separately. However, if energy intake is the outcome of primary interest, this is the behavior that should be measured.

12.3.3.1 Hot red pepper

Early Japanese work with capsaicin indicated that its inclusion in a breakfast meal (10 g red pepper, 30 mg capsaicin) did not alter energy intake at the subsequent meal ($n = 13$F), though the macronutrient distribution of the lunch meal was altered (Yoshioka *et al.*, 1999). In contrast, a Canadian study ($n = 10$M) indicated serving an appetizer (6 g red pepper, 18 mg capsaicin) just prior to lunch did result in lower energy intake at an *ad libitum* lunch meal and snack (–189 kcal (1 kcal = 4.184 kJ); Yoshioka *et al.*, 1999). A later study by the same group added red pepper to a soup and noted only a trend for reduced energy intake for Japanese males ($n = 16$) at the highest dose (2.8 mg capsaicin and 55 000 SHU) (Yoshioka *et al.*, 2004). With exposures at a dose of 0.9 g red pepper (3.8 mg capsaicin and 80 000 SHU) over 2 days, a significant reduction in daily energy intake (about –125 kcal) was observed when the spice was added to tomato juice or ingested in capsules swallowed with tomato juice. However, the effect was greater with oral exposure (Westerterp-Plantenga *et al.*, 2005).

Another trial involving Danish adults ($n = 17$M, 23F) using a lower, palatable dose of capsaicin reported no effect of capsaicin intake on energy or gram weight of food intake (Reinbach *et al.*, 2010). Reflecting the importance of customary exposure effects on participant responses, another study conducted in the United States ($n = 10$M, 3F regular spicy food users; $n = 4$M, 8F non-users) observed a reduction of energy intake (–66 kcal) at a challenge meal presented 4.5 hours after a meal containing capsaicin, but only among those with low customary exposure (Ludy and Mattes, 2011). Other Dutch work (n = 8M, 7F), noted only a trend for lower intake associated with capsaicin exposure, revealed it was present just under conditions of energy balance (−143 kcal) and not during energy deficit (Janssens *et al.*, 2014). Thus, there is limited evidence that oral exposure to capsaicin reduces energy intake under conditions that maintain food palatability.

12.3.4 Increased energy intake

As noted previously, there are limited data indicating pungent spices can be used to increase energy intake by enhancing the appeal of foods. In one American trial (n = 16M, 18F), herb-flavored dip was used to promote intake of vegetables among preschoolers. The addition of various spices led to fewer rejections and greater gram intake of vegetables (Savage *et al.*, 2013). One Chinese pilot study in hospitalized elderly individuals (n = 7M, 14F) observed higher energy intake with flavors and spices including ginger, garlic, sesame oil, oyster sauce, spiced soy sauce, and soybean paste compared with unflavored foods (Henry *et al.*, 2003). However, energy intake was still well below energy needs. Other work has revealed that the use of spices may lead to variable effects on food palatability, while energy intake remains unchanged (Reinbach *et al.*, 2010; Markey *et al.*, 2011; Gregersen *et al.*, 2013).

Spices have also been used to compensate for reduced food appeal when selected ingredients linked to health are purposefully modified. In one American trial (n = 47M, 101F), spices were added to enhance the appeal of reduced-fat foods in adults (Peters *et al.*, 2014). The foods with added spice were rated as more palatable and comparable to the full-fat versions. It is assumed this would lead to greater energy intake, but this was not measured. A similar approach has been attempted in a British study (n = 69M, 79F) with reduced sodium foods (Ghawi *et al.*, 2014). Here again, over time, the spices successfully maintained the appeal of a low-sodium soup, though acute energy intake was not altered. In female swine tested while transitioning between diets – a time when energy intake often drops – the addition of "hot flavored spices" led to marked individual differences in acceptance/preference with no overall change in energy intake (Clouard and Val-Laillet, 2014). Whether this strategy will work in the long-term with animals that like particular spices will require further study.

Another approach to spice use in maintaining the palatability of foods modified for some therapeutic use is to add selected spices that are commonly associated with the other flavor-active component to compensate for the latter's elimination. For example, vanilla is often associated with sweetness, so will the addition of vanilla permit a reduction of sugar without loss of appeal? Preliminary evidence suggests there may be some efficacy to this approach (Blank and Mattes, 1990), but further testing with chemesthetically active spices over nutritionally relevant timeframes is required to ascertain their value.

An additional variation in the use of spices to enhance intake entails adding them to foods to increase sensory variety and, as a result, dietary acceptability. Monotony can lead to reduced intake (Zandstra *et al.*, 2000), so where it is deemed desirable to promote intake of a single food or limited set of foods, it may be possible to add spices to combat monotony effects. For example, there is a considerable literature suggesting that nuts provide an array of health benefits, but to realize them, this limited class of foods must be eaten regularly (Jones *et al.*, 2014b). While a plausible hypothesis, in a 12-week trial comparing peanut intake between groups provided with a single flavor (honey roasted, salted, spicy, or unsalted) or a combination of three flavors, no effect of the flavor compounds on energy intake was observed (Jones *et al.*, 2014a). Whether this was due to a property of peanuts that renders them resistant to monotony effects, the trial duration was too short to measure differential responses, or the diversity of specific flavor compounds was not optimal, could not be determined and warrants further investigation.

12.3.4.1 Cinnamon

Energy intake has been monitored in a limited number of studies exploring the effects of cinnamon exposure. In one American trial, participants were provided with Cinnulin PF® (a water-soluble cinnamon extract) at a dose of 500 mg (equivalent to 10 g whole cinnamon powder and containing at least 1% of the putative active polyphenol polymer; n = 8M, 4F) or placebo capsules (n = 3M, 7F) twice per day. There was a trend for the Cinnulin-treated group to increase daily energy intake over 12 weeks compared with placebo treatment (+241 versus −93 kcal day^{-1}, respectively) (Ziegenfuss *et al.*, 2006).

Taken together the data provide very limited support for a positive influence of pungent spices on energy intake. Ultimately, it is more likely that efficacy will depend on their contribution to palatability rather than a metabolic effect. The most promising use may be their addition to foods with lower acceptability due to purposeful manipulation (e.g., salt, sugar, or fat reduction) for some intended purpose. There are currently no data implicating spice use with positive energy balance.

12.4 Thermogenesis

Consumption of pungent spices is often linked to enhanced thermogenesis (i.e., augmented energy expenditure and/or substrate oxidation). While hot red pepper (capsaicin) is the best studied of the pungent spices, limited data exist on the thermogenic effects of black pepper (piperine), ginger (gingerols, shogaols, and zingerone), and mustard (allyl isothiocyanate). Allyl isothiocyanate is an organosulfur compound, while capsaicin, piperine, and gingerols all have similar chemical structures with an aromatic ring and an alkyl side-chain with carbonyl function (Astrup *et al.*, 2010) (Fig. 12.1). Similarity in chemical structure may contribute to similarities in thermogenic effects, as the intensity of pungency is affected by alterations in the alkyl side-chain (e.g., chain length, degree of unsaturation, alkyl end grouping) or amide group position near the polar aromatic end (DeMan, 2013). The leading explanation for the thermogenic effects of these spices is stimulation of the sympathetic nervous system by the binding of pungent compounds to thermosensitive ion receptors, namely TRPV1 (transient receptor potential cation channel, subfamily V, member 1) and TRPA1 (transient receptor potential cation channel, subfamily A, member 1; Eldershaw *et al.*, 1994, Okumura *et al.*, 2010). Through this mechanism, the pungent principles of the spices stimulate the secretion of catecholamines from the adrenal *medulla* (Kawada *et al.*, 1988; Yoshioka *et al.*, 1995).

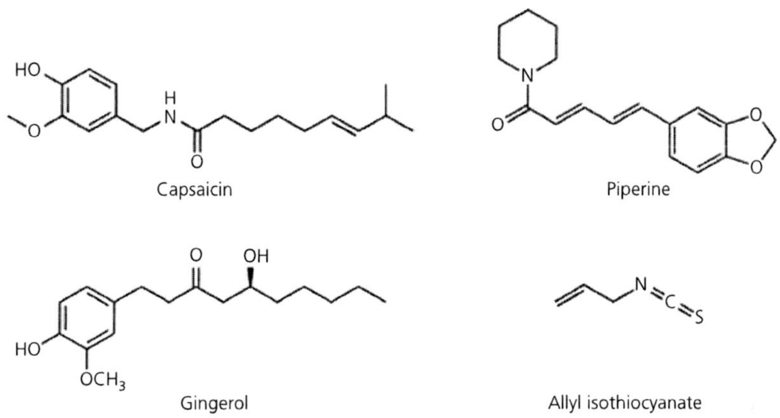

Fig. 12.1 Chemical structures of capsaicin, piperine, and gingerol have an aromatic ring and alkyl side-chain. Allyl isothiocyanate is an organosulfur compound.

Previous investigations indicate that oral sensory exposure, compared with gastrointestinal exposure via encapsulated powders, augments thermogenesis (Westerterp-Plantenga *et al.*, 2005; Ludy and Mattes, 2011). Some investigations suggest that thermogenic effects are greatest with palatable flavors (LeBlanc and Brondel, 1985), while others suggest that unpleasant flavors produce the greatest effects (Ludy and Mattes, 2011).

12.4.1 Hot red peppers (capsaicin)

Capsaicin is the best studied of the pungent compounds and exerts thermogenic effects by activating TRPV1 receptors, commonly termed "capsaicin receptors" (Caterina *et al.*, 1997). Numerous human studies have demonstrated that capsaicin increases energy expenditure (Yoshioka *et al.*, 1995; Yoshioka *et al.*, 1998; Matsumoto *et al.*, 2000; Lejeune *et al.*, 2003; Ludy and Mattes, 2011; Chaiyata *et al.*, 2003) and substrate oxidation (Yoshioka *et al.*, 1995; Yoshioka *et al.*, 1998; Lim *et al.*, 1997; Matsumoto *et al.*, 2000; Lejeune *et al.*, 2003; Ludy and Mattes, 2011; Shin and Moritani, 2007). For example, in an American study where participants (n = 14M, 11F) were fed 1 g cayenne red pepper (2 mg capsaicin) mixed into a tomato soup lunch, 4.5 hour postprandial energy expenditure increased by 10 kcal and fat oxidation was augmented (Ludy and Mattes, 2011). Similarly, in Dutch participants (n = 23M, 68F) resting energy expenditure was increased 119 kcal day^{-1} and fat oxidation was enhanced during a 3-month weight maintenance period following a 5–10% weight loss with daily ingestion of 135 mg capsaicin in encapsulated supplements (Lejeune *et al.*, 2003). Additionally, in a Thai investigation where healthy participants (n = 12F) consumed a glucose drink mixed with 5 g fresh chili pepper (3.5 mg capsaicin), postprandial energy expenditure was increased 12.1% at 5 minutes, 4.7% at 15 minutes, and 5.2% at 30 minutes after capsaicin ingestion (Chaiyata *et al.*, 2003).

However, there have been conflicting reports. For example, most studies indicate an augmentation of fat oxidation (Yoshioka *et al.*, 1998; Lejeune *et al.*, 2003; Shin and Moritani, 2007; Ludy *et al.*, 2012; Ludy and Mattes, 2011), while fewer studies suggest increases in carbohydrate oxidation (Yoshioka *et al.*, 1995; Lim *et al.*, 1997; Matsumoto *et al.*, 2000). Other studies find no thermogenic effects (Ahuja *et al.*, 2007; Smeets and Westerterp-Plantenga, 2009), or reductions in postprandial energy expenditure (Ahuja *et al.*, 2006). A few plausible explanations for these mixed results may be considered.

One explanation is that body composition moderates thermogenic effects. All studies demonstrating increased thermogenesis following capsaicin consumption were conducted in lean individuals (Yoshioka *et al.*, 1995; Yoshioka *et al.*, 1998; Matsumoto *et al.*, 2000; Lejeune *et al.*, 2003; Ludy and Mattes, 2011), or did not report body composition (Chaiyata *et al.*, 2003). In contrast, an Australian randomized crossover trial where participants (n = 14M, 22F) were instructed to consume a 30 g chili blend (55% cayenne red pepper) daily for 4 weeks, demonstrated no effect on two-hour postprandial diet-induced thermogenesis or substrate oxidation for the entire sample. However, energy expenditure was reduced among overweight and obese individuals (Ahuja *et al.*, 2006; Ahuja *et al.*, 2007). Likewise, when Japanese women with no long-term history of spicy food use consumed 3 mg capsaicin in a yellow curry sauce at breakfast, energy expenditure increased in lean (n = 8F), but not in overweight or obese (n = 8F) participants (Matsumoto *et al.*, 2000). Similar thermogenic effects have been noted with caffeinated coffee. Whereas caffeine stimulates thermogenesis in both lean and obese individuals, the magnitude is greater for lean individuals with larger increases in energy expenditure (Bracco *et al.*, 1995) and enhanced fat oxidation (Acheson *et al.*, 1980).

Alternatively, the dose may mediate the thermogenic response. A meta-analysis showed that higher doses (135–150 mg capsaicin, equivalent to 45–50 g cayenne red pepper or 13–15 jalapeño peppers per day) provide the greatest thermogenic effects (Ludy *et al.*, 2012). In line with this finding, a Dutch randomized crossover trial where participants (n = 11M, 19F) were served a low-dose (1 g) cayenne red pepper lunch containing pasta, sausage, and tomato sauce showed no effect on 3.5-hour postprandial energy expenditure compared with a pepper-free lunch (Smeets and Westerterp-Plantenga, 2009). A caveat is that large doses require compliance. This may have contributed to the lack of significant thermogenic differences in "naïve or infrequent consumers" who were supposed to consume 16.5 g cayenne red pepper (33 mg capsaicin) per day (Ahuja *et al.*, 2006; Ahuja *et al.*, 2007). Similar compliance issues have been noted with the bitter flavors of catechins and caffeine in tea (Hursel *et al.*, 2009).

12.4.2 Black pepper (piperine)

Piperine, an alkaloid, is the primary chemical compound responsible for the pungency of black pepper. It has the dual role of activating two thermosensitive receptors, TRPV1 (McNamara *et al.*, 2005) and TRPA1 (Okumura *et al.*, 2010). Piperine stimulates TRPV1 more strongly than TRPA1, but not as strongly as capsaicin or allyl isothiocyanate (present in mustard) (Okumura *et al.*, 2010). However, piperine is better absorbed than capsaicin (97% versus 80%, respectively) (Ganesh Bhat and Chandrasekhara, 1986; Iwai *et al.*, 2003). It is the most commonly consumed spice worldwide and is increasingly being consumed to stimulate metabolism (Szallasi, 2005); however, the reports of its effects on thermogenesis are sparse and contradictory.

When piperine was infused into the hind limb of male Wistar rats, it stimulated oxygen uptake in a dose-dependent manner (Eldershaw *et al.*, 1994). In another study of male Wistar rats, administration of piperine led to increased catecholamine secretion, particularly epinephrine, which is suggestive of thermogenic enhancement by the sympathetic nervous system (Kawada *et al.*, 1988). In contrast, a study of male Swiss albino mice demonstrated reduced thyroid hormone concentrations – both triiodothyronine (T3) and thyroxin (T4) – after 15 days of daily oral piperine administration, which suggests decreased thermogenesis (Panda and Kar, 2003).

To date, only two human studies have investigated the effects of piperine on energy expenditure and thermogenesis. The first was a Danish randomized cross-over trial where participants ($n = 22$ healthy young males who liked spicy food) ingested a brunch meal with 1.3 g black pepper divided between shredded beetroot and scrambled eggs. Four-hour postprandial diet-induced thermogenesis and substrate oxidation did not vary for the piperine meal compared with a no-pepper control (Gregersen *et al.*, 2013). The second was an American randomized cross-over trial where participants ($n = 17$ overweight post-menopausal females) were fed 1.5 g black pepper (68.6 mg piperine) divided equally between three meals while undergoing whole room indirect calorimetry for 24 hours. Similar to the other human investigation, piperine did not alter energy expenditure or substrate oxidation compared with a no-pepper control (O'Connor *et al.*, 2013).

12.4.3 Ginger (gingerols, shogaols, and zingerone)

The main chemical compounds responsible for the irritancy of ginger are the gingerols, contained in fresh ginger, and the shogaols and zingerone, which are both dehydration and degradation products of the gingerols (Dedov *et al.*, 2002). Collectively, these compounds activate TRPV1 receptors and increase the secretion of epinephrine, suggesting a role for the sympathetic nervous system involvement in ginger's thermogenic effects (Liu and Simon, 1996; Dedov *et al.*, 2002; Iwasaki *et al.*, 2006; Kawada *et al.*, 1988). When extracts of fresh and dried ginger were infused into the hind limb of male Wistar rats, oxygen consumption was stimulated. Gingerol displayed the greatest thermogenic effects, while shogaols demonstrated a more modest augmentation of energy expenditure. Notably, high concentrations of ginger extracts inhibited oxygen consumption (Eldershaw *et al.*, 1992).

There are three human investigations related to the thermogenic effects of ginger. In a British investigation, participants ($n = 3$M, 5F) were served a breakfast prepared with 30 g fresh, grated ginger. This had no effect on postprandial diet-induced thermogenesis compared with a ginger-free breakfast (Henry and Piggott 1987). In a Danish randomized crossover trial, participants ($n = 22$ healthy young males who liked spicy food) ingested a brunch meal with 20 g finely chopped ginger mixed into stewed apples. Four-hour postprandial diet-induced thermogenesis and substrate oxidation did not vary for the ginger meal compared with a no-ginger control (Gregersen *et al.*, 2013). In contrast, an American investigation provided participants ($n = 10$ overweight men) with a hot ginger beverage containing 2 g dried ginger powder along with breakfast. The ginger beverage yielded a 43 kcal day^{-1} increase in post-prandial thermogenesis relative to hot water (Mansour *et al.*, 2012). The varied findings may be explained by the form of ginger administration. Dried ginger, not fresh, contains shogaols, which increase ginger's pungency (Suekawa *et al.*, 1984).

12.4.4 Mustard (allyl isothiocyanate)

Allyl isothiocyanate, an organosulfur compound, is responsible for the pungency of mustard. Like the pungent compounds in black pepper and ginger, allyl isothiocyanate binds to both TRPV1 (Ohta *et al.*, 2007; Everaerts *et al.*, 2011) and TRPA1 receptors (Jordt *et al.*, 2004; Iwasaki *et al.*, 2008). In a study of female Sprague-Dawley rats, intramuscular injection of allyl isothiocyanate augmented brown adipose tissue activity (Yoshida *et al.*, 1988). However, a separate study failed to demonstrate catecholamine stimulation when allyl isothiocyanate was infused into the femoral vein of male Wistar rats (Kawada *et al.*, 1988).

Reports of the thermogenic effects of mustard are sparse. Two human investigations were identified. In a Danish randomized crossover trial, participants ($n = 22$ healthy young males who liked spicy food) ingested a brunch meal with 21 g Dijon mustard mixed into shredded beetroot. Four-hour postprandial diet-induced thermogenesis tended to increase by 14% following the mustard meal, compared with a no-mustard control (14.1 versus 12.4 kcal h^{-1}, respectively; $p = 0.08$) (Gregersen *et al.*, 2013). In a British investigation, participants ($n = 12$) were served a breakfast prepared with 3 g mustard sauce and 3 g chili sauce. Three-hour postprandial diet-induced thermogenesis following the spiced breakfast increased 25%, compared with a spice-free breakfast (153 versus 128% resting metabolic rate, respectively; $p < 0.01$) (Henry and Emery, 1986). However, it is not possible to distinguish the effects of capsaicin (chili) and allyl isothiocyanate (mustard). This is particularly limiting since a subsequent investigation demonstrated cross-desensitization between mustard oil and capsaicin (Simons *et al.*, 2003). Additionally, previous studies suggest that a synergism may exist between bioactive ingredients where combinations produce greater thermogenic effects than single ingredients. For example, a Swiss investigation ($n = 10$ healthy males) showed that the effects of green tea on energy expenditure and fat oxidation were more than could be explained by caffeine or catechins alone (Dulloo *et al.*, 1999).

Collectively, the thermogenic data presented in this section provide modest support for the effects of pungent spices on energy balance. With hot red pepper, the most pronounced enhancements in energy expenditure and fat oxidation occur at higher doses, among lean individuals, and with oral sensory exposure. While human data on the thermogenic effects of other spices are sparse, form (e.g., dried versus fresh) and combination (i.e., single versus multiple spices) may play important roles. Through the drying process, spices (e.g., ginger) may acquire greater pungency. Additional effects may occur when combining spices (e.g., red pepper and mustard). The roles of customary exposure and palatability in mediating thermogenic effects require further investigation.

12.5 Body weight

Few published studies have been of sufficient duration to measure changes of body weight related to the use of pungent spices. Most have hypothesized anorexigenic and thermogenic effects and/or inhibition of pancreatic enzyme activity, augmented metabolic rate, and prevention of adipocyte differentiation resulting in weight loss

(Westerterp-Plantenga *et al.*, 2006; Astrup *et al.*, 2010; Kazemipoor *et al.*, 2012; González-Castejón and Rodriguez-Casado, 2011). There are also reports of various pungent spices, especially capsaicin, diminishing (Nopanitaya, 1973) or facilitating energy absorption (Platel *et al.*, 2002; Pradeep *et al.*, 1991). Capsaicin and piperine have been reported to stimulate the activity of enzymes at the intestinal brush border membrane and to increase microvilli length thereby increasing the absorptive surface (Prakash and Srinivasan, 2010). This is desirable for nutrient absorption, but increased permeability for ions and macromolecules could also pose a risk for food allergies and intolerance (Jensen-Jarolim *et al.*, 1998).

Despite the various mechanisms described, results of spice use to moderate body weight or fat mass have generally not been supportive. In one 8-week trial with Wistar rats, the addition of clove, ginger, red pepper, or black pepper to Yaji (a Nigerian meat sauce) had no significant effect on body weight (Akpamu *et al.*, 2011). Similarly, clove exerted no effect on body weight in adult albino rats over a 90-day trial (Agbaje *et al.*, 2009). In contrast, weight gain has been reported in Wistar rats following the ingestion of black pepper for 35 (Hassan *et al.*, 2010) and 55 days (Mbongue *et al.*, 2005). In one French human trial, administration of Satiereal® (a saffron extract described earlier), via capsules, led to a progressive loss of body weight (–1 kg) over an 8-week period compared with a placebo treatment (Gout *et al.*, 2010). Whether this would be modified if participants had ingested the spice in a manner that provided oral stimulation has not been examined, nor has the sustainability of the effect. Another American trial provided a mixture of chromium picolinate, inulin, capsicum, L-phenylalanine, and other lipotropic compounds to adults (n = 3M, 53F) on an energy-restricted diet and activity regimen (Hoeger *et al.*, 1998). No effect was observed relative to the control group (n = 9M, 58F).

The efficacy of capsaicin to aid in maintenance of reduced body weight has also been tested. Following a 4-week very low energy diet that did or did not include 135 mg day^{-1} capsaicin, Dutch participants (n = 23M, 68F) were monitored for three months (Lejeune *et al.*, 2003). While fat oxidation was elevated in the capsaicin treated group (n = 12M, 30F), there was no difference from the placebo (n = 11M, 38F) in body weight. Thus, despite mechanistic, appetitive, and acute feeding studies results providing evidence of the plausibility of spices to moderate body weight, the very limited evidence from longer-term trials has not yielded the expected results.

12.6 Individual variability

Individual variability in responses to different spices, doses, and modes of delivery can possibly hamper identification of the potential health effects of spice ingestion. Increasing knowledge of the factors contributing to variability should permit experimental control and better isolation of useful properties. Dose is an obvious critical methodological factor that must be reconciled to compare findings across trials. However, this is not straightforward since the sensory impact of the dose varies according to the food in which it is delivered and the prior experience of the consumer with the spice. Nearly all the work on individual differences has focused on capsaicin. Findings on whether individuals with prior experience with capsaicin rate its burn lower than those without experience are mixed (Stevenson and Prescott, 1994; Ludy and Mattes, 2012; Byrnes and Hayes, 2013). However, users rate spicy food as more pleasant than non-users (Ludy and Mattes, 2012). The time required to induce a hedonic shift for the burn or sting of capsaicin or to recover from it has not been determined, but one British trial noted an increase in acceptability over a 5-week period (Stevenson and Yeomans, 1995). Individuals who consume foods containing capsaicin report a higher liking for spicy foods (e.g., horseradish, mustard, and wasabi) and cuisines (e.g., Chinese, Indian, and Mexican) (Ludy and Mattes, 2012). This is specific to spicy foods as users and non-users do not differ in hedonic ratings for non-spicy items such as apples, bread, chicken, lettuce, milk, and rice (Ludy and Mattes, 2012). Prior dietary experience with capsaicin may help reconcile discrepant responses between individuals or within individuals over time.

Capsaicin may also modify the experience of other tastes present in a food, especially sourness, bitterness, and umami (Simons *et al.*, 2002). There is a direct dose–response association, but how this may interact with prior experience with capsaicin is not well characterized. Because capsaicin specifically alters the perception of some tastes and odors, it may alter the flavor of foods (i.e., the totality of the sensory experience) and would be expected to modify acceptability ratings. The masking effect does not alter flavor identification and occurs equally in users and non-users (Lawless *et al.*, 1985), but the hedonic impact is not comparably altered (Stevenson and Yeomans, 1993). In one British trial, non-users expressed no strong hedonic impression to the high concentrations rated negatively by users whereas strong doses were comparably disliked (Stevenson and Yeomans, 1993).

The timing of stimulus presentation is also critical as a pause of only 2.5–5.0 minutes between stimulations may allow the onset of desensitization and lower burn intensity ratings (Green, 1989; Green and Rentmeister-Bryant, 1998). Such pauses are likely under customary dining conditions so they may lead to underestimations of the contribution of capsaicin to the irritancy and acceptability of foods when it occurs. The reverse effect, sensitization (i.e., growth of sensation with repeated stimulation), does not appear to be functional during food ingestion (Prescott, 1999) (see Chapter 6 for more discussion of sensation and desensitization). The composition of the food containing capsaicin also alters the burn it elicits, with stronger ratings for lower concentrations in water than in lipid vehicles (Lawless *et al.*, 2000). Personality traits are also linked to capsaicin intake. There is a positive association between consumption of foods containing capsaicin and sensation seeking and sensitivity to reward (Ludy and Mattes, 2012; Byrnes and Hayes, 2013) and for women only, extroversion (Ludy and Mattes, 2012) (see Chapter 2 for additional discussion).

12.7 Conclusion

Pungent spices contribute many potential beneficial health properties (Fig. 12.2). Although the energy balance effects of spices are small in magnitude, they should not be discounted. Nutritionally relevant outcomes are present at modest doses, enhanced with oral exposure, and are found in traditional ethnic recipes. At these low doses, and even when several times greater than typical human intake, the

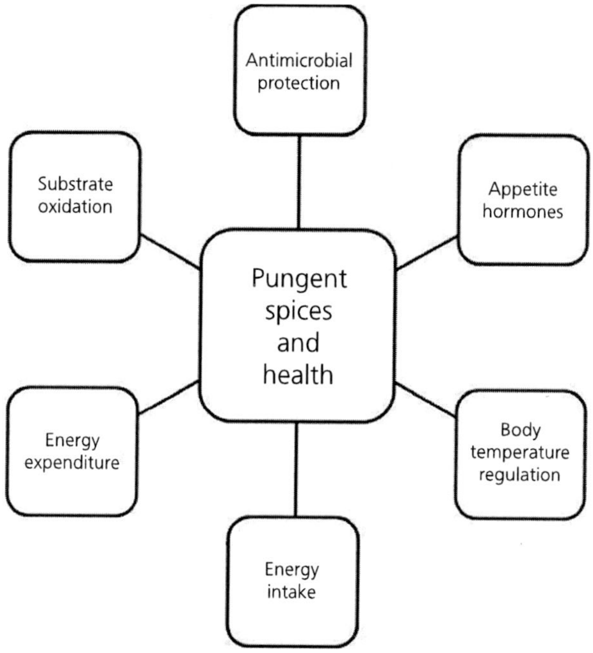

Fig. 12.2 Summary of proposed benefits of pungent spices on health.

potential for adverse effects is low (Srinivasan, 2005). Collectively, this indicates that pungent spices are unlikely to incite harm, may aid in food safety, and hold a potentially therapeutic role in weight management.

References

Acheson, K., Zahorska-Markiewicz, B., Pittet, P., Anantharaman, K., and Jequier, E. 1980. Caffeine and coffee: their influence on metabolic rate and substrate utilization in normal weight and obese individuals. *American Journal of Clinical Nutrition*, **33** (5), 989–997.

Agbaje, E., Adeneye, A., and Daramola, A. 2009. Biochemical and toxicological studies of aqueous extract of Syzigium aromaticum (l.) Merr. and perry (myrtaceae) in rodents. *African Journal of Traditional, Complementary and Alternative Medicines*, **6** (3), 241–254.

Ahuja, K.D.K., Robertson, I.K., Geraghty, D.P., and Ball, M.J. 2006. Effects of chili consumption on postprandial glucose, insulin, and energy metabolism. *American Journal of Clinical Nutrition*, **84** (1), 63–69.

Ahuja, K.D.K., Robertson, I.K., Geraghty, D.P., and Ball, M.J. 2007. The effect of 4-week chilli supplementation on metabolic and arterial function in humans. *European Journal of Clinical Nutrition*, **61** (3), 326–333.

Akpamu, U., Nwaopara, A., Izunya, A., *et al.* 2011. A comparative study on weight changes in rats fed with diet containing *Yaji*, *Yaji*-additives and *Yaji*-spices. *Biology and Medicine*, **3** (5), 6–15.

Astrup, A., Kristensen, M., Gregersen, N.T., *et al.* 2010. Can bioactive foods affect obesity? *Annals of the New York Academy of Sciences*, **1190** (1), 25–41.

Best, R.L., and Appleton, K.M. 2011. Comparable increases in energy, protein and fat intakes following the addition of seasonings and sauces to an older person's meal. *Appetite*, **56** (1), 179–182.

Billing, J., and Sherman, P.W. 1998. Antimicrobial functions of spices: why some like it hot. *Quarterly Review of Biology*, **73** (1), 3–49.

Blank, D.M., and Mattes, R.D. 1990. Sugar and spice: similarities and sensory attributes. *Nursing Research*, **39** (5), 290–293.

Bracco, D., Ferrarra, J.M., Arnaud, M.J., Jequier, E., and Schutz, Y. 1995. Effects of caffeine on energy metabolism, heart rate, and methylxanthine metabolism in lean and obese women. *The American Journal of Physiology*, **269** (4), E671–E678.

Byrnes, N.K., and Hayes, J.E. 2013. Personality factors predict spicy food liking and intake. *Food Quality and Preference*, **28** (1), 213–221.

Caterina, M.J., Schumacher, M.A., Tominaga, M., *et al.* 1997. The capsaicin receptor: a heat-activated ion channel in the pain pathway. *Nature*, **389** (6653), 783–784.

Chaiyata, P., Puttadechakum, S., and Komindr, S. 2003. Effect of chili pepper (*Capsicum frutescens*) ingestion on plasma glucose response and metabolic rate in Thai women. *Journal of the Medical Association of Thailand*, **86** (9), 854–860.

Clouard, C., and Val-Laillet, D. 2014. Impact of sensory feed additives on feed intake, feed preferences, and growth of female piglets during the early postweaning period. *Journal of Animal Science*, **92** (5), 2133–2140.

de Graaf, C., Polet, P., and van Staveren, W.A. 1994. Sensory perception and pleasantness of food flavors in elderly subjects. *Journal of Gerontology*, **49** (3), 93–99.

Dedov, V.N., Tran, V.H., Duke, C.C., *et al.* 2002. Gingerols: a novel class of vanilloid receptor (VR1) agonists. *British Journal of Pharmacology*, **137** (6), 793–798.

DeMan, J.M. 2013. *Principles of Food Chemistry*, New York: Springer.

Dressler, H., and Smith, C. 2013. Food choice, eating behavior, and food liking differs between lean/normal and overweight/obese, low-income women. *Appetite*, **65** (1), 145–152.

Dulloo, A.G., Duret, C., Rohrer, D., *et al.* 1999. Efficacy of a green tea extract rich in catechin polyphenols and caffeine in increasing 24-h energy expenditure and fat oxidation in humans. *The American Journal of Clinical Nutrition*, **70** (6), 1040–1045.

Edwards, S.J., Montgomery, I.M., Colquhoun, E.Q., Jordan, J.E., and Clark, M.G. 1992. Spicy meal disturbs sleep: an effect of thermoregulation? *International Journal of Psychophysiology*, **13** (2), 97–100.

Eldershaw, T.P., Colquhoun, E.Q., Bennett, K.L., Dora, K.A., and Clark, M.G. 1994. Resiniferatoxin and piperine: capsaicin-like stimulators of oxygen uptake in the perfused rat hindlimb. *Life Sciences*, **55** (5), 389–397.

Eldershaw, T.P.D., Colquhoun, E.Q., Dora, K.A., Peng, Z.C., and Clark, M.G. 1992. Pungent principles of ginger (*Zingiber officinale*) are thermogenic in the perfused rat hindlimb. *International Journal of Obesity*, **16** (10), 755–763.

Essed, N.H., van Staveren, W.A., Kok, F.J., and de Graaf, C. 2007. No effect of 16 weeks flavor enhancement on dietary intake and nutritional status of nursing home elderly. *Appetite*, **48** (1), 29–36.

Everaerts, W., Gees, M., Alpizar, Y.A., *et al.* 2011. The capsaicin receptor TRPV1 is a crucial mediator of the noxious effects of mustard oil. *Current Biology*, **21** (4), 316–321.

Flint, A., Gregersen, N.T., Gluud, L.L., *et al.* 2007. Associations between postprandial insulin and blood glucose responses, appetite sensations and energy intake in normal weight and overweight individuals: a meta-analysis of test meal studies. *British Journal of Nutrition*, **98** (1), 17–25.

Ganesh Bhat, B., and Chandrasekhara, N. 1986. Studies on the metabolism of piperine: absorption, tissue distribution and excretion of urinary conjugates in rats. *Toxicology*, **40** (1), 83–92.

Ghawi, S.K., Rowland, I., and Methven, L. 2014. Enhancing consumer liking of low salt tomato soup over repeated exposure by herb and spice seasonings. *Appetite*, **81** (1), 20–29.

Glanz, K., and Basil, M. 1998. Why Americans eat what they do: Taste, nutrition, cost, convenience, and weight control concerns. *Journal of the American Dietetic Association*, **98** (10), 1118–1126.

González-Castejón, M., and Rodriguez-Casado, A. 2011. Dietary phytochemicals and their potential effects on obesity: a review. *Pharmacological Research*, **64** (5), 438–455.

Gout, B., Bourges, C., and Paineau-Dubreuil, S. 2010. Satiereal, a *Crocus sativus* L extract, reduces snacking and increases satiety in a randomized placebo-controlled study of mildly overweight, health women. *Nutrition Research*, **30** (5), 305–313.

Green, B.G. 1989. Capsaicin sensitization and desensitization on the tongue produced by brief exposures to a low concentration. *Neuroscience Letters*, **107** (1–3), 173–178.

Green, B.G., and Rentmeister-Bryant, H. 1998. Temporal characteristics of capsaicin desensitization and stimulus-induced recovery in the oral cavity. *Physiology & Behavior*, **65** (1), 141–149.

Gregersen, N.T., Belza, A., Jensen, M., *et al.* 2013. Acute effects of mustard, horseradish, black pepper and ginger on energy expenditure, appetite, ad libitum energy intake and energy balance in human subjects. *British Journal of Nutrition*, **109** (3), 556–563.

Griep, M.I., Mets, T.F., and Massart, D.L. 2000. Effects of flavour amplification of Quorn® and yoghurt on food preference and consumption in relation to age, BMI and odour perception. *British Journal of Nutrition*, **83** (2), 105–113.

Hachiya, S., Kawabata, F., Ohnuki, K., *et al.* 2007. Effects of CH-19 Sweet, a non-pungent cultivar of red pepper, on sympathetic nervous activity, body temperature, heart rate, and blood pressure in humans.. *Bioscience, Biotechnology, and Biochemistry*, **71** (3), 671–676.

Hameed, S., Dhillo, W., and Bloom, S. 2009. Gut hormones and appetite control. *Oral Diseases*, **15** (1), 18–26.

Hassan, D., Ogah, D., Yusuf, N., Musa, M., and Saidu, G. 2010. The effect of acute and chronic (short and long term) oral administrations of black pepper (*Piper guineense*) aqueous extract on the body weight and haematological values of albino-Wistar rat. *Journal of Medicinal Plants*, **4** (12), 1122–1125.

Haxton, H.A. 1948. Gustatory sweating. *Brain: A Journal of Neurology*, **71** (1), 16–25.

Henry, C.J., and Emery, B. 1986. Effect of spiced food on metabolic rate. *Human Nutrition – Clinical Nutrition*, **40** (2), 165–168.

Henry, C.J., and Piggott, S.M. 1987. Effect of ginger on metabolic rate. *Human Nutrition – Clinical Nutrition*, **41** (1), 89–92.

Henry, C.J.K., Woo, J., Lightowler, H.J., *et al.* 2003. Use of natural food flavours to increase food and nutrient intakes in hospitalized elderly in Hong Kong. *International Journal of Food Sciences and Nutrition*, **54** (4), 321–327.

Hlebowicz, J., Darwiche, G., Bjorgell, O., and Almer, L.O. 2007. Effect of cinnamon on postprandial blood glucose, gastric emptying, and satiety in healthy subjects. *American Journal of Clinical Nutrition*, **85** (6), 1552–1556.

Hoeger, W.W., Harris, C., Long, E.M., and Hopkins, D.R. 1998. Four-week supplementation with a natural dietary compound produces favorable changes in body composition. *Advances in Therapy*, **15** (5), 305–314.

Hu, M.L., Rayner, C.K., Wu, K.L., *et al.* 2011. Effect of ginger on gastric motility and symptoms of functional dyspepsia. *World Journal of Gastroenterology*, **17** (1), 105–110.

Hursel, R., Viechtbauer, W., and Westerterp-Plantenga, M. 2009. The effects of green tea on weight loss and weight maintenance: a meta-analysis. *International Journal of Obesity*, **33** (9), 956–961.

International Food Information Council Foundation. 2014, June 24, 2014 last update, *Food & Health Survey: The Pulse of America's Diet: From Beliefs to Behaviors*, http://www.foodinsight.org/articles/2014-food-and-health-survey (accessed 7 August, 2014).

Iwai, L., Yazawa, W., and Watanabe, T. 2003. Roles as metabolic regulators of the non nutrients, capsaicin and capsiate, supplemented to diets. *Proceedings of the Japan Academy*, **79** (7), 207–212.

Iwasaki, Y., Morita, A., Iwasawa, T., *et al.* 2006. A nonpungent component of steamed ginger—[10]-shogaol—increases adrenaline secretion via the activation of TRPV1. *Nutritional Neuroscience*, **9** (3–4), 169–178.

Iwasaki, Y., Tanabe, M., Kobata, K., and Watanabe, T. 2008. TRPA1 agonists—allyl isothiocyanate and cinnamaldehyde—induce adrenaline secretion. *Bioscience, Biotechnology, and Biochemistry*, **72** (10), 2608–2614.

Janssens, P.L., Hursel, R., and Westerterp-Plantenga, M.S. 2014. Capsaicin increases sensation of fullness in energy balance, and decreases desire to eat after dinner in negative energy balance. *Appetite*, **77** (1), 46–51.

Jensen-Jarolim, E., Gajdzik, L., Haberl, I., *et al.* 1998. Hot spices influence permeability of human intestinal epithelial monolayers. *Journal of Nutrition*, **128** (3), 577–581.

Jones, J.B., Provost, M., Keaver, L., *et al.* 2014a. Effects of daily consumption of one or varied peanut flavors on acceptance and intake. *Appetite*, **82** (1), 208–212.

Jones, J.B., Provost, M., Keaver, L., *et al.* 2014b. A randomized trial on the effects of flavorings on the health benefits of daily peanut consumption. *American Journal of Clinical Nutrition*, **99** (3), 490–496.

Jordt, S., Bautista, D.M., Chuang, H., *et al.* 2004. Mustard oils and cannabinoids excite sensory nerve fibres through the TRP channel ANKTM1. *Nature*, **427** (6971), 260–265.

Kawada, T., Sakabe, S., Watanabe, T., Yamamoto, M., and Iwai, K. 1988. Some pungent principles of spices cause the adrenal medulla to secrete catecholamine in anesthetized rats. *Proceedings of the Society for Experimental Biology and Medicine*, **188** (2), 229–233.

Kazemipoor, M., Radzi, C.W.J.W.M., Cordell, G.A., and Yaze, I. 2012. Potential of traditional medicinal plants for treating obesity: a review. *International Conference on Nutrition and Food Sciences*, **39** (1), 1–6.

Khan, A., Safdar, M., Ali Khan, M.M., Khattak, K.N., and Anderson, R.A. 2003. Cinnamon improves glucose and lipids of people with type 2 diabetes. *Diabetes Care*, **26** (12), 3215–3218.

Khan, R., Khan, Z., and Shah, S. 2010. Cinnamon may reduce glucose, lipid and cholesterol level in type 2 diabetic individuals. *Pakistan Journal of Nutrition*, **9** (5), 430–433.

Koskinen, S., Kälviäinen, N., and Tuorila, H. 2003. Flavor enhancement as a tool for increasing pleasantness and intake of a snack product among the elderly. *Appetite*, **41** (1), 87–96.

Lawless, H., Rozin, P., and Shenker, J. 1985. Effects of oral capsaicin on gustatory, olfactory and irritant sensations and flavor identification in humans who regularly or rarely consume chili pepper. *Chemical Senses*, **10** (4), 579–589.

Lawless, H.T., Hartono, C., and Hernandez, S. 2000. Thresholds and suprathreshold intensity functions for capsaicin in oil and aqueous based carriers. *Journal of Sensory Studies*, **15** (4), 437–477.

LeBlanc, J., and Brondel, L. 1985. Role of palatability on meal-induced thermogenesis in human subjects. *American Journal of Physiology - Endocrinology and Metabolism*, **248** (3), 333–336.

Lee, T.S. 1954. Physiological gustatory sweating in a warm climate. *Journal of Physiology*, **124** (3), 528–542.

Lejeune, M.P.G.M., Kovacs, E.M.R., and Westerterp-Plantenga, M.S. 2003. Effect of capsaicin on substrate oxidation and weight maintenance after modest body-weight loss in human subjects. *British Journal of Nutrition*, **90** (3), 651–659.

Lim, K., Yoshioka, M., Kikuzato, S., *et al.* 1997. Dietary red pepper ingestion increases carbohydrate oxidation at rest and during exercise in runners. *Medicine & Science in Sports & Exercise*, **29** (3), 355–361.

Liu, L., and Simon, S.A. 1996. Similarities and differences in the currents activated by capsaicin, piperine, and zingerone in rat trigeminal ganglion cells. *Journal of Neurophysiology*, **76** (3), 1858–1869.

Lopez-Carrillo, L., Lopez-Cervantes, M., Robles-Diaz, G., *et al.* 2003. Capsaicin consumption, Helicobacter pylori positivity and gastric cancer in Mexico. *International Journal of Cancer*, **106** (2), 277–282.

Ludy, M.J. 2013. Some like it hot: characteristics underlying spicy food preference and aversion. *FASEB Journal*, **27**, 1035.17.

Ludy, M.J., and Mattes, R.D. 2011. The effects of hedonically acceptable red pepper doses on thermogenesis and appetite. *Physiology & Behavior*, **102** (3), 251–258.

Ludy, M.J., and Mattes, R.D. 2012. Comparison of sensory, physiological, personality, and cultural attributes in regular spicy food users and non-users. *Appetite*, **58** (1), 19–27.

Ludy, M.J., Moore, G.E., and Mattes, R.D. 2012. The effects of capsaicin and capsiate on energy balance: critical review and meta-analyses of studies in humans. *Chemical Senses*, **37** (2), 103–121.

Mang, B., Wolters, M., Schmitt, B., *et al.* 2006. Effects of a cinnamon extract on plasma glucose, HbA1c, and serum lipids in diabetes mellitus type 2. *European Journal of Clinical Investigation*, **36** (5), 340–344.

Mansour, M.S., Ni, Y., Roberts, A.L., *et al.* 2012. Ginger consumption enhances the thermic effect of food and promotes feelings of satiety without affecting metabolic and hormonal parameters in overweight men: A pilot study. *Metabolism*, **61** (10), 1347–1352.

Markey, O., McClean, C.M., Medlow, P., *et al.* 2011. Effect of cinnamon on gastric emptying, arterial stiffness, postprandial lipemia, glycemia, and appetite responses to high-fat breakfast. *Cardiovascular Diabetology*, **10** (9), 78.

Masic, U., and Yeomans, M.R. 2014. Umami flavor enhances appetite but also increases satiety. *American Journal of Clinical Nutrition*, **100** (2), 532–538.

Mathey, M.F., Siebelink, E., de Graaf, C., and Van Staveren, W.A. 2001. Flavor enhancement of food improves dietary intake and nutritional status of elderly nursing home residents. *The Journals of Gerontology Series A: Biological Sciences and Medical Sciences*, **56** (4), M200–M205.

Matsumoto, T., Miyawaki, C., Ue, H., *et al.* 2000. Effects of capsaicin–containing yellow curry sauce on sympathetic nervous system activity and diet-induced thermogenesis in lean and obese young women. *Journal of Nutritional Science and Vitaminology*, **46** (6), 309–315.

Mattes, R.D. 2012. Spices and energy balance. *Physiology & Behavior*, **107** (4), 584–590.

Mbongue, F., Kamtchouing, P., Essame, O., *et al.* 2005. Effect of the aqueous extract of dry fruits of Piper guineense on the reproductive function of adult male rats. *Indian Journal of Pharmacology*, **37** (1), 30–32.

McNamara, F.N., Randall, A., and Gunthorpe, M.J. 2005. Effects of piperine, the pungent component of black pepper, at the human vanilloid receptor (TRPV1). *British Journal of Pharmacology*, **144** (6), 781–790.

Nabhan, G.P. 2004. *Why Some Like it Hot: Food, Genes, and Cultural Diversity*, Washington, DC: Shearwater Books.

Nopanitaya, W. 1973. Long term effects of capsaicin on fat absorption and the growth of the rat. *Growth*, **37** (3), 269–279.

O'Connor, A., Corbin, K.D., Nieman, D.C., and Swick, A.G. 2013. A randomized, controlled trial to assess short-term black pepper consumption on 24-hour energy expenditure and substrate utilization. *Functional Foods in Health and Disease*, **3** (10), 377–388.

Ohta, T., Imagawa, T., and Ito, S. 2007. Novel agonistic action of mustard oil on recombinant and endogenous porcine transient receptor potential V1 (pTRPV1) channels. *Biochemical Pharmacology*, **73** (10), 1646–1656.

Okumura, Y., Narukawa, M., Iwasaki, Y., *et al.* 2010. Activation of TRPV1 and TRPA1 by black pepper components. *Bioscience, Biotechnology, and Biochemistry*, **74** (5), 1068–1072.

Osaka, T., Kobayashi, A., Lee, T.H., *et al.* 2000. Lack of integrative control of heat production and heat loss after capsaicin administration. *Pflügers Archiv - European Journal of Physiology*, **440** (3), 440–445.

Panda, S., and Kar, A. 2003. Piperine lowers the serum concentrations of thyroid hormones, glucose, and hepatic 5′D activity in adult male mice. *Hormone and Metabolic Research*, **9** (35), 523–526.

Peters, J.C., Polsky, S., Stark, R., Zhaoxing, P., and Hill, J.O. 2014. The influence of herbs and spices on overall liking of reduced fat food. *Appetite*, **79** (1), 183–188.

Platel, K., Rao, A., Saraswathi, G., and Srinivasan, K. 2002. Digestive stimulant action of three Indian spice mixes in experimental rats. *Food/Nahrung*, **46** (6), 394–398.

Pradeep, K.U., Geervani, P., and Eggum, B. 1991. Influence of spices on utilization of sorghum and chickpea protein. *Plant Foods for Human Nutrition*, **41** (3), 269–276.

Prakash, U.N. and Srinivasan, K. 2010. Beneficial influence of dietary spices on the ultrastructure and fluidity of the intestinal brush border in rats. *British Journal of Nutrition*, **104** (1), 31–39.

Prescott, J. 1999. The generalizability of capsaicin sensitization and desensitization. *Physiology & Behavior*, **66** (5), 741–749.

Reinagel, M. 2012, Today's contemporary spice cabinet. *Food & Nutrition Magazine*, winter 2012, http://www.foodandnutrition.org/Winter-2012/Todays-Contemporary-Spice-Cabinet/ (accessed 6 October 2015).

Reinbach, H.C., Smeets, A., Martinussen, T., Møller, P., and Westerterp-Plantenga, M.S. 2009. Effects of capsaicin, green tea and CH-19 sweet pepper on appetite and energy intake in humans in negative and positive energy balance. *Clinical Nutrition*, **28** (3), 260–265.

Reinbach, H.C., Martinussen, T., and Møller, P. 2010. Effects of hot spices on energy intake, appetite, and sensory specific desires in humans. *Food Quality and Preference*, **21** (6), 655–661.

Rozin, P., and Schiller, D. 1980. The nature and acquisition of a preference for chili pepper by humans. *Motivation and Emotion*, **4** (1), 77–101.

Savage, J.S., Peterson, J., Marini, M., BordiJr, P.L., and Birch, L.L. 2013. The addition of a plain or herb-flavored reduced-fat dip is associated with improved preschoolers' intake of vegetables. *Journal of the Academy of Nutrition and Dietetics*, **113** (8), 1090–1095.

Schiffman, S.S. 1993. Perception of taste and smell in elderly persons. *Critical Reviews in Food Science & Nutrition*, **33** (1), 17–26.

Schiffman, S.S., and Warwick, Z.S. 1988. Flavor enhancement of foods for the elderly can reverse anorexia. *Neurobiology of Aging*, **9** (1), 24–26.

Sherman, P.W., and Billing, J. 1999. Darwinian Gastronomy: Why We Use Spices Spices taste good because they are good for us. *Bioscience*, **49** (6), 453–463.

Sherman, P.W., and Hash, G.A. 2001. Why vegetable recipes are not very spicy. *Evolution and Human Behavior*, **22** (3), 147–163.

Shin, K.O., and Moritani, T. 2007. Alterations of autonomic nervous activity and energy metabolism by capsaicin ingestion during aerobic exercise in healthy men. *Journal of Nutritional Science and Vitaminology*, **53** (2), 124–132.

Simons, C.T., O'Mahony, M., and Carstens, E. 2002. Taste suppression following lingual capsaicin pre-treatment in humans. *Chemical Senses*, **27** (4), 353–365.

Simons, C.T., Carstens, M.I., and Carstens, E. 2003. Oral irritation by mustard oil: self-desensitization and cross-desensitization with capsaicin. *Chemical Senses*, **28** (6), 459–465.

Skulas-Ray, A. C., Kris-Etherton, P. M., Teeter, D. L., *et al.* 2011. A high antioxidant spice blend attenuates postprandial insulin and triglyceride responses and increases some plasma measures of antioxidant activity in healthy, overweight men. *Journal of Nutrition*, **141** (8), 1451–1457.

Sloan, A.E. 2014. What, when and where America eats. *Food Technology*, **68** (1), 20–31.

Smeets, A., and Westerterp-Plantenga, M. 2009. The acute effects of a lunch containing capsaicin on energy and substrate utilisation, hormones, and satiety. *European Journal of Clinical Nutrition*, **48** (4), 229–234.

Smeets, A.J., Janssens, P.L., and Westerterp-Plantenga, M.S. 2013. Addition of capsaicin and exchange of carbohydrate with protein counteract energy intake restriction effects on fullness and energy expenditure. *Journal of Nutrition*, **143** (4), 442–447.

Sockett, P. 1995. The epidemiology and costs of diseases of public health significance, in relation to meat and meat products. *Journal of Food Safety*, **15** (2), 91–112.

Solomon, T., and Blannin, A. 2007. Effects of short-term cinnamon ingestion on in vivo glucose tolerance. *Diabetes, Obesity and Metabolism*, **9** (6), 895–901.

Solomon, T.P., and Blannin, A.K. 2009. Changes in glucose tolerance and insulin sensitivity following 2 weeks of daily cinnamon ingestion in healthy humans. *European Journal of Applied Physiology*, **105** (6), 969–976.

Srinivasan, K. 2005. Spices as influencers of body metabolism: an overview of three decades of research. *Food Research International*, **38** (1), 77–86.

Stevenson, R.J., and Yeomans, M.R. 1993. Differences in ratings of intensity and pleasantness for the capsaicin burn between chili likers and non-likers: implications for liking development. *Chemical Senses*, **18** (5), 471–482.

Stevenson, R.J., and Prescott, J. 1994. The effects of prior experience with capsaicin on ratings of its burn. *Chemical Senses*, **19** (6), 651–656.

Stevenson, R.J., and Yeomans, M.R. 1995. Does exposure enhance liking for the chilli burn? *Appetite*, **24** (2), 107–120.

Suekawa, M., Ishige, A., Yuasa, K., Sudo, K., Aburada, M., and Hosoya, E. 1984. Pharmacological studies on ginger. I. Pharmacological actions of pungent constitutents, (6)-gingerol and (6)-shogaol. *Journal of Pharmacobio-Dynamics*, **7** (11), 836–848.

Szallasi, A., and Blumberg, P.M. 1999. Vanilloid (Capsaicin) receptors and mechanisms. *Pharmacological Reviews*, **51** (2), 159–212.

Szallasi, A. 2005. Piperine: researchers discover new flavor in an ancient spice. *Trends in Pharmacological Sciences*, **26** (9), 437–439.

Tepper, B.J., and Trail, A.C. 1998. Taste or health: A study on consumer acceptance of corn chips. *Food Quality and Preference*, **9** (4), 267–272.

Ueki, S., Miyoshi, M., Shido, O., Hasegawa, J., and Watanabe, T. 2008. Systemic administration of [6]-gingerol, a pungent constituent of ginger, induces hypothermia in rats via an inhibitory effect on metabolic rate. *European Journal of Pharmacology*, **584** (1), 87–92.

Vanschoonbeek, K., Thomassen, B.J., Senden, J.M., Wodzig, W.K., and van Loon, L.J. 2006. Cinnamon supplementation does not improve glycemic control in postmenopausal type 2 diabetes patients. *Journal of Nutrition*, **136** (4), 977–980.

Wadikar, D., and Premavalli, K. 2011. Appetizer administration stimulates food consumption, weight gain and leptin levels in male Wistar rats. *Appetite*, **57** (1), 131–133.

Westerterp-Plantenga, M.S., Smeets, A., and Lejeune, M.P.G. 2005. Sensory and gastrointestinal satiety effects of capsaicin on food intake. *International Journal of Obesity*, **29** (6), 682–688.

Westerterp-Plantenga, M.S., Diepvens, K., Joosen, A.M.C.P., Bérubé-Parent, S., and Tremblay, A. 2006. Metabolic effects of spices, teas, and caffeine. *Physiology & Behavior*, **89** (1), 85–91.

Whiting, S., Derbyshire, E., and Tiwari, B. 2012. Capsaicinoids and capsinoids. A potential role for weight management? *A systematic review of the evidence. Appetite*, **59** (2), 341–348.

Wu, K.L., Rayner, C.K., Chuah, S.K., *et al.* 2008. Effects of ginger on gastric emptying and motility in healthy humans. *European Journal of Gastroenterology & Hepatology*, **20** (5), 436–440.

Yoshida, T., Yoshioka, K., Wakabayashi, Y., Nishioka, H., and Kondo, M. 1988. Effects of capsaicin and isothiocyanate on thermogenesis of interscapular brown adipose tissue in rats. *Journal of Nutritional Science and Vitaminology*, **34** (6), 587–594.

Yoshioka, M., Imanaga, M., Ueyama, H., *et al.* 2004. Maximum tolerable dose of red pepper decreases fat intake independently of spicy sensation in the mouth. *British Journal of Nutrition*, **91** (6), 991–995.

Yoshioka, M., Lim, K., Kikuzato, S., *et al.* 1995. Effects of red-pepper diet on the energy metabolism in men. *Journal of Nutritional Science and Vitaminology*, **41** (6), 647–656.

Yoshioka, M., St-Pierre, S., Drapeau, V., *et al.* 1999. Effects of red pepper on appetite and energy intake. *British Journal of Nutrition*, **82** (2), 115–123.

Yoshioka, M., St-Pierre, S., Suzuki, M., and Tremblay, A. 1998. Effects of red pepper added to high-fat and high-carbohydrate meals on energy metabolism and substrate utilization in Japanese women. *British Journal of Nutrition*, **80** (6), 503–510.

Zandstra, E., de Graaf, C., and Van Trijp, H. 2000. Effects of variety and repeated in-home consumption on product acceptance. *Appetite*, **35** (2), 113–119.

Ziegenfuss, T.N., Hofheins, J.E., Mendel, R.W., Landis, J., and Anderson, R.A. 2006. Effects of a water-soluble cinnamon extract on body composition and features of the metabolic syndrome in pre-diabetic men and women. *Journal of the International Society of Sports Nutrition*, **3** (2), 45–53.

CHAPTER 13

On food and chemesthesis – food science and culinary perspectives

Christopher R. Loss[1] and Ali Bouzari[2]
[1] *The Culinary Institute of America, Hyde Park, NY, USA*
[2] *Pilot R+D, Healdsburg, CA, USA*

13.1 Introduction: putting chemesthesis in the context of flavor

The fields of food science and culinary arts are beginning to merge (Barham *et al.*, 2010; Cheng *et al.*, 2011; Loss, 2014) and to focus attention on flavor as a multimodal (Delwiche, 2004; Auvray and Spence, 2008) and contextual based (Meiselman, 1996, 2013) perception. At the same time, sensory scientists are exploring the temporal nature of flavor (Piggott, 2000) and how its progression over the course of a meal or dish contributes to consumer experience at the table (Reinbach *et al.*, 2007; Dinnella *et al.*, 2012). Product and menu developers are beginning to leverage the insights and empirical understanding that culinary professionals have of ingredient functionality, flavor, and the consumer's dining experiences (Gustafsson, 2004; Meiselman, 2009; Vega and Ubbink, 2008). Ingredients containing chemesthetic agents clearly contribute to flavor perception (Green, 1996), but a better understanding of the role of ingredients and their chemesthetic properties is needed from the combined perspective of food scientists and chefs. This combined perspective is referred to as the "culinary science" perspective.

In order to discuss the contributions of chemesthesis to flavor, we need to have a definition and better understanding of what flavor is. Delwiche (2004) describes flavor as an integration or confluence of all of the sensory experiences when eating, including irritation or pain. A *well liked* flavor is perhaps best described by Lawless (2000) as a "balance" between harmony and contrast and between the new and the familiar. An integral and influential component of flavor "balance" are the herbs, spices, fruits, and vegetables that provide cultural "signatures" of regional cuisines around the world (Rozin and Rozin, 2005). These ingredients are the major contributors of chemesthetic properties in food.

When considering flavor we also need to be aware that it is not a "snapshot" in time, but more of a movie that unfolds while we eat (T. Acree, personal communication, 5 April, 2008). The temporal nature of the perception of the onset and decay of chemesthesis are unique compared with that of taste, aroma, and the physical/tactile aspects of food, and one more reason that a consideration of their contribution to food flavor is key.

In this chapter we focus on the use of ingredients in the kitchen and at the product development bench that contribute chemesthetic properties, as well as the role of these ingredients in global cuisines, and their influence on flavor perception. We begin by providing some background on the origins and applications of chemesthetic agents in the kitchen and at the lab bench. We discuss the connections between the regional cultivation of herbs and spices and the origins of global cuisines. Then we describe the diverse ingredient sources of chemesthetic agents, and describe how culinary techniques can influence the quality of chemesthetic ingredients and the dishes they are incorporated into. Methods used by chefs to extract, concentrate, and infuse chemesthetic agents into foods will be described. Culinary techniques and traditions will be compared with the current understanding of the physicochemical properties of chemesthetic agents. We will also explore the applications of chemesthetic ingredients in cuisines from around the world, including modern, and *avant-garde* applications.

13.2 Historical and cultural context for the use of chemesthetic ingredients in foods

The plant materials that contribute burning, cooling, and tingling to our foods have always been a part of our environment and have influenced cuisines as well as medical practices across a diversity of cultures (Freedman, 2007; Nilius and Appendino, 2013). Herbs and spices, and the flavors they help to create, have driven exploration and discovery, and contributed significantly to globalization and commerce from the Middle Ages on. The goals of the expeditions of Marco Polo and Christopher Columbus were to find and bring back exotic and flavorful spices that in many cases were equal to their weight in gold (Parry, 1953). Today, the common peppercorn (*Piper nigrum*) is ubiquitous in western kitchens and dining tables, yet it was once an exotic spice that significantly influenced globalization.

In *Naturalis Historia* (77 AD), Pliny the Elder provides descriptions of the use of ginger, pepper, and cinnamon (amongst other spices) as remedies for common ailments. In the late 1700s, Chef Antonin Careme emphasized the importance of using spices in moderation so as not to overpower the flavor of other ingredients (Freedman, 2007), and in his seminal 1907 *Le Guide Culinaire,* Auguste Escoffier (1907) divided seasonings into descriptive categories that contained some of the common chemesthetic qualities. For example, "hot" included peppercorns and cayennes; pungent condiments were those that included onions, chives, shallots, and horseradish.

13.2.1 Cultural connections to chemesthetic agents

Specific chemesthetic ingredients or their combinations have become synonymous with regional and global cuisines (Raghavan, 2007). Research in ethnobotany (Denes *et al.,* 2012), evolutionary biology (Sherman and Billings, 1999), and psychology

(Rozin and Rozin, 2005) have revealed intimate connections between the ingredients people cultivate and cook with (especially those containing chemesthetic properties), and the flavor profiles, nutritional content, and microbial quality of those foods. Regional Mexican cuisines more often than not include chilis and their burning capsaicinoids in their staple and iconic dishes (moles, salsas, and menudo). Huajiao (also known as "Sichuan pepper") and its tingly qualities are synonymous with the cuisine of the Sichuan province in China (e.g., ma po tofu, Sichuan hot pot, and Chengdu wan tons). Likewise, the cooling of mint found in raita and mint chutneys is closely associated with Pakistani and Indian cuisine.

13.2.2 History of use of chemesthetic agents in prepared foods and food service

Spices were being added to sauces and soups for commercial production as far back as 1300 (Freedman, 2007). Spice blends referred to as "blanch powders" containing chemesthetic ingredients such as ginger, nutmeg, and cinnamon were sold as early as 1400. Spice companies today provide blends of spices in ratios commonly found in regional global cuisines (e.g., McCormick's line of spice blends, or "ethnic food kits" that are available via the internet (The Spice Kit Co. Ethnic Foods Across the Kitchens of America https://www.kickstarter.com/projects/133033989/cook-easy-ethnic-meals-at-home). This adds value for consumers who might be unfamiliar with the cuisine style but want to experiment and explore new flavor profiles in their own kitchens.

Chemesthetic agents have become a component in the product developer's tool-box as a way to create unique and engaging flavor profiles that convey a sense of *authenticity* for regional cuisines around the globe. Product and menu developers are leveraging consumer awareness of global cuisines and incorporating chemesthetic ingredients, such as different varieties of chilis, including aleppos from Turkey, gochujang from Korea, and the aji amarillo from Peru (Nachay, 2013). A variety of sauces, condiments, and spice blends (see Table 13.1) with distinct chemesthetic properties from regions around the world, have found their way into professional kitchens, both fine dining and multi-unit operation restaurants (Sloan, 2011; Raghavan, 2004). Chemesthetic agents are also used to create novel or "extreme" sensory experiences, such as atomic hot wings (using capsaicin), salt and vinegar potato chips (acidic bite of sodium diacetate), chili infused vodkas (capsaicin), and cooling/tingling gums, and other candies (menthol and hydroxy-α sanshool).

13.2.3 Chemesthetics and health

As Hippocrates has noted, "Your nourishment ought to be your remedies and your medicaments shall be your food". Throughout history the line between health tincture and culinary recipes using chemesthetic containing ingredients has been somewhat blurry (Freedman, 2007), and Wright (2005) has described "flavorists" (i.e., flavor chemists) as a cross between chef and pharmacist. Ingredients containing chemesthetic agents have been used to address health and wellness issues for thousands of years. As evidenced by this book, research on the topic is ongoing and of great interest to both academics and industry. Of particular interest is the use of

Table 13.1 Common spice mixtures/pastes used in global cuisines (adapted from McGee, 2004; Raghavan, 2007).

Name of spice mixture/paste	Region	Components
Cajun blend	Southern US (Louisiana)	Sweet paprika, ground celery, garlic, black pepper, onion, cayenne red pepper, caraway, Greek oregano, cumin, dill seed, mace, cardamom, turmeric, basil, marjoram, thyme, rosemary, bay leaf
Chermoula	Morocco	Onion, garlic, coriander leaf, chili, cumin, black pepper, saffron
Chimmichurri	Argentina	Jalapeño, onion, parsley, oregano, black pepper, lemon juice, vinegar
Creole sauce blend	Southern US (Louisiana)	Onions, garlic, bay leaf, thyme, parsley, black pepper, red cayenne pepper, green onions, lemon juice, oregano
Dukkah	Egypt	Coriander, cumin, sesame seeds, black peppercorns, mint leaves, roasted hazelnuts, roasted chickpeas
Five spice	China	Star anise, Sichuan pepper, cassia, clove, fennel
Garam masala	India	Cumin, coriander, cardamom, black pepper, clove, mace, cinnamon
Green curry paste	Thailand	Thai basil, lemongrass, green cayenne chili, cilantro, shallot, galangal, coriander seeds, black pepper, shrimp paste, coconut milk, kaffir lime leaf
Hot pepper blends	West Africa	Black peppercorns, white peppercorns, green peppercorns, pink peppercorns, allspice, cubeb, ginger, red cayenne pepper, grains of paradise, dried habaneros
Panch phoran	India	Cumin, fennel, nigella, fenugreek, mustard
Pesto	Italy	Sweet basil, pine nut, parmesan and Pecorino Romano cheese, garlic, olive oil
Piri-piri	South Africa	Garlic, onions, ground red piri piri pepper, hot paprika, bay, parsley, black pepper, olive oil, lemon juice
Quatre épices	France	Black pepper, nutmeg, clove, cinnamon
Ras el hanout	Morocco	Cardamom, cassia, mace, clove, cumin, chili, rose petals
Recado rojo	Mexico	Annatto, Mexican oregano, cumin, clove, cinnamon, black pepper, allspice, garlic, salt
Shichimi	Japan	Sansho ("green huajiao"), mustard, poppy seed, sesame seed, mustard, dried mandarin peel
Tabasco® sauce	North America	Cayenne pepper, Tabasco peppers, vinegar, salt
Tahini	Middle East	Sesame seeds, sesame oil, garlic, lemon juice
Za'atar	Middle East	Marjoram, oregano, thyme, sesame, sumac
Zhug	Middle East	Cumin, coriander, garlic, chili

chemesthetic containing ingredients and their antioxidant properties to minimize the risk of cardiovascular disease, cancer, and other diseases (Srinivasan, 2014).

13.3 Sources of chemesthetic agents in the kitchen and at the product development lab bench

There are a variety of useful references for the chef and food technologist looking to incorporate chemesthetic-containing ingredients into their recipes and new consumer packaged goods (CPGs) (McGee, 2004; Wright, 2005; Raghavan, 2007; Baser and Buchbauer, 2010; Norman, 2002). Although herbs and spices come to mind as the most commonly used ingredients with chemesthetic properties, they are not the only sources. Ingredients can be found throughout the kitchen and from all food groups that contain chemesthetic agents, excluding perhaps fresh meats and fresh dairy products. Fruits, vegetables, fermented dairy products, fermented meats, fermented vegetables, wine, beer, soft drinks, and carbonated water all contribute to the chemesthetic quality of foods. In the product development setting, food technologists have typically used oil, water or ethanol extracts, steam distillates, high pressure CO_2 extracts, and dry spice blends for a more consistent flavor profile typically required in CPGs.

13.3.1 Herbs

Herbs are defined as fresh or dried plant leaves that are used to contribute flavor to food or beverages, by creating "complexity" and enhancing liking for the overall flavor (McGee, 2004). Well-trained and experienced chefs distinguish themselves by being able to effectively utilize herbs and leverage their distinct sensory properties without overpowering a dish. Fresh herbs contribute to the aromatic quality of a dish, but most fresh herbs used in the professional kitchen also contain compounds with some chemesthetic properties. McGee (2004) has thoroughly reviewed the compounds that contribute to the flavor of the most commonly used herbs in the kitchen. Some of the sensory descriptors related to chemesthetic qualities include warming, pungent, penetrating, anise, and "distinctive". Some herbs contribute more than one chemesthetic quality. For example, bay leaves are described as both "warm" (supposedly due to the presence of methyl eugenol), and "penetrating" (due to the presence of cineole). Tarragon can be described as "warm" due to its estragole content, and anise, due to anethole. Spearmint is perceived as both "warm" (myrcene) and penetrating (cineole).

13.3.2 Spices

Spices are the dried seeds, berries, roots, or bark of plants (McGee, 2004), and the most common ingredients used in the kitchen when trying to generate pronounced heating, cooling, and tingling. Removal of moisture from spices has a variable effect on volatile and non-volatile chemesthetic agents that remain (Raghvan, 2007). Eugenol in clove is highly volatile, whereas gingerol (in ginger) and capsaininoids (in chilis) are less volatile and more likely to become concentrated. Drying will also liberate volatile aroma compounds, which will impact the intensity of the overall flavor balance of the ingredient. It is possible that suprathreshold levels of volatile compounds might suppress other stimuli, including chemesthetic qualities, and so dehydrated herbs and spices might be perceived as more intense chemesthetically speaking. Chefs commonly use chemesthetic-containing spices for pickling, marinades, condiments, rubs, and finishing seasonings.

Some of the more common chemesthetic contributing spices include anise, asafoetida, fennel seed, dill seed, clove (see Chapter 8), and black pepper (see later). Anise contains anethole, which contributes a cooling sensation in spice blends. Recent work suggests that asafoetida contains alkenyldisulfides that contribute a pungent quality similar to garlic and onion (Shokoohinia *et al.*, 2013). Asafoetida is commonly used in South Indian pickle preparations enhancing flavor quality of fruits and vegetables. Fennel seeds and dill also contribute a cooling quality to dishes due to their carvone content. In garlic dill pickles, the cooling quality of the dill might "balance" or suppresses the heat of the garlic and other TRPV1 agonists in the pickling spice blend.

Cabbage constitutes a major family of spices that contribute pungent burn and cooling, with particular effect in the nasal cavity. Chopped cabbage itself has a sulfurous aroma, originating from dimethyl disulfide, which is enhanced when cooked, and might be considered a mild irritant by certain consumer segments. Enzyme activated sulfur compounds such as isothiocyanates are largely responsible for the highly aromatic and pungent qualities of wasabi, mustards, horseradish, and radishes in general.

Black, brown, and yellow mustards are versatile ingredients that contribute chemesthetic qualities as well as viscosity and emulsification of sauces and condiments. These factors collectively contribute to the overall flavor profile of prepared vinaigrette dressings, for example, that are used to enhance the flavor of vegetables and whole grains. Prepared mustards from whole seeds are typically made with vinegar and or alcohol to control the enzymatic activity and hence slow the release of pungent compounds (McGee, 2004). Whole mustard seeds are also included in pickling brines used to preserve a wide variety of vegetables (CIA, 2013).

Peppercorns are perhaps one of the most ubiquitous chemesthetic contributors in the kitchen and at the table, providing the bite or pungency of piperine (also see Chapter 6). It is used as a topical seasoning for finished dishes, in pickle recipes, whole or cracked in sauces and in forcemeats, amongst other preparations. It is typically used in recipes (like salt) "to taste", and despite its prevalence, has rarely been the subject of sensory studies in food.

13.3.3 Fruits

A variety of fruits are used to contribute cooling, burning, numbing, and astringent sensations to foods. Some of these fruits include: chilis (heating), lime (cooling), lemon (cooling), mangos and mango powder (cooling), ice apple (cooling), persimmon (astringent), unripe bananas (astringent), and fresh tengjiao (numbing/tingling). Fruits will often be incorporated into Indian chutneys, pickles, and salads to create a dynamic flavor profile. Fresh tengjioa, similar to huajiao (with a slightly milder numbing–tingling effect) is added to soups and stews as whole fresh (green) berries to contribute aroma, and provide a slight tingly effect. Although there is little evidence to support the cooling effects of citrus fruits via a chemesthetic mechanism, chefs combine lemon or lime with "hot" dishes to "cool" or perhaps balance the intensity of capsaicin. Although real food systems have been minimally explored through scientifically rigorous sensory studies, mixture suppression may play a role in this perceived cooling. This is an area of research that would benefit from the combined effort of chefs and sensory scientists. In pozole, a traditional Mexican stew/soup, chilis and limes provide heat and cooling, respectively. Mango powder, which has a sweet and sour quality, lends a cooling quality to chutneys and curries, and is also applied by sprinkling on the surface of foods.

13.3.4 Vegetables

Some of the common vegetables that contribute chemesthetic qualities to food include fennel bulbs (versus the dried seed), radishes, turnips, mustard greens, and watercress. These ingredients are commonly incorporated raw into fresh salads, or used as a garnish on top of soups, stews, and meat dishes. They contribute a contrast in the overall flavor profile, and act as "islands of flavor" that enhance the sensory experience. In menu development, pungent greens such as arugula (also called rocket) will often be combined with "bright" citrus fruits to balance the bite and bitterness of some of these vegetables. Sautéing, steaming, or simmering in water are also common techniques used to control the intensity of the pungency of mustard ingredients.

13.3.5 Fermented foods

Fermented foods can contribute chemesthetic properties in a variety of ways: (1) due to the primary ingredient being fermented, (2) from the herbs or spices that are part of the fermentation recipe, (3) as a result of the fermentation byproducts (the hydrolytic or enzymatic breakdown of lipids, proteins or carbohydrates), or (4) the production of carbon dioxide or alcohol. In many cases a combination of these factors are at play. The umami, sour, and salty qualities of fermented foods are developed during the production process, and contribute "balance" and complexity to the overall flavor quality.

Hot peppers and pepper-based sauces are fermented to develop flavor complexity, and to preserve, and otherwise optimize food safety (Katz, 2012). Kimchi is primarily comprised of cabbage that has a mild pungent quality (Talavera-Bianchi *et al.*, 2010) and garlic that is combined with chili, ginger, or onions (Katz, 2012) and then fermented. Carbon dioxide is also generated during the kimchi fermentation process contributing another dimension to the chemesthetic profile of the product. Kimchi typically is used to lend its hot and pungent qualities to foods such as rice, eggs, or meats that are relatively mild in flavor. Relishes, salsas, and chutneys that contain peppers and garlic are also commonly fermented.

Fermented dairy products can be used to contribute a perceived cooling quality or contrast to spicy foods. In some cases this might just be a physical cooling; alternatively, the lactic acid from fermentation may be cooling in the same way citrus fruits cool or balance spicy dishes. More research is needed to explore these dynamic interactions. Although the mechanism is not well understood, tangy, lightly fermented products such as crème fraîche, sour cream, and yogurt, commonly accompany "hot" and spicy foods. Turkish Tulum, an aged goat cheese, can become "pungent" due to the breakdown of lipids and proteins (Hayaloglu, 2007). However, Mexican Añejo cheese has been described as "astringent" and "pungent" (Hernandez-Morales, 2010), Italian provolone is often described as "sharp" (CIA, 2011) and La Serena cheese from Spain as "pungent" (Carbonell *et al.*, 2002).

Fermented meats will often contain herbs and spices that lend chemesthetic qualities. These meats and sausages, such as salami, contain a variety of spices with chemesthetic properties, including garlic, coriander seeds, fennel seeds, and paprika (CIA, 2012).

Naturally fermented soft drinks, which are beginning to gain popularity, contain carbon dioxide (CO_2) produced by lactic acid bacteria, creating a tingling sensation.

Beer and sparkling wine or champagne contain CO_2 and alcohol, which can contribute tingle and burn, respectively (depending upon the concentration) (see Chapters 6 and 8). Carbonated alcoholic beverages are often paired with high fat or fried foods, creating a pleasing overall mouthfeel. The CO_2 and alcohol are perceived to "cut through" the fats and oils, perhaps "cleansing" and "resetting" the palate, both physically and chemically.

13.3.6 Extracts and dry blends

At the product development bench, food technologists use herbs, spices, fruits, and vegetables that have chemesthetic qualities in a variety of forms including dried, frozen, and extracts. Spices can be used whole or ground into powder, and fresh herbs in pastes or purees (Raghavan, 2004). Each of these forms has pros and cons for various applications.

Culinary professionals prefer to work with fresh herbs whenever possible because the concentration and balance of sensory attributes (including appearance) is optimal. However, the quality of fresh herbs has a limited shelf life. Fresh herbs quickly degrade under standard processing conditions, are often too expensive, and are frequently unavailable in the volume needed for commercial packaged goods.

Dried herbs have improved stability and can be more affordable, but their flavor profile is altered from the original due to concentration effects and oxidation of the essential oils that often contain the chemesthetic compounds of interest. Spices are already dried, so their flavor profile is familiar in that form (e.g., cinnamon, mace, nutmeg, and allspice). Grinding of spices can be done cryogenically which will help retain the native quality of the flavor constituents. Dried spice blends are commonly used in a variety of commercial applications for sausages, breads, topical applications for chicken and fish, and as seasonings for sauces. These blends can be supplied by spice companies or flavor houses and cover a variety of cuisines styles including Moroccan, Brazilian, Thai, South Indian, Mexican, Vietnamese, and Jamaican (Raghavan, 2007).

Freeze-dried and individually quick frozen (IQF) herbs, fruits, vegetables, and other "aromatics" are good alternatives to their fresh counterparts, especially for product developers designing menu items for multi-unit food service operations. These ingredients retain their aroma/taste/chemesthetic balance better than dried herbs, however, they are more expensive, and are not an "identity match" to the fresh product.

As reviewed in Chapter 3, extracts from herbs, spices, and other ingredients containing chemesthetic properties can be obtained through steam distillation (Starmans and Nijhuis, 1996), solvent extraction (Starmans and Nujhuis, 1996), or super-critical CO_2 extraction (Sovilj *et al.*, 2011; Lang and Wai, 2001). Enzymes are also utilized to help liberate flavor-contributing compounds from spices (Sowbhagya and Chitra, 2010).

13.3.7 Plant breeding

Another approach to generating and controlling the chemesthetic and overall flavor quality of foods is by the selective breeding of fruits and vegetables (Saha *et al.*, 2010; Eggink *et al.*, 2014; Stewart *et al.*, 2007). Tanaka *et al.* (2014) describe the application of genotyping and selective breeding to develop a milder pepper containing lower capsaicin levels compared with existing breeds. Chefs and plant breeders have begun to work together to customize and enhance overall flavor of fruits and vegetables (see Hubbard, 2010).

13.4 Culinary techniques and chemesthetic agents

13.4.1 Incorporation

Chemesthetic experiences are a part of nearly every facet of the contemporary culinary landscape. Chefs employ two general methods for incorporating chemesthetic ingredients into completed dishes: extraction of the chemesthetic component from the physical matrix of the source ingredient, and utilization of the source ingredient as a whole.

13.4.1.1 Extraction

The separation of chemesthetic components from the food matrix in culinary applications is primarily limited to the simple process of chemical and physical extraction. The most common manifestation of this process is in steeping of ingredients in a liquid extraction medium. The three basic categories of extraction media readily available in the kitchen are lipid-based, ethanol-based, and water-based solvents. The chemical nature of the solvent will also determine the co-migration of other solutes such as pigments, taste and aroma compounds that also contribute to the flavor quality of the final dish (McGee, 2004).

13.4.1.1.1 Lipid-based

Fats and oils with chemesthetic properties are easily prepared by steeping ingredients in lipids (Norman, 2002; CIA, 2007). Chemesthetic extracts can be prepared with saturated or unsaturated lipids depending on the desired final application, and on the functionality of the delivery mechanism. In products where solid or semisolid lipids are the desired vector for delivering chemesthetic agents, such as charcuterie, compound butters, and laminated doughs, saturated lipids such as lard, tallow, duck fat, butter, and coconut oil are used. Unsaturated lipids such as olive, canola, grapeseed, and peanut oils are used more frequently when a pourable consistency is desired. Marinades, salad dressings, and simple infused oils are all examples of this technique.

13.4.1.1.2 Ethanol-based

Ethanol-based chemesthetic extracts are used primarily in cocktails. As a solvent, ethanol is highly effective at extracting a wide variety of taste and aroma compounds. This ability to extract solutes of varying polarity allows ethanol-based extracts to carry most chemesthetic compounds. While the flavor of raw alcohol is rarely incorporated into cooked foods, ethanol-based extracts of chemesthetic ingredients have become very popular in beverages. Common examples include the mint in a Mojito, the ginger in a Dark and Stormy, or any of the various chili and pepper-infused cocktails that have emerged in the past few years along with the rise of the "mixologist" (Birch, 2010) as a member of culinary brigade. Here are a few examples of mixed drinks with pronounced chemesthetic quality (in addition to the alcohol):

- "Pancho Villa": tequila, cayenne, grapefruit, and red sea salt dusted rim (Luxembourg Café, New York, NY)
- "L'Ete Chez Lux": gin, ginger, watermelon, lime (Luxembourg Café, New York, NY)
- "Whammy": bourbon, lemon, mint, black pepper, peaches (Freeman's, New York, NY)

13.4.1.1.3 Water-based

Most chemesthetic compounds are weakly soluble in water. Extraction of chemesthetic agents into aqueous media is primarily applied in situations when the extraction medium is also the finished product. Stocks, soups, and sauces simmered with chemesthetic ingredients are some of the most prevalent examples of this technique. These extracts are usually less intense than extracts using ethanol or lipid-based media, and most contemporary soups and sauces have some amount of lipid component to aid in the extraction.

13.4.1.2 Whole-ingredient incorporation

The primary benefit to the incorporation of the source ingredient into a dish rather than an extract of the chemesthetic components is the inclusion of a more complete sampling of the aroma and taste compounds found in the source ingredient. This approach also provides the chef with the ability to leverage the ingredient functionality of the source material and an opportunity to incorporate textural contrasts.

Mustards used in traditional mayonnaise, for example, are most often incorporated in the form of powdered seeds, rather than infused oils or other extracts. This is because the mustard seed itself contains both hydrocolloids and surface-active emulsifiers, which aid in the formation of a stable emulsion. Chilis are often ground into pastes for use in marinades, sauces, and soups because of the added viscosity afforded by the plant particulates (McGee, 2004). The pulverized plant tissue can aid a marinade in clinging to a piece of meat or give sauces and soups a thick, substantial mouthfeel without the dilution of flavor inherent to other neutral-flavored thickeners such as starch. In the case of fermented condiments featuring chilis, ginger, and so on, the fermentable sugars contained within the ingredient act as a fertile substrate for microorganisms, allowing the food matrix to be transformed while preserving the chemesthetic compounds inherent in the main ingredient.

Textural contrast (Hyde and Witherly, 1993) can also be a large motivator for the inclusion of whole ingredients over extracts. Spices, ginger, horseradish, and other chemesthetic ingredients can be used to provide brittle, crunchy textures to act as counterpoints to soft purees, soups, and meat dishes. Keeping chemesthetic ingredients whole also allows for a less homogenous dispersion of the chemesthetic experience. Rather than infusing ginger into an ice cream base or jalapeños into a ceviche marinade, chefs can incorporate whole chunks of ginger into the finished ice cream or can garnish a ceviche with slices of fresh jalapeño. This introduces a more temporal quality to the chemesthetic experience, providing stimuli in isolated, concentrated bursts rather than a continuous, low roar.

13.4.2 Impact of culinary technique on intensity

Culinary techniques can affect the intensity of the chemesthetic experience. Drying, one of the simplest culinary techniques, can be implemented to increase the concentration of non-volatile chemesthetic components, although loss of volatile chemesthetic compounds may result.

Enzymes are responsible for the evolution of several chemesthetic compounds, most notably the lachrymators in the mustard family and in alliums. Chefs exploit this behavior in various ways, from activating enzymes through physical damage of plant tissue and changes in water activity to denaturing them with heat and changes in pH.

The volatile chemesthetic compounds are all susceptible to loss of potency through outgassing due to heating. The loss of volatile compounds is desirable in some cases to temper the intensity of the chemesthetic experience, but there are many instances when chefs seek to retain as much of the compound within the food as possible. In these cases, chefs will coarsely grind spices and rough chop fresh plant foods in order to minimize surface area and exposure to the atmosphere. Many preparations such as fried and roasted foods are garnished with spices, sauces, and pastes at the end of cooking to preserve the quality of non-volatile chemesthetic compounds.

Fermentation is a frequently used alternative to heat-based processing methods that can aid in the preservation of chemesthetic compounds. For prolonged ferments, such as long-fermented, capsaicin-based hot sauces, the extended fermentation time can result in the gradual degradation of chemesthetic compounds, yielding flavorful products with mildly chemesthetic properties.

13.5 Applications of chemesthetic agents in the food industry

Research indicates that the perception of stimuli that contribute to flavor is highly variable due to genetic (Dinehart *et al.*, 2006; Keller *et al.*, 2007; Hayes *et al.*, 2013) and contextual factors (Meiselman, 1996). Flavor is also a temporal and dynamic perception that evolves over time in the oral cavity. The contribution of chemesthetic ingredients to the overall liking of flavors of food are significant in this respect because they impact flavor balance and often have unique and pronounced temporal effects on overall flavor perception (also see Chapter 6). It might be argued that this is in part why the average consumer avoids using chemesthetic agents in cooking, because they are difficult to control. These factors play an important role in developing "craveability" and providing new, authentic, and highly liked "experiences" for consumers in restaurants and for prepared foods. Evaluation of flavor profiles containing chemesthetic agents must be considered in the context of specific consumer segments, which should be based on psychographics such as extreme eaters, food neophilics, and variety seekers (see Chapter 2).

13.5.1 Chemesthetic agents in global cuisines

Cuisines from around the world harness specific combinations of chemesthetic ingredients that are reflected in common recipes (Rozin and Rozin, 2005) and spice blends (Table 13.1). There are, of course, many regional cuisines that share common spices. Cumin and chili are shared by the Middle East and Morocco. Chilis are a common source of "heat" in many warm climate cuisines. Jamaican curries and Northern Thai curries share many common spices with pronounced chemesthetic properties, including coriander seeds, cumin, limes, chilis, tamarind, cilantro, and coconut. Chefs consider coconut to play the role of a solvent/diluent that can balance or mute certain qualities (Danhi, 2008). The ratio of ingredients and how they are combined and transformed by culinary techniques are what ultimately influences the flavor quality that identifies a particular cuisine (Danhi, 2008).

A common "flavor theme" across a variety of cuisines and cultures is to combine ingredients containing chemesthetic agents that have contrasting qualities. A popular example is the compound Sichuan flavor principle of "má là", where "má" refers to the numbing–tingling of Sichuan peppers, and "là" refers to the heat and burn of chilis. According to chefs, this combination produces "flavor waves of alternating burning and numbing that stimulates appetite and enhances enjoyment" (Shirley Cheng, personal communication). Dishes such as "Water boiled pork" and "Tengiao sliced beef" contain both tingling and burning qualities. Traditional Sichuan banquets are designed to follow this pattern of "flavor waves" throughout the entire meal. In one instance, we surveyed diners (primarily chefs and sensory scientists) throughout their meal to measure this perceived rise and fall of flavor intensity (Fig. 13.1).

The western Indian dish, panipuri, also combines heating and cooling to enhance the enjoyment of a potato based snack type food. The potato or lentil filling contains chili for heat, balanced with tamarind (acidic, considered cooling and slightly astringent). The entire deep fried ball of spiced potato is dipped into a chutney that is thinned with lime juice, and contains garlic, cilantro (coriander

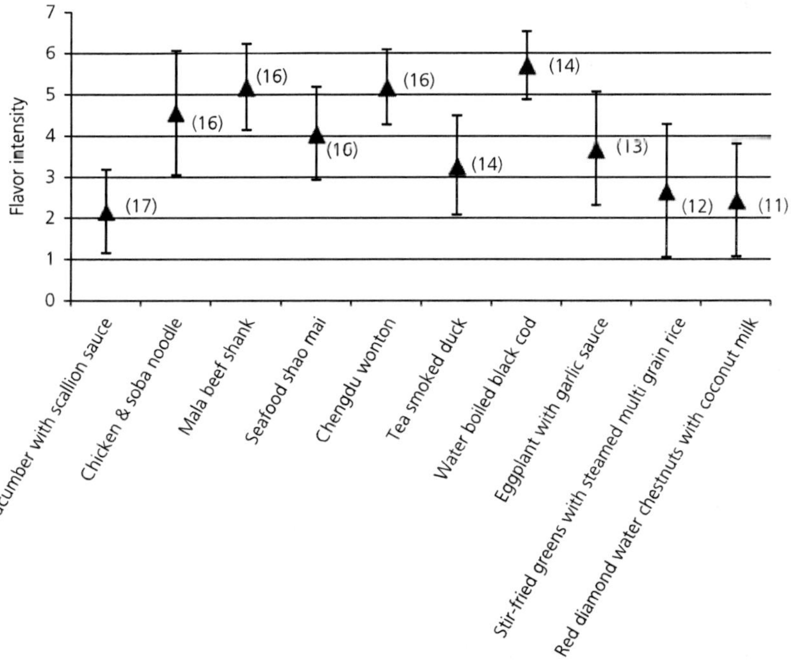

Fig. 13.1 Flavor intensity throughout a ten-course Sichuan banquet served in the order presented from left to right along the *x*-axis. Errors bars are for number of diners, in parentheses.

leaf), and chili. This complex balance of heating, astringency and cooling due to the citrus (combined with the physical textural contrasts) makes for a very "craveable" sensory experience.

Salsa tamulada is a condiment common in Mexico that is a simple blend of ground dry roasted habanero, bitter orange juice, and salt. The flavor profile is considered to be extremely spicy with floral and fruity aromas. The bitter and acidic orange juice is believed to balance and slightly cool the heat of the chilis.

Katz (2012) describes "Salgam suyu", a pungent Turkish beverage made from fermented carrots and turnips that is paired with "raki", an anise flavored liquor. This is another example of combining cooling and pungent qualities together to create an overall flavor balance.

Pastry chefs are also interested in combining contrasting and balanced chemesthetic experiences. For example, in Mexico, chili is combined with lemon sorbet. In the Sonoran region of Mexico, the extremely hot chiltepin peppers (as common as salt and pepper on tables in that region) are often sprinkled on top of ice cream.

13.5.2 Creating "craveable" culinary experiences with chemesthetic agents

Chefs and product developers want to create "craveable" products that keep consumers engaged and satisfied without becoming boring or monotonous. We like contrasting and dynamic textures and sensory experiences in general (Lawless, 2000). Chemesthetic agents such as capsaicin in chilis are used in foods to create a progression of flavor (both gradual onset and increase of burn, and slow drop off) that keeps the consumers' senses piqued. Consider, for example, the flavor of a bag of plain salted pretzels, versus a bag of salted pretzels laced with mustard powder or dipped in prepared mustard, or plain potato chips versus Sriracha flavored chips. Chicharrones (fried fat) are very crispy and desirable from a textural standpoint, but when coated with chili and lime powder they take on an entirely new and somewhat irresistible quality. However, if the diner is not anticipating this experience, or has had no or little previous experience with these sensory stimuli, it can generate a negative hedonic experience. Careful wording on menus and package labels is critical for this reason, providing an important role for marketing and consumer behavior scientists in new product development.

Cooling effects of certain ingredients can also be used to "cleanse the palate" at the end of a multicourse meal. For example, frozen fennel meringue has been combined with granny smith (tart) gelée and crystalized anise hyssop leaf in a plated dessert (Migoya, 2012).

All consumers are neophobic to some degree, and some more so than others (Meiselman *et al.*, 2010). Chemesthetic agents can significantly decrease liking scores amongst neophobic consumer segments. One approach is to introduce chemesthetic ingredients into familiar foods, and balance the new with familiar (Lawless, 2000). When we combined a low concentration of a mà lá seasoning blend with a sodium reduced French fry seasoning mix, the liking scores were not significantly altered even though the saltiness scores went down (compared with a typical salt level found in fast foods; Cheng and Loss, 2012). Customer comments for the mà lá seasoned fries included "refreshing", "hot", "tingling", "satisfying", and "savory". It was also observed that French fries with low levels of the mà lá spice blend had a less oxidized, old-oil aroma compared with French fries that did not contain the mà lá blend.

On the other hand, chemesthetic agents are used to create novel sensory experiences for the adventurous or extreme eater (Veeck, 2010). "Atomic wings" and "five alarm" chilis are popular amongst a certain segment of consumers (see Chapter 2). Pop-rocks and red hot candies also create unique and intense experiences.

13.5.3 Food safety and preservation

Herbs, spices, and other ingredients and dishes with chemesthetic qualities have antimicrobial properties and can extend the shelf life and improve the safety of foods. This topic has been reviewed from both food science (Tajkarimi *et al.*, 2010) and evolutionary biology perspectives (Sherman and Billings, 1999). Research Chefs and Food Technologists are driven by increasing consumer demand to "clean up" food product labels, by removing "artificial" preservatives, and incorporating natural antimicrobials. Many herbs, spices, and other vegetables that contribute distinct and desirable flavors also bring along with them antimicrobial properties. Clove, cinnamon, garlic, rosemary, and oregano have pronounced antimicrobial properties. Ginger, chilis, and cumin (Tajkarimi *et al.*, 2010) and assofoetida (Kavoosi *et al.*, 2013)have also been noted to have similar characteristics . Culinary Scientists and Chefs will need to better understand how to create well-liked and "craveable" dishes using these ingredients (perhaps adding value by connecting them to "authentic" flavor profiles) while conferring sufficient antimicrobial properties.

A survey of meat-based recipes in cookbooks, specifically the herbs, spices, and other aromatic vegetables they contained, revealed a correlation between the usage of antimicrobial spices with the region in which they were grown (Sherman and Billings, 1999). Warmer regions that posed a greater risk of foodborne disease or food spoilage contained a greater number of antimicrobial spices that also happen to contribute chemesthetic qualities. For example, onion, ginger, garlic, chilis, pepper, lime, and coriander were most common in regions such as Thailand, The Philippines, and India. It should be noted that although certain herbs and spices (typically their extracts) have pronounced antimicrobial properties, the whole ingredients (fresh or dried) can carry their own spoilage and pathogenic microbial populations (Banerjee and Sarkar, 2003), which may have been introduced and allowed to develop during post-harvest processing.

13.5.4 Modern applications of chemesthetic agents in fine dining

The rise of culinary science as a discipline has informed the progression of *avant-garde* cuisine. The incorporation of new technologies and methods of food preparation have allowed chefs to incorporate new experiences based on scientific insight (Myhrvold *et al.*, 2011). The *avant-garde* mindset also allows chefs to break traditional barriers of flavor combinations, incorporating chemesthetic ingredients in ways seldom seen before.

Modern drying technology allows chefs to concentrate the flavors of different ingredients by removing water at lower temperatures, thus preventing some of the degradation of chemesthetic compounds that is traditionally seen in drying. These techniques include new dehydrators with more precise control of temperatures, as well as freeze-drying technology.

In addition to drying food, the rotary evaporator has been incorporated into modern kitchens to allow for preparation of various low-temperature distillates and concentrates. This provides chefs with some interesting possibilities, such as separating non-volatile chemesthetic compounds, for example, capsaicin, from volatile fractions. Chefs have thus been able to complement foods with the aroma of habanero, for example, without any of the expected burn, yielding a surprising and impactful sensory experience.

One of the limiting factors in using chemesthetic agents in culinary settings is the time required for simple diffusion to allow chemesthetic compounds to permeate most foods. *Sous-vide* technology allows for this process to be expedited. By sealing food with chemesthetic materials in vacuum-sealed plastic pouches and cooking at precise temperatures in water baths, chefs gain a greater level of control over several aspects of their medium. Many modern chefs have developed protocols for the extraction of chemesthetic ingredients through specific temperature–time combinations. These protocols are optimized through trial and error to allow maximum efficiency in extraction through elevated temperatures while avoiding degradation of flavor and preventing outgassing in the sealed environment. A related technique is the process of vacuum infusion (Myrhvold *et al.*, 2011). Solid food ingredients, including fruits, vegetables, doughs, and meats, can be placed in a liquid containing chemesthetic ingredients and exposed to strong vacuum. This voids gas from intercellular spaces in the tissue. When atmospheric pressure is reintroduced, it forces chemesthetic liquid to saturate the tissues.

Pressure-cooking has also revolutionized the culinary landscape. One of the largest applications of pressure-cooking to chemesthetic ingredients is the ability to soften plant tissues that are normally simply pulverized or extracted, allowing for more possibilities in the realm of whole-ingredient applications. Mustard seeds, for instance, can be pressure-cooked whole, causing the seed coat to soften and expand, creating a texture very similar to the popping sensation of caviar, while delivering contained bursts of chemesthetic experience. Pressure cookers also allow for the retention of volatile compounds.

Liquid nitrogen was first incorporated into kitchens as a means of freezing desserts, such as ice cream, quickly (Myrhvold *et al.*, 2011). Recently, the potential of liquid nitrogen to aid in the efficient grinding and pulverization of chemesthetic ingredients has been realized in professional kitchens. Many chemesthetic compounds are volatile, and the heat and high surface area created by grinding often result in the loss of these components. Incorporation of liquid nitrogen into the grinding apparatus causes the product to be both more brittle, allowing finer grinding, and to remain cool, reducing the extent of volatile off-gassing that occurs.

All of these modern applications of chemesthetic ingredients have allowed chefs to present chemesthetic stimuli in innovative and often provocative ways. The continued evolution of chemesthetic agents in the professional kitchen is one of the keys to training diners to accept chemesthetic agents in the wider food culture.

References

Auvray M. and Spence, C. 2008. The multisensory perception of flavor. *Consciousness and Cognition*, **17** (3), 1016–1031.

Banerjee, M. and Sarkar, P.K. 2003. Microbiological quality of some retail spices in India. *Food Research International*, **36** (5), 469–474.

Barham, P., Skibsted, L.H., Bredie, W.L.P., *et al.* 2010. Molecular Gastronomy: A New Emerging Scientific Discipline. *Chemical Reviews*, **110** (4), 2313–2365.

Baser, K.H.C. and Buchbauer, G. (eds.). 2010. *Handbook of Essential Oils; Science Technology, and Applications*, Boca Raton, FL: CRC Press.

Birch, H. 2010. Shaken, not stirred. *Chemistry World*, **7** (12), 34–37.

Carbonell, M., Nunez, M., and Fernandez-Garcia, E. 2002. Evolution of the volatile compounds of ewe raw milk cheese during ripening; Correlation with flavour characteristics. *Lait*, **82** (6), 683–698.

CIA. 2007. *Hors d'Oeuvre at Home with the Culinary Institute of America; Essential Techniques and Recipes for Creating Great Small Bites*, Hoboken: John Wiley & Sons, Inc.

CIA. 2011. *The Professional Chef*, 9th edn, Hoboken: John Wiley & Sons, Inc.

CIA. 2012. *Garde Manger: The Art and Craft of the Cold Kitchen*, 4th edn, Hoboken: John Wiley & Sons, Inc.

CIA. 2013. *Preserving: Putting Up the Season's Bounty*, New York: Houghton Mifflin Harcourt.

Cheng, M., Ogbeide, G.A., and Hamouz, F.L. 2011. The development of culinary arts and food science into a new academic discipline. *Journal of Culinary Science and Technology*, **9** (1), 17–26.

Cheng, S. and Loss, C. 2012. Consumer acceptance of the mala flavour profile and applications for sodium reduction strategies. Poster presented at Eurosense, Bern, Switzerland.

Danhi, R. 2008. *Southeast Asian Flavors; Adventures in Cooking the Foods of Thailand Vietnam Malaysia and Singapore*, El Segundo, CA: Mortar and Press.

Delwiche, J. 2004. The impact of perceptual interactions on perceived flavor. *Food Quality and Preference*, **15**, 137–146.

Denes, A., Papp, N., Babai, D., Czucz, B., and Monar, Z. 2012. Wild plants used for food by Hungarian ethnic groups living in the Carpathian Basin. *ACTA Societatis Botanicorum Poloniae*, **81** (4), 381–396.

Dinehart, M.E., Hayes, J.E., Bartoshuk, L.M., Lanier, S.L., and Duffy, V.B. 2006. Bitter taste markers explain variability in vegetable sweetness, bitterness, and intake. *Physiology and Behavior*, **87** (2), 304–313.

Dinella, C., Masi, C., Zoboli, G., and Monteleone, E. 2012. Sensory functionality of extra-virgin olive oil in vegetable foods assessed by temporal dominance of sensations and descriptive analysis. *Food Quality and Preference*, **26**, 141–150.

Eggink, P.M., Tikunov, Y., Maliepaard, C., *et al.* 2014. Capturing flavors from *Capsicum baccatum* by introgression in sweet pepper. *Theoretical and Applied Genetics*, **127** (2), 373–390.

Escoffier, A. 1907. *A Guide to Modern Cookery*, London: William Heinemann.

Freedman, P. (ed.). 2007. *Food; The History of Taste*, Berkeley, CA: University of California Press.

Green, B.G. 1996. Chemesthesis: pungency as a component of flavor. *Trends in Food Science and Technology*, **7** (12), 415–420.

Gustafsson, I. 2004. Culinary arts and meal science – a new scientific research discipline. *Food Service Technology*, **4** (1), 9–20.

Hayaloglu, A.A., Fox, P.F., and Guven, M. 2007. Cheeses of Turkey 1. Varieties ripened in goat skinned bags. *Lait*, **87** (2), 79–95.

Hayes, J.E., Feeney, E.L., and Allen A.L. 2013. Do polymorphisms in chemosensory genes matter for human ingesitve behavior? *Food Quality and Preference*, **30** (2), 202–216.

Hernandez-Morales, C., Hernandez-Montes, A., and Aguirre-Mandujano, E. 2010. Physicochemical, microbiological, textural, and sensory characteristics of Mexican Anejo cheese. *International Journal of Dairy Technology*, **63** (4), 552–560.

Hubbard, K. 2011. *Chef Dan Barber: Why Plant Breeders Matter*. Organic Seed Alliance [online], Available http://blog.seedalliance.org/2011/06/23/chef-dan-barber-why-plant-breeders-matter/ (accessed 23 June, 2011).

Hyde, R.J. and Witherly, S.A. 1993. Dynamic contrast – A sensory contribution to palatability. *Appetite*, **21** (1), 1–16.
Katz, A.E. 2012. *The Art of Fermentation*, White River Junction, VT: Chelsea Green Publishing.
Kavoosi, G., Tafsiry, A., Ebdam, A.A., and Rowshan, V. 2013. Evaluation of antioxidant and antimicrobial activities of essential oils from *Carum copticum* seed and *Ferula assafoetida* latex. *Journal of Food Science*, **78** (2), T356–T361.
Keller, A., Hanyi, Z., Qiuyi, C., and Vosshall, L.B. 2007. Genetic variation in a human odorant receptor alters odour perception. *Nature*, **449**, 468–472.
Lang, Q.Y. and Wai, C.M. 2001. Supercritical fluid extraction in herbal and natural product studies – a practical review. *Talanta*, **53** (4), 771–782.
Lawless, H.T. 2000. Sensory combinations in the meal, in *Dimensions of the Meal; The Science, Culture, Business, and Art of Eating* (ed. H.L. Meiselman), Gaithersburg, MD: Aspen Publication, pp. 92–106.
Loss, C.R. 2014. Enhancing science education and developing a cross-disciplinary research program through the culinary medium. Presented at the Trends in Cooking Science Symposium, The American Chemical Society, San Francisco, CA.
McGee, H. 2004. *On Food and Cooking; The Science and Lore of the Kitchen*, New York: Scribner.
Meiselman, H.L. 1996. The contextual basis for food acceptance, food choice and food intake: the food, the situation, and the individual, in *Food Choice Acceptance and Consumption* (eds. H.L. Meiselman and H.J.H. MacFie), New York: Springer, US, pp. 239–263.
Meiselman, H.L. (ed.). 2009. *Meals in Science and Practice; Interdisciplinary Research and Business Applications*, Cambridge, UK: Woodhead Publishing.
Meiselman, H., King, S., and Gillette, M. 2010. The demographics of neophobia in a large commercial US sample. *Food Quality and Preference*, **21** (7), 893–897.
Meiselman, H.L. 2013. The future in sensory/consumer research: evolving to a better science. *Food Quality and Preference*, **27** (2), 208–214.
Migoya, F. 2102. *The Elements of Dessert*, Hoboken: John Wiley & Sons, Inc.
Myhrvold, N., Young, C., and Bilet, M. 2011. *Modernist Cuisine: The Art and Science of Cooking*, Bellvue WA: The Cooking Lab.
Nachay, K. 2013. Firing up product development. *Food Technology*, **67** (11), 25–32.
Nilius, B. and Appendino, G. 2013. Spices: the savory and beneficial science of pungency. *Reviews in Physiology Biochemistry and Pharmacology*, **164**, 1–76.
Norman, J. 2002. *Herbs and Spices; The Cooks Reference*, New York, NY: DK Publishing.
Parry, G. 1953. *The Story of Spices*, New York, NY: Chemical Publishing Company.
Piggott, R.J. 2000. Dynamism in flavour science and sensory methodology. *Food Research International*, **33**, 191–197.
Raghavan, S. 2004. Developing ethnic foods and ethnic flair with spices. *Food Technology*, **58** (8), 35–42.
Raghavan, S. 2007. *Spices, Seasonings, and Flavorings*, 2nd edn, Boca Raton, FL: CRC Press.
Reinbach, H.C., Meinert, L., Ballabio, D., *et al.* 2007. Interactions between oral burn, meat flavor and texture in chili spiced pork patties evaluated by time intensity. *Food Quality and Preference*, **18** (6), 909–919.
Rozin, E. and Rozin, P. 2005. Culinary themes and variations, in *The Taste Culture Reader; Experiencing Food and Drink* (ed. C. Korsmeyer), New York: Berg, pp. 34–41.
Saha, S., Hedau, N.K., Kumar, S., Mahajan, V., and Gupta, H.S. 2010. Variability in hot pepper for phytochemicals offers promising tools in plant-breeding programmes. *ACTA Agriculturae Scandinavica, Section B - Soil and Plant Science*, **60** (3), 227–234.
Sherman, P. and Billing, J. 1999. Darwinian gastronomy: Why we use spices: Spices taste good because they are good for us. *BioScience*, **49** (6), 453–463.
Shokoohinia, Y., Cianese G., Apendino, G., *et al.* 2013. Some like it pungent and vile. TRPA1 as a molecular target for malodorous vinyl disulfides from asafoetida. *Fitoterapia*, **90**, 247–251.
Sloan, E. 2011. Spices, seasonings come on strong. *Food Technology*, **65** (5), 20.
Sovilj, M.N., Nikolovski, B.G., and Spasojevic, M.D. 2011. Critical review of supercritical fluid extraction of selected spice plant materials. *Macedonian Journal of Chemistry and Chemical Engineering*, **30** (2), 197–220.

Sowbhagya, H.B. and Chitra, V.N. 2010. Enzyme-assisted extraction of flavorings and colorants form plant material. *Critical Reviews in Food Science and Nutrition*, **50** (2), 146–161.

Starmans, D.A.J. and Nijhuis, H.H. 1996. Extraction of secondary metabolites from plant material: A review. *Trends in Food Science and Technology*, **7** (6), 191–197.

Srinivasan, K. 2014. Antioxidant potential of spices and their active constituents. *Critical Reviews in Food Science and Nutrition*, **54** (3), 352–372.

Stewart, C., Mazourek, M., Stellari, G.M., Oconnell, M., and Jahn, M. 2007. Genetic control of pungency in *C. chinense* via the *Pun1* locus. *Journal of Experimental Botany*, **58** (5), 979–991.

Talavera-Bianchi, M., Chambers, E., and Chambers, D.H. 2010. Lexicon to describe the flavor of fresh leafy vegetables. *Journal of Sensory Studies*, **25**, 163–183.

Tajkarimi, M.M., Ibrahim, S.A., and Cliver, D.O. 2010. Animicrobial herb and spice compounds in foods. *Food Control*, **21** (9) 1199–1218.

Tanaka, Y., Yoneda, H., Hosokawa, M., Miwa, T., and Yazawa, S. 2014. Application of marker-assisted selection in breeding of a new fresh pepper cultivar (*Capsicum annuum*) containing capsinoids, low-pungent capsaicinoid analogs. *Scientia Horticulturae*, **165**, 242–245.

Veeck, A. 2010. Encounters with extreme foods: neophilic/neophobic tendencies and novel foods. *Journal of Food Products Marketing*, **16** (2), 246–260.

Vega, C., Ubbink, J. 2008. Molecular gastronomy: food fad or science supporting innovative cuisine? *Trends in Food Science and Technology*, **19** (7), 372–382.

Wright, J. 2005. *Flavor Creation*, Carol Stream, IL: Allured Publishing Corporation.

CHAPTER 14
Overview of chemesthesis with a look to the future

E. Carstens
Department of Neurobiology, Physiology and Behavior, University of California Davis, Davis, CA, USA

14.1 Introduction

The term chemesthesis was first introduced in 1990 by Green *et al.* (1990) and was defined as the chemical sensibility of the skin and mucous membranes. Irritation, the "common chemical sense" (Parker, 1912), and for food "the trigeminal sense", are terms that have also been used to describe thermal, irritant, and mechanical sensations elicited by chemical stimuli. Chemesthetic qualities are evoked by chemical stimuli that activate sensory receptors involved in signaling pain (nociceptors), temperature (warm or cool thermoreceptors), or touch (mechanoreceptors). Descriptors commonly used to describe chemesthetic sensory sub-qualities include numbing, burning, stinging, warming, cooling, and tingle. Examples of chemesthetic sensory qualities are the pungent burn of chilis containing capsaicin, the cool sensation elicited by menthol, and the tingling or buzzing sensation elicited by sanshools. Cutaneous and mucosal chemosensitivity presumably warns the organism of the presence of chemical irritants and initiates protective reflexes or more complex physiological and behavioral responses to expel, remove, or otherwise avoid the offending stimulus. Examples of protective responses elicited by chemesthetic stimulation include coughing or sneezing to clear the airways or nasal sinus, tearing to clear the ocular surface, salivation in the oral cavity, and scratching or rubbing the skin. Green (2012) recently coined the term "chemofensor complex" to describe the chemical sensitivity of body tissues that collectively provide protection from hostile environmental stimuli. As noted in Chapter 1, it is now known that chemosensitive molecular receptors are expressed not only by epithelial and nerve cells on the body surface, but also by cells of internal organs such as the liver, gastrointestinal tract, pancreas, and brain (Green, 2012). Some cells communicate with nerves or the immune system, while others do not and presumably participate in "non-sensory" (e.g., humoral) signaling. Thus, both sensory and non-sensory chemoreception contributes to a general chemofensor complex that serves to protect the organism against environmental hazards.

Chemicals that elicit chemesthetic sensations act at molecular receptors and/or ion channels expressed in epithelial cells and/or sensory nerve endings innervating the skin or mucosa. Many of these chemosensitive molecular receptors are members of the family of thermosensitive transient receptor potential (TRP) ion channels (see Chapter 4). Thus, TRPV1 responds to temperatures at or above the heat pain threshold, as well as to capsaicin (Caterina *et al.*, 1997). TRPA1 was originally reported to respond to intense cold (Story *et al.*, 2003) and to mustard oil and many other irritant chemicals (Jordt *et al.*, 2004). TRPM8 responds to menthol and cooling in the innocuous range (McKemy *et al.*, 2002; Peier *et al.*, 2002a), whereas TRPV3 responds to innocuous warming and clove and oregano oils (Peier *et al.*, 2002b; Moqrich *et al.*, 2005; Xu *et al.*, 2006). TRPV4 also responds to innocuous warming, hypertonicity (Güler *et al.*, 2002) and chemical agonists, including GSK1016790A and 4-α phorbol 12,13-didecanoate (4-αPDD) (Vincent and Duncton, 2011). Presumably, TRPM8 is expressed by sensory nerves that signal cooling (cold fibers), TRPV3 and TRPV4 by sensory nerves that signal warming (warm fibers), and TRPV1 and TRPA1 by nerves that signal pain (nociceptors). Thermosensitive TRP channels thus provide a satisfactory explanation for the thermal nature of chemesthetic sensory qualities, such as the cool sensation elicited by menthol or the burn elicited by capsaicin. Other chemesthetic sensations such as itch or tingle involve additional types of molecular receptors and ion channels. This chapter provides an update on chemesthesis and the underlying neural mechanisms, with an emphasis on open questions and potential future approaches to address them.

14.2 Peripheral innervation of oral, ocular, and nasal mucosa and skin

Oral, ocular, and nasal chemesthesis is mediated by nerve fibers traveling in the corresponding branch of the trigeminal nerve, with cell bodies located in the trigeminal ganglion (TG). Cutaneous chemesthesis is mediated by sensory nerve fibers whose cell bodies are located in the dorsal root ganglia (DRG) for the lower body, or within the TG for facial and scalp skin. The ocular surface and nasal sinus are more sensitive to airborne irritants compared with the oral mucosa and skin. The cornea is innervated by the ciliary nerves, whose peripheral nerve endings terminate within a few microns of the ocular surface where they are in contact with the mucin/glycocalyx layer (Rosenthal and Boorsook, 2012). Similarly, peripheral branches of ethmoidal nerve endings innervating the nasal sinus terminate very close to the mucosal surface, and detection thresholds for ocular and nasal irritation by volatile organic compounds are similar (Cometto-Muniz and Cain, 1995). In contrast, trigeminal innervation of the tongue involves paragemmal nerve endings that terminate beneath or around the outside of taste buds, with occasional intragemmal fibers that penetrate into the lumen of the taste bud (Whitehead and Kachele, 1994). Chemicals dissolved in saliva need to traverse tight epithelial junctions of the tongue surface, or pass into the interior of taste buds, in order to contact paragemmal or intragemmal trigeminal nerve fibers, presumably accounting for the higher detection levels for oral compared with ocular or nasal chemesthesis.

In the skin, myelinated nerve fibers innervate several types of mechanoreceptive end organs including hair follicle and delta-hair receptors (hairy skin only), Merkel cells, Meissner corpuscles, Ruffini endings, Pacinian corpuscles, mechanical nociceptors, and others (see Le Pichon and Chesler, 2014 for recent review). With the exception of tingle (discussed further later), most chemesthetic sensory qualities appear to be mediated largely via activation of unmyelinated thermoreceptors or nociceptors. Unmyelinated free nerve endings may be generally divided into peptidergic and non-peptidergic classes. Based on stimulus-evoked responses, unymelinated fibers can be subclassified as warm or cold fibers responsive to innocuous warming or cooling, respectively, or as nociceptors that respond to noxious thermal (hot and cold) stimuli and irritant chemicals. The response selectivity of a sensory receptor presumably reflects its expression pattern of transduction molecules. Free nerve endings densely innervate the skin, with peripheral terminals reaching the strata granulosum and spinosum of the epidermis just beneath the skin surface (Le Pichon and Chesler, 2014; Zylka *et al.*, 2005). The outermost skin layer (stratum corneum) forms a barrier to chemical diffusion that helps to account for the lower chemosensitivity of the skin compared with oral or nasal mucosa.

Unmyelinated trigeminal fibers project centrally to terminate primarily in the ipsilateral trigeminal subnucleus caudalis (Vc), where there is an inverted somatotopic map of the head such that the oral cavity is located dorsomedially (Carstens *et al.*, 1995), the nasal sinus laterally (Anton *et al.*, 1991), and the ocular surface ventrolaterally (Bereiter *et al.*, 2002). Neurons in dorsomedial and ventrolateral Vc respond to a wide variety of irritant chemical stimuli applied to the tongue or ocular surface, respectively (Carstens *et al.*, 1998; Zanotto *et al.*, 2007, 2008; Akiyama *et al.*, 2010; Klein *et al.*, 2011a, d, 2014). This can be explained to a large extent by the broad chemical tuning of trigeminal afferents projecting to Vc. Calcium imaging studies have revealed that many individual TG cells respond to multiple chemicals including menthol, capsaicin, allyl isothiocyanate (AITC, the pungent chemical in mustard oil), cinnamic aldehyde, eugenol, carvacrol, and others (Akiyama *et al.*, 2010; Klein *et al.*, 2011a–d, 2014). Similarly, many DRG cells (Akiyama *et al.*, 2012, 2013) as well as higher-order neurons in the spinal cord respond to multiple irritant chemical stimuli (Carstens, 1997; Jinks and Carstens, 2002; Akiyama and Carstens, 2013).

14.3 TRPV1

TRPV1, formerly called the vanilloid receptor-1 (VR-1), was the first of the thermosensitive TRP channels to be sequenced and cloned (Caterina *et al.*, 1997). TRPV1 responds to heat at or near the thermal pain threshold, as well as to capsaicin, protons, and other chemicals including piperine, zingerone, 6-shogaol, and gingerol (Liu and Simon, 1996; Iwasaki *et al.*, 2006; Riera *et al.*, 2009) (Fig. 14.1, top). TRPV1 is expressed in C-fiber polymodal nociceptors that signal heat pain. As discussed in Chapter 6, topical application of capsaicin to the tongue or facial skin of humans elicited irritation that usually increased with repeated application at short interstimulus intervals (sensitization), followed after a hiatus of >2.5 minutes by desensitization,

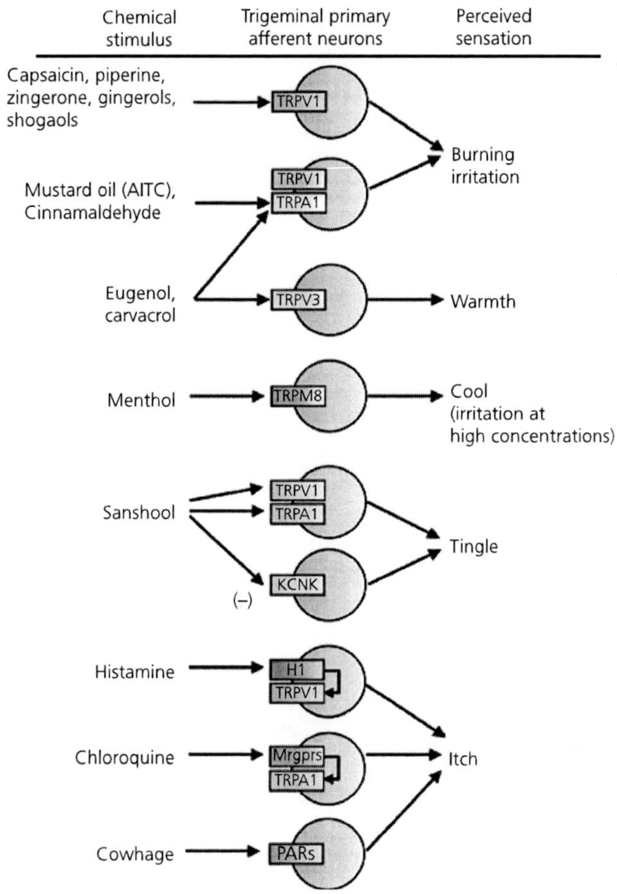

Fig. 14.1 Summary diagram of neural mechanisms involved in chemesthesis. The circles represent trigeminal primary sensory neurons. Boxes indicate receptors expressed by the sensory neuron. On the left side, arrows point to receptors that are activated by the indicated chemicals (except for sanshool, which inhibits KCNK, the two-pore potassium channel subtypes KCNK3, KCNK9, and KCNK18). On the right side, arrows point to the sensory qualities elicited by the chemicals. Histamine activates the H1 histamine receptor that in turn activates TRPV1 (indicated by an arrow), which is co-expressed by the sensory neuron. Chloroquine activates Mrgprs that in turn activate TRPA1, which is co-expressed by the same neuron.

and then in turn by progressively increasing irritation when reapplied at short interstimulus intervals (stimulus-induced recovery) (Green, 1998). The sensory properties of capsaicin are reflected in the responses of rodent Vc neurons, which exhibited sensitization, desensitization, and stimulus-induced recovery (Dessirier *et al.*, 2000a). TRPV1 is also activated by acidic stimulation, and it is noteworthy that citric acid elicited a temporally sensitizing pattern of oral irritation in humans (Dessirier *et al.*, 2000b), and Vc neuronal responses in rats (Sudo *et al.*, 2002).

Capsaicin cross-desensitized oral irritation elicited by a variety of chemicals including AITC, cinnamaldehyde, nicotine, carbonation, menthol, and others (Green, 1991; Cliff and Green, 1996; Dessirier *et al.*, 1997; Dessirier *et al.*, 2001a, b;

Simons *et al.*, 2003b). Capsaicin enhanced thermally-evoked heat pain in humans (Albin *et al.*, 2008) and produced thermal hyperalgesia in rats (Gilchrist *et al.*, 1996), effects that may be mediated peripherally by capsaicin enhancement of heat transduction by TRPV1 (Guenther *et al.*, 1999). The molecular mechanism underlying this interaction is currently not understood. Thermal sensitivity of TRPV1 is imparted by amino acid residues at extracellular sites on the pore loop region (Kim *et al.*, 2013), whereas capsaicin binding sites are intracellular on transmembrane segment 3 (see Bevan *et al.*, 2014, for recent review). Given recent spectacular advances in visualizing the molecular structure of TRPV1 using methods such as electron-cryomicroscopy (Liao *et al.*, 2013; Cao *et al.*, 2013), one may anticipate in the relatively near future a progressively clearer if not complete understanding of the molecular mechanism of thermotransduction by TRPV1 and how it is modulated by capsaicin and other chemicals.

Capsaicin, particularly at higher concentrations (330 μM), induced desensitization that persisted for several days (Carstens *et al.*, 2007; Karrer and Bartoshuk, 1991). There may be multiple mechanisms underlying acute and long-term desensitization by capsaicin, such as an immediate calcium- and calcineurin-dependent modulation of TRPV1 (Cholewinski *et al.*, 1993; Docherty *et al.*, 1996), neurotransmitter depletion, and inhibition of voltage-sensitive sodium channels (Liu *et al.*, 2001), as well as longer-term degeneration of intradermal C-fibers (Nolano *et al.*, 1999; see O'Niell *et al.*, 2012, for recent review).

The aversiveness of capsaicin, which achieves high concentrations in the seeds of pepper plants, may confer an evolutionary advantage. Many avian species are insensitive to capsaicin (Mason *et al.*, 1991) due to expression of a capsaicin-insensitive variant of TRPV1 (Caterina *et al.*, 1997). This capsaicin-insensitivity allows birds to swallow and disperse pepper plant seeds, whereas mammals that would normally chew the seeds are repelled by the capsaicin burn and thus avoid the pepper plants (Tewksbury and Nabhan, 2001).

Topical application of capsaicin to the skin was reported to elicit itch as the dominant sensation (Green and Flammer, 1989; Green, 1990; Green and Shaffer, 1993). Focal application of capsaicin via cowhage spicules also elicited itch accompanied by burning and pricking/stinging sensations (Sikand *et al.*, 2009, 2011b). In this regard, it is noteworthy that TRPV1 is necessary for histamine-evoked itch sensation. Rodent scratching behavior elicited by intradermal injection of histamine was nearly abolished in knockout mice lacking TRPV1 (Imamachi *et al.*, 2009).

These data suggest critical roles for TRPV1 in both burning pain and itch sensations. This is further supported by the following intriguing observation. In wildtype mice, intradermal cheek injection of capsaicin elicits forelimb wiping indicative of pain. In TRPV1 global knockout mice, TRPV1 was selectively expressed back into sensory neurons expressing Mas-related G-protein-coupled receptor (Mrgpr) A3, a molecular receptor mediating chloroquine-evoked itch (see later). In these animals, cheek injection of capsaicin elicited hindlimb scratching indicative of itch, indicating that the MrgprA3-expressing sensory neurons elicit itch regardless of their mode of activation (Han *et al.*, 2013). A question to be answered by future studies is whether irritation and histamine-mediated itch are signaled by separate or overlapping populations of sensory neurons expressing TRPV1, and how these two sensations are discriminated (see later). Possible approaches to these questions may involve the recently described method to selectively silence sensory neurons via infiltration of a small local anesthetic molecule through open TRP channels (Roberson *et al.*, 2013), or optogenetic approaches. For example, by selectively expressing channelrhodopsin in Nav1.8-expressing sensory neurons, illumination of the skin with blue light elicited nocifensive behavioral responses (Daou *et al.*, 2013). It is not inconceivable that TRPV1- expressing neurons, or the channel itself, can be rendered light-sensitive by incorporation of channelrhodopsin in order to investigate behavioral itch- and pain-related consequences elicited by non-invasive optical stimulation. Itch as a chemesthetic sense is discussed further in later sections.

14.4 TRPA1

TRPA1 (formerly ANKTM1) was originally reported to respond to intense cooling (Story *et al.*, 2003) as well as irritant chemicals including mustard oil (Jordt *et al.*, 2004). TRPA1 is co-expressed with TRPV1 in a high percentage of sensory neurons (Story *et al.*, 2003). The cold-sensitivity of TRPA1 is controversial and remains an open question (for review, see Caspani and Heppenstall, 2009). Several studies confirmed the sensitivity of TRPA1 to a variety of irritant chemicals including cinnamaldehyde, allicin from garlic, nicotine, oleocanthal (from extra-virgin olive oil), ibuprofen, and others (Bandell *et al.*, 2004; Bautista *et al.*, 2005, 2006; Macpherson *et al.*, 2005, 2007; Gerhold and Bautista, 2008; Talavaera *et al.*, 2009; Peyrot des Gachons *et al.*, 2011) (Fig. 14.1). In humans, allyl isothiocyanate (AITC; from mustard oil), cinnamaldehyde, and nicotine elicited oral irritation that progressively declined with repeated applications (desensitization) (Dessirier *et al.*, 1997; Prescott and Swain-Campbell, 2000; Simons *et al.*, 2003b; Klein *et al.*, 2011b). The mechanism underlying desensitization may involve covalent binding of TRPA1 agonists to cysteine residues of TRPA1 to prevent further activation (Macpherson *et al.*, 2007), and/or intracellular calcium-dependent (Wang *et al.*, 2008) or calcium-independent (Akopian *et al.*, 2007) inactivation of TRPA1, and/or trafficking of TRPA1 from the cell membrane (Schmidt *et al.*, 2009).

AITC and capsaicin (Simons *et al.*, 2003b), and cinnamaldehyde and menthol (Zanotto *et al.*, 2008; Klein *et al.*, 2011b) exhibited mutual cross-desensitization. Both AITC and cinnamaldehyde weakly and briefly enhanced lingual heat pain in humans (Albin *et al.*, 2008) and cinnamaldehyde produced thermal hyperalgesia in rodents (Tsagareli *et al.*, 2010). These sensory and behavioral effects are reflected in the temporally desensitizing responses of spinal and Vc neurons to AITC, cinnamaldehyde, and nicotine, and enhancement of their responses to noxious heat (Dessirier *et al.*, 2000a; Simons *et al.*, 2004; Merrill *et al.*, 2008; Sawyer *et al.*, 2009a). The mechanism by which TRPA1 agonists enhance heat sensitivity speculatively involves an interaction with TRPV1 co-expressed in the same sensory neurons, but is currently unknown and requires future investigation. Moreover, future studies are needed to determine the role of TRPA1 in cold pain sensation, and how the central nervous system discriminates between cold- and heat-pain sensations based on input from nociceptors that co-express both TRPV1 and TRPA1.

TRPA1 expressed in sensory nerves innervating the airway epithelia was recently reported to be activated by noxious chemicals, such as chlorine and reactive oxygen species in polluted air (Bessac *et al.*, 2008). Activation of these sensory afferents triggers protective reflex responses, such as sneezing and cough to expel the irritant chemicals from the airways.

Interestingly, topical application of cinnamaldehyde to the skin elicits burning pain as well as itch in most subjects, but does not elicit any cold sensation (Namer *et al.*, 2005). Non-histaminergic itch, such as the itch elicited by the antimalarial drug, chloroquine, requires TRPA1. Scratching behavior elicited by chloroquine or bovine adrenal medulla protein 8-22 (BAM8-22), which act at Mrgpr receptors independent of histamine (see more later), is virtually abolished in knockout mice lacking TRPA1 (Wilson *et al.*, 2011). As noted previously, future studies are needed to unravel how the central nervous system differentiates between itch and pain sensations, when both are signaled by TRPA1-expressing sensory neurons.

14.5 TRPV3, TRPV4, and warming

TRPV3 is expressed in keratinocytes and sensory neurons and is activated by innocuous warming into the noxious range (Peier *et al.*, 2002b; Smith *et al.*, 2002; Xu *et al.*, 2002; Chung *et al.*, 2004; Nilius *et al.*, 2014). TRPV4 is also activated by innocuous warmth up to approximately 42 °C, at which point its conductance plateaus or begins to decline (Güler *et al.*, 2002; Chung *et al.*, 2003). The properties of TRPV3 and TRPV4 appear to match those of two types of warm fibers, one exhibiting an increasing response from innocuous warmth into the noxious range, and the other exhibiting increased firing through the innocuous warm range with a decline in firing at temperatures approaching the pain threshold (reviewed in Green, 2004). Knockout mice lacking TRPV3 exhibited deficits in thermal preference behavior for innocuous warm and noxious heat stimuli (Moqrich *et al.*, 2005), and keratinocytes overexpressing TRPV3 exhibited increased responses to heat (Huang *et al.*, 2008). However, a subsequent study reported no deficit in thermal preference or noxious heat sensitivity in knockout mice lacking both TRPV3 and TRPV4 (Huang *et al.*, 2011), calling into question the roles of these TRP channels in signaling innocuous warmth and heat pain. A very recent study generated TRPV3/TRPV1 double knockout mice, which exhibited a thermal deficit within a narrow (48–50 °C) temperature range (Marics *et al.*, 2014). These workers suggested a cooperative interaction between TRPV3 in skin cells and TRPV1 in sensory nerve endings for thermal sensitivity to moderate heat, as well as additional mechanisms for heat sensation (Marics *et al.*, 2014). An alternative hypothesis holds that TRPV1 and TRMP8 are sufficient for detection of innocuous warming and cooling (Pogorzala *et al.*, 2013).

Eugenol (from clove oil) and carvacrol (from oregano oil) are considered agonists of TRPV3 (Xu *et al.*, 2006), although eugenol also activates TRPA1 and TRPV1 (Bandell *et al.*, 2004) and carvacrol activates TRPA1 (Bandell *et al.*, 2004; Xu *et al.*, 2006; de la Roche *et al.*, 2013) (Fig. 14.1). Both chemicals elicited oral irritation that was described as having qualities of tingling, warming, numbing, cooling, burning, and stinging (Cliff and Heymann, 1992; Klein *et al.*, 2013). Eugenol- and carvacrol-evoked irritation exhibited desensitization, and both agents briefly enhanced innocuous warmth and heat pain sensations in humans (Klein *et al.*, 2013). Similarly, lingual application of eugenol and carvacrol activated rat Vc neurons in a temporally desensitizing pattern and briefly enhanced their responses to noxious heat (Klein *et al.*, 2014). The numbing quality is of interest, particularly given that eugenol also possesses a local anesthetic action via inhibition of voltage-sensitive ion channels (Lee *et al.*, 2005; Park *et al.*, 2006). Speculatively, eugenol may produce localized regions of anesthesia in addition to activating skin cells and/or nerve endings, resulting in a complex sensory input that is interpreted as numbing.

To investigate the peripheral effects of these chemicals, we used calcium imaging of trigeminal ganglion (TG) and DRG cells; 8–30% of these cells were activated by eugenol and carvacrol, and 2–5% were activated by farnesyl pyrophosphate (Klein *et al.*, 2014), a putative selective TRPV3 agonist (Bang *et al.*, 2010). Moreover, subpopulations of TG and DRG cells were sensitive to innocuous warming and noxious heat. Eugenol and carvacrol enhanced responses to both innocuous warming as well as noxious heat (Klein *et al.*, 2014), consistent with psychophysical data (Klein *et al.*, 2013). These data suggest that some sensory neurons express TRPV3, and may be involved in signaling innocuous warmth (i.e., expressed in warm fibers) as well as heat pain (i.e., expressed in nociceptors). It is reiterated that TRPV3 and TRPV4 are highly expressed in keratinocytes. In an intriguing study employing calcium imaging of co-cultured keratinocytes and DRG neurons, it was shown that heating activated the keratinocytes prior to the DRG neurons (Mandadi *et al.*, 2009). Moreover, responses of DRG neurons to heating were blocked by the purinergic receptor antagonist PPADS, implicating ATP as a neuromodulator released from keratinocytes to activate sensory neurons (Mandadi *et al.*, 2009). Altogether, the evidence indicates that both keratinocytes and sensory neurons express TRPV3 and can transduce innocuous warm temperatures. The data also point to a peripheral site of chemical modulation of thermal responsiveness of warm fibers and nociceptors. Like TRPV1, amino acid residues essential for thermal sensitivity of TRPV3 were located extracellularly in the pore loop domain (Kim *et al.*, 2013). The mechanism by which eugenol and carvacrol enhance the thermosensitivity of TRPV3 is unknown and awaits future studies. Another open question is the contribution of TRPV3 to heat pain, given that it responds to innocuous warmth as well as noxious heat (Peier *et al.*, 2002b) and is co-expressed with TRPV1 in sensory neurons (Smith *et al.*, 2002; discussed in Green, 2004).

14.6 TRPM8 and cold

TRPM8 was originally shown to respond to innocuous cooling as well as to menthol (McKemy *et al.*, 2002; Peier *et al.*, 2002a) (Fig. 14.1). Knockout mice lacking TRPM8 exhibited reduced sensitivity to innocuous cool temperatures in thermal preference tests (Bautista *et al.*, 2007; Colburn *et al.*, 2007; Dhaka *et al.*, 2007). However, evidence for a role of TRPM8 in noxious cold sensitivity was inconclusive. In a recent study, adult mice with ablation of TRPM8-expressing neurons exhibited an almost complete insensitivity to both innocuous and noxious cold stimuli (Knowlton *et al.*, 2013), supporting a role for TRPM8 in signaling decreasing temperatures into the noxious range. Nevertheless, the dual role of TRPM8 in innocuous cooling and noxious cold sensing, and the possibility of additional cold transduction mechanisms, are topics that need further elucidation (for recent review, see McKemy, 2013).

In humans, menthol elicited a cooling sensation at low concentrations, but irritation at higher concentrations in a manner that declined across repeated applications (Cliff and Green, 1994, 1996; Dessirier *et al.*, 2001a). Similarly, responses of Vc neurons exhibited a temporally desensitizing pattern of firing in response to lingual application of menthol (Zanotto *et al.*, 2007). Menthol-evoked cooling is presumably due to expression of TRPM8 in cold fibers. Menthol is also co-expressed with TRPV1 in sensory neurons (McKemy *et al.*, 2002; Reid *et al.*, 2002; Viana *et al.*, 2002; Xing *et al.*, 2006) that are presumably nociceptors. Thus, menthol activation of TRPM8 expressed in nociceptors represents a plausible mechanism underlying menthol irritation. Menthol enhanced oral cold sensation (Albin *et al.*, 2008) and cold-evoked responses of Vc neurons (Zanotto *et al.*, 2007), an effect that may be explained by the ability of menthol to enhance gating of TRPM8 by cooling (McKemy *et al.*, 2002). Consistent with these findings, topical hindpaw application of menthol increased cold sensitivity in a rat thermal preference assay, although higher concentrations of menthol reduced cold sensitivity, which is consistent with an analgesic action (Klein *et al.*, 2010). Menthol cross-desensitized nicotine-evoked oral irritation (Dessirier *et al.*, 2001b) and either cross-sensitized or desensitized capsaicin-evoked oral irritation depending on timing (Cliff and Green, 1996; Green and McAuliffe, 2000). Menthol and cinnamaldehyde also exhibited mutual cross-desensitization of oral irritation that might be explained by a peripheral interaction at intraoral trigeminal nerve endings (Klein *et al.*, 2011b).

14.7 Tingle

Tingle, buzzing, or numbing are terms that are often used to describe the sensation elicited by sanshools from Szechuan peppers of the genus Xanthoxylum (Sugai *et al.*, 2005), or by carbonated beverages (see Chapter 8). Lingual application of an analog of hydroxy-α sanshool, isobutylalkylamide (IBA), elicited a sensation of tingle that decreased across repeated applications and was not cross-desensitized by capsaicin or AITC (Albin and Simons, 2010). Hydroxy-α sanshool excited cells expressing TRPV1 and TRPA1 (Koo *et al.*, 2007; Riera *et al.*, 2009; Klein *et al.*, 2011d), and, importantly, inhibited the two-pore potassium channel (KCNK) subfamilies KCNK3, KCNK9, and KCNK18 to activate sensory neurons (Bautista *et al.*, 2008) (Fig. 14.1). Hydroxy-α sanshool activated a variety of primary afferent fiber types, including cold receptors, mechanoreceptors, nociceptors, and mechanically insensitive fibers (Bryant and Mezine, 1999; Lennertz *et al.*, 2010). Activation of rapidly adapting low-threshold mechanoreceptors by sanshools appears particularly relevant to the mechanical quality of tingle sensation. While inhibition of potassium channels provides a satisfactory explanation for the ability of sanshools to activate a variety of primary afferent fiber types, more work is needed to clarify the additional role of thermosensitive TRP channels. Intradermal injection of IBA activated both wide dynamic range and low-threshold spinal dorsal horn neurons

(Sawyer *et al.*, 2009b). Sanshool activation of multiple types of primary afferent and spinal neurons may give rise to an abnormal, mixed barrage of sensory input to the central nervous system that elicits the tingling, buzzing paraesthesia. The mechanism underlying numbing is more obscure, and may involve localized regions of anesthesia as suggested earlier for eugenol.

Carbonated beverages elicit an oral irritant sensation that is thought to be mediated not by the mechanical impact of bursting bubbles, but rather by conversion of the CO_2 in the bubbles into carbonic acid. This is supported by the observations that tingle still persists tens of seconds after the carbonated beverage has been expectorated (Green, 1992), and that drugs such as acetazolamide or dorzolamide, which prevent carbonic anhydrase from converting CO_2 into carbonic acid, reduce carbonated water-evoked irritation (Simons *et al.*, 1999; Dessirier *et al.*, 2000c) and activation of trigeminal primary afferents (Komai and Bryant, 1993) and Vc neurons (Simons *et al.*, 1999). Activation of primary sensory neurons by carbonic acid may be mediated by amiloride-sensitive acid-sensing ion channels expressed in trigeminal nociceptors (Dessirier *et al.*, 2001c). In addition, ingestion of carbonated water under hyperbaric conditions that prevented formation of bubbles elicited oral irritation, which was equivalent to the irritation elicited under normobaric conditions with bubble formation (Wise *et al.*, 2013).

Tingle, buzzing, and numbing are thus paraesthetic sensations that can be elicited by different chemicals such as sanshools and carbonic acid, and appear to involve the activation of multiple primary sensory neuron subtypes by a general blockade of potassium channels accompanied by inhibition of TRP channels. The central mechanisms underlying this enigmatic sensory quality remain an open question.

14.8 NaCl

NaCl evokes a salt taste sensation, but can simultaneously elicit irritation at concentrations as low as 0.5 M (Green and Gelhard, 1989; Gilmore and Green, 1993; Dessirier *et al.*, 2001b). Oral irritation elicited by a high NaCl concentration exhibited sensitization and cross-sensitized capsaicin-evoked irritation (Dessirier *et al.*, 2001b), a pattern also observed for NaCl-evoked activation of Vc neurons (Sudo *et al.*, 2002). NaCl-evoked irritation was significantly attenuated by amiloride, suggesting possible involvement of epithelial sodium channels expressed in lingual nociceptors (Dessirier *et al.*, 2001b).

14.9 Itch

Itch is often described as an unpleasant sensation associated with the desire to scratch. It may be considered to be a chemesthetic sensory quality since it is usually elicited by exogenous or endogenous chemical stimuli. Among the long list of chemicals that can elicit an itch sensation are the biogenic amines histamine and serotonin, neuropeptides such as endothelin-1 and substance P, Mrgpr subtypes

A3, C11, and D (see later), proteases, lipid mediators, cytokines, bile acids, oxidative stress, and agonists of toll-like receptors (for reviews, see Hägermark, 1992; Akiyama and Carstens, 2013).

Current evidence indicates the existence of at least two itch-signaling pathways, histaminergic and non-histaminergic (see Akiyama and Carstens, 2013 for recent review). The prototypical itch mediator in human skin is histamine, an endogenous substance in cutaneous mast cells, white blood cells, as well as in neurons where it functions as a neurotransmitter. Cutaneous application of histamine elicits a wheal, flare, and itch sensation largely via activation of mechanically insensitive C-fiber pruriceptors (Schmelz *et al.*, 1997; Namer *et al.*, 2008). Histamine excites pruriceptors by acting at the H1 (and also H4) histamine receptor linked to phospholipase Cβ3 and TRPV1 (Imamachi *et al.*, 2009) (Fig. 14.1). In this regard, it is interesting that topical skin application of the TRPV1 agonist capsaicin initially elicits an itch sensation followed later by burning/stinging (Green and Shaffer, 1993). Moreover, localized application by insertion of a capsaicin-laden spicule elicits both itch and burning/stinging sensations (LaMotte *et al.*, 2009; Sikand *et al.*, 2009, 2011b).

In contrast, non-histaminergic itch-inducers such as cowhage elicit itch unaccompanied by flare (Johanek *et al.*, 2007) via activation of mechanosensitive C-fiber (Namer *et al.*, 2008) and A-delta nociceptors (Ringkamp *et al.*, 2011). Cowhage spicules contain proteases that act at protease-activated receptor (PAR) subtypes PAR2 and PAR4 (Reddy *et al.*, 2008). In addition, a variety of other molecular receptors have been recently implicated in non-histaminergic itch, notably the Mas-related G-protein-coupled receptor (Mrgpr) subtypes MrgprA3 that signals itch elicited by the antimalarial drug chloroquine (Liu *et al.*, 2009), MrgprC11 that signals itch elicited by BAM8-22 (Sikand *et al.*, 2011a), and MrgprD that signals β-alanine-evoked itch (Liu *et al.*, 2012). Itch elicited by chloroquine and BAM8-22 involves activation of pruriceptors that co-express TRPA1 (Wilson *et al.*, 2011) (Fig. 14.1). In primates, cowhage and histamine excited largely non-overlapping populations of spinothalamic tract neurons (Davidson *et al.*, 2007, 2012), whereas this distinction appears to be more blurred in rodents (Akiyama and Carstens, 2013). Importantly, all pruritogen-responsive neurons additionally responded to capsaicin and AITC. It is thus noteworthy that chemesthetic qualities of irritation (burning/stinging, pricking) and itch involve some of the same molecular players. Future studies are needed to determine how these different chemesthetic qualities are processed and discriminated by the central nervous system.

14.10 Interactions between chemesthesis and taste

Chemesthetic sensations play an important role in food and beverage flavor and consumer acceptance, as judged by the popularity of spicy food, carbonated drinks, and mentholated oral hygiene products. The orbitofrontal cortex is thought to play an integrative role in flavor perception; neurons with gustatory and olfactory input also respond to oral capsaicin (Rolls *et al.*, 2003). While spices are often claimed to "enhance" food flavor, there appears to be a mutually inhibitory interaction between chemesthesis and taste. Capsaicin reduced the perceived intensity of certain taste qualities, particularly sucrose-evoked sweetness (Simons *et al.*, 2002), and both capsaicin and nicotine suppressed tastant-evoked responses of gustatory neurons in the nucleus of the solitary tract (Simons *et al.*, 2003a, 2006). It was recently reported that carbonation reduced brain activation elicited by sucrose in humans (Di Salle *et al.*, 2013). Conversely, tastants also appear to have an inhibitory effect on chemesthesis. Certain tastants, particularly sucrose, suppressed the oral burning sensation elicited by capsaicin (Stevens and Lawless, 1986). It was recently reported that lingual application of sodium chloride and monosodium glutamate, but not sucrose, suppressed responses of rat Vc neurons to noxious heat and chemical (pentanoic acid) stimuli (Boucher *et al.*, 2013). Sucrose elicits analgesia in newborns (Anseloni *et al.*, 2002), and consumption of chocolate by adult rats induced analgesia (Foo and Mason, 2009). These effects are thought to be mediated by activation of central pain-inhibitory pathways rather than by a peripheral gustatory–trigeminal interaction. This brief synopsis indicates that there are intriguing interactions between chemesthesis and taste that have great importance to food palatability and acceptance. Clearly, more research is needed to expand our understanding of these interactions and the underlying neural mechanisms.

14.11 Summary and conclusions

Figure 14.1 provides a summary schematic diagram of the molecular receptors and ion channels expressed in sensory neurons that are activated by various chemicals to elicit chemesthetic sensory qualities.

References

Akiyama, T., Carstens, M.I., and Carstens, E. 2010. Facial injections of pruritogens and algogens excite partly overlapping populations of primary and second order trigeminal neurons in mice. *Journal of Physiology,* **104** (5), 2442–2450.

Akiyama, T., Tominaga, M., Davoodi, A., *et al.* 2012. Cross-sensitization of histamine-independent itch in mouse primary sensory neurons. *Neuroscience,* **226**, 305–312.

Akiyama, T. and Carstens, E. 2013. Neural processing of itch. *Neuroscience,* **250**, 697–714.

Akiyama, T., Tominaga, M., Davoodi, A., *et al.* 2013. Roles for substance P and gastrin-releasing peptide as neurotransmitters released by primary afferent pruriceptors. *Journal of Physiology,* **109** (3), 742–748.

Akopian, A.N., Ruparel, N.B., Jeske, N.A., and Hargreaves, K.M. 2007. Transient receptor potential TRPA1 channel desensitization in sensory neurons is agonist dependent and regulated by TRPV1-directed internalization. *Journal of Physiology,* **583**, 175–193.

Albin, K.C., Carstens, M.I., and Carstens, E. 2008. Modulation of oral heat and cold pain by irritant chemicals. *Chemical Senses,* **33** (1), 3–15.

Albin, K.C. and Simons, C.T. 2010. Psychophysical evaluation of a sanshool derivative (alkylamide) and the elucidation of mechanisms subserving tingle. *PLoS One,* **5** (3), e9520.

Anseloni, V.C., Weng, H.R., Terayama, R., *et al.* 2002. Age-dependency of analgesia elicited by intraoral sucrose in acute and persistent pain models. *Pain,* **97** (1–2), 93–103.

Anton, F., Herdegen, T., Peppel, P., and Leah, J.D. 1991. c-FOS-like immunoreactivity in rat brainstem neurons following noxious chemical stimulation of the nasal mucosa. *Neuroscience,* **41** (2–3), 629–641.

Bandell, M., Story, G.M., Hwang, S.W., *et al.* 2004. Noxious cold ion channel TRPA1 is activated by pungent compounds and bradykinin. *Neuron,* **41** (6), 849–857.

Bang, S., Yoo, S.,Yang, T.J., Cho, H., and Hwang, S.W. 2010. Farnesyl pyrophosphate is a novel pain- producing molecular via specific activation of TRPV3. *Journal of Biological Chemistry,* **285** (25), 19362–19371.

Bautista, D.M., Movahed, P., Hinman, A., *et al.* 2005. Pungent products from garlic activate the sensory ion channel TRPA1. *Proceedings of the National Academy of Sciences, USA,* **102** (34), 12248–12252.

Bautista, D.M., Jordt, S.E., Nikai, T., *et al.* 2006. TRPA1 mediates the inflammatory actions of environmental irritants and proalgesic agents. *Cell,* **124** (6), 1269–1282.

Bautista, D.M., Siemens, J., Glazer, J.M., *et al.* 2007. The menthol receptor TRPM8 is the principal detector of environmental cold. *Nature,* **448** (7150), 204–208.

Bautista, D.M., Sigal, Y.M., Milstein, A.D., *et al.* 2008. Pungent agents from Szechuan peppers excite sensory neurons by inhibiting two-pore potassium channels. *Nature Neuroscience,* **11** (7), 772–729.

Bereiter, D.A., Bereiter, D.F., and Hirata, H. 2002. Topical cannabinoid agonist, WIN55, 212-2, reduces cornea-evoked trigeminal brainstem activity in the rat. *Pain,* **99** (3), 547–556.

Bessac, B.F., Sivula, M., von Hehn, C.A., *et al.* 2008. TRPA1 is a major oxidant sensor in murine airway sensory neurons. *Journal of Clinical Investigation,* **118** (5), 1899–1910.

Bevan, S., Quallo, T., and Andersson, D.A. 2014. TRPV1. *Handbook of Experimental Pharmacology,* **222**, 207–245.

Boucher, Y., Felizardo, R., Klein, A.H., Carstens, M.I., and Carstens, E. 2013. Gustatory modulation of the responses of trigeminal subnucleus caudalis neurons to noxious stimulation of the tongue in rats. *European Journal of Neuroscience,* **38** (6), 2812–2822.

Bryant, B.P. and Mezine, I. 1999. Alkylamides that produce tingling paresthesia activate tactile and thermal trigeminal neurons. *Brain Research,* **842** (2), 452–460.

Cao, E., Liao, M., Cheng, Y., and Julius, D. 2013. TRPV1 structures in distinct conformations reveal activation mechanisms. *Nature,* **504** (7478), 113–118.

Caspani, O. and Heppenstall, P.A. 2009. TRPA1 and cold transduction: an unresolved issue? *Journal of General Physiology,* **133** (3), 245–249.

Carstens, E. 1997. Responses of rat spinal dorsal horn neurons to intracutaneous microinjection of histamine, capsaicin, and other irritants. *Journal of Physiology,* **77** (5), 2499–2514.

Carstens, E., Albin K.C., Simons, C.T., and Carstens, M.I. 2007. Time course of self-desensitization of oral irritation by nicotine and capsaicin. *Chemical Senses,* **232** (9), 811–816.

Carstens, E., Kuenzler, N., and Handwerker, H.O. 1998. Activation of neurons in rat trigeminal subnucleus caudalis by different irritant chemicals applied to oral or ocular mucosa. *Journal of Physiology,* **80** (2), 465–492.

Carstens, E., Saxe, I., and Ralph, R. 1995. Brainstem neurons expressing c-Fos immunoreactivity following irritant chemical stimulation of the rat's tongue. *Neuroscience,* **69** (3), 939–953.

Caterina, M.J., Schumacher, M.A., Tominaga, M., *et al.* 1997. The capsaicin receptor: a heat-activated ion channel in the pain pathway. *Nature,* **389** (6653), 816–824.

Cholewinski, A., Burgess, G.M., and Bevan, S. 1993. The role of calcium in capsaicin-induced desensitization in rat cultured dorsal root ganglion neurons. *Neuroscience,* **55** 1015–1023.

Chung, M.K., Lee, H., and Caterina, M.J. 2003. Warm temperatures activate TRPV4 in mouse 308 keratinocytes. *Journal of Biological Chemistry,* **278** (34), 32037–32046.

Chung, M.K., Lee, H., Mizuno, A., Suzuki, M., and Caterina, M.J. 2004. TRPV3 and TRPV4 mediate warmth-evoked currents in primary mouse keratinocytes. *Journal of Biological Chemistry,* **279** (20), 21569–21575.

Cliff, M.A. and Green, B.G. 1994. Sensory irritation and coolness produced by menthol: evidence for selective desensitization of irritation. *Physiology & Behavior,* **56**, 1021–1029.

Cliff, M.A. and Green, B.G. 1996. Sensitization and desensitization to capsaicin and menthol in the oral cavity: interactions and individual differences. *Physiology & Behavior,* **59** (3), 487–494.

Cliff, M. and Heymann, H. 1992. Descriptive analysis of oral pungency. *Journal of Sensory Studies,* **7**, 279–290.

Colburn, R.W., Lubin, M.L., Stone D.J. Jr, *et al.* 2007. Attenuated cold sensitivity in TRPM8 null mice. *Neuron,* **54** (3), 379–386.

Cometto-Muñiz, J.E. and Cain, W.S. 1995. Relative sensitivity of the ocular trigeminal, nasal trigeminal and olfactory systems to airborne chemicals. *Chemical Senses,* **20** (2), 191–198.

Daou, I., Tuttle, A.H., Longo, G., *et al.* 2013. Remote optogenetic activation and sensitization of pain pathways in freely moving mice. *Journal of Neuroscience,* **33** (47), 18631–18640.

Davidson, S., Zhang, X., Khasabov, S.G., *et al.* 2012. Pruriceptive spinothalamic tract neurons: physiological properties and projection targets in the primate. *Journal of Physiology*, **108** (6), 1711–1723.

Davidson, S., Zhang, X., Yoon, C.H., *et al.* 2007. The itch- producing agents histamine and cowhage activate separate populations of primate spinothalamic tract neurons. *Journal of Neuroscience*, **27** (37), 10007–10014.

de la Roche, J., Eberhardt, M.J., Klinger, AB, *et al.* 2013. The molecular basis for species-specific activation of human TRPA1 protein by protons involves poorly conserved residues within transmembrane domains 5 and 6. *Journal of Biological Chemistry*, **288** (28), 20280–20292.

Dessirier, J.M., O'Mahony, M., and Carstens, E. 1997. Oral irritant effects of nicotine: psychophysical evidence for decreased sensation following repeated application and lack of cross-desensitization to capsaicin. *Chemical Senses*, **22** (5), 483–492.

Dessirier, J.M., Simons, C.T., Sudo, M., Sudo, S., and Carstens, E. 2000a. Sensitization, desensitization and stimulus-induced recovery of trigeminal neuronal responses to oral capsaicin and nicotine. *Journal of Physiology*, **84** (4), 1851–1862.

Dessirier, J.M., O'Mahony, M., Iodi-Carstens, M., and Carstens, E. 2000b. Sensory properties of citric acid: psychophysical evidence for sensitization, self-desensitization, cross-desensitization and cross-stimulus-induced recovery following capsaicin. *Chemical Senses*, **25** (6), 769–780.

Dessirier, J.M., Simons, C.T., Carstens, M.I., O'Mahony, M, and Carstens, E. 2000c. Psychophysical and neurobiological evidence that the oral sensation elicited by carbonated water is of chemogenic origin. *Chemical Senses*, **25** (3), 277–284.

Dessirier, J.-M., O'Mahony, M., and Carstens, E. 2001a. Oral irritant properties of menthol: sensitizing and desensitizing effects of repeated application and cross-desensitization to nicotine. *Physiology & Behavior*, **73**, 25–36.

Dessirier, J.M., O'Mahony, M., Iodi-Carstens, M., Yao E., Carstens, E. 2001b. Oral irritation by sodium chloride: sensitization, self-desensitization, and cross-sensitization to capsaicin. *Physiology & Behavior*, **72** (3), 317–324.

Dessirier, J.M., Simons, C.T., O'Mahony, M., and Carstens, E. 2001c. The oral sensation of carbonated water: cross-desensitization by capsaicin and potentiation by amiloride. *Chemical Senses*, **26** (6), 639–643.

Dhaka, A., Murray, A.N., Mathur, J., *et al.* 2007. TRPM8 is required for cold sensation in mice. *Neuron*, **54**, 371–378.

Di Salle, F., Cantone, E., Savarese, M.F., *et al.* 2013. Effect of carbonation on brain processing of sweet stimuli in humans. *Gastroenterology*, **145** (3), 537–9.e3.

Docherty, R.J., Yeats, J.C., Bevan, S., and Boddeke H.W. 1996. Inhibition of calcineurin inhibits the desensitization of capsaicin-evoked currents in cultured dorsal root ganglion neurones from adult rats. *Pflugers Archiv*, **431**, 828–837.

Foo, H. and Mason, P. 2009. Analgesia accompanying food consumption requires ingestion of hedonic foods. *Journal of Neuroscience*, **29** (41), 13053–13062.

Gerhold, K.A. and Bautista, D.M. 2008. TRPA1: irritant detector of the airways. *Journal of Physiology*, **586** (14), 3303.

Gilchrist, H.D., Allard, B.L., and Simone, D.A. 1996. Enhanced withdrawal responses to heat and mechanical stimuli following intraplantar injection of capsaicin in rats. *Pain*, **67** (1), 179–188.

Gilmore, M.M. and Green, B.G. 1993. Sensory irritation and taste produced by NaCl and citric acid: effects of capsaicin desensitization. *Chemical Senses*, **18**, 257–272.

Green, B.G. 1990. Spatial summation of chemical irritation and itch produced by topical application of capsaicin. *Perception & Psychophysics*, **48** (1), 12–18.

Green, B.G. 1991. Capsaicin cross-desensitization on the tongue: psychophysical evidence that oral chemical irritation is mediated by more than one sensory pathway. *Chemical Senses*, **16**, 675–689.

Green, B.G. 1992. The effects of temperature and concentration on the perceived intensity and quality of carbonation. *Chemical Senses*, **17**, 435–450.

Green, B.G. 1998. Capsaicin desensitization and stimulus-induced recovery on facial compared to lingual skin. *Physiology & Behavior*, **65** (3), 517–523.

Green, B.G. 2004. Temperature perception and nociception. *Journal of Neurobiology*, **61** (1), 13–29.

Green, B.G. 2012. Chemesthesis and the chemical senses as components of a "chemofensor complex". *Chemical Senses*, **37** (3), 201–206.

Green, B.G. and Flammer, L.J. 1989. Localization of chemical stimulation: capsaicin on hairy skin. *Somatosensory & Motor Research*, **6** (5-6), 553–566.

Green, B.G. and Gelhard, B. 1989. Salt as an oral irritant. *Chemical Senses*, **14**, 259–271.

Green, B.G., Mason, J.R., and Kare, M.R. (eds.) 1990. Irritation, in *Chemical Senses*, vol. 2. New York: Marcel-Dekker, p. iv.

Green, B.G. and McAuliffe, B.L. 2000. Menthol desensitization of capsaicin irritation. Evidence of a short-term anti-nociceptive effect. *Physiology & Behavior*, **68** (5), 631–639.

Green, B.G. and Shaffer, G.S. 1993. The sensory response to capsaicin during repeated topical exposures: differential effects on sensations of itching and pungency. *Pain*, **53** (3), 323–334.

Güler, A.D., Lee, H., Iida, T., *et al.* 2002. Heat-evoked activation of the ion channel, TRPV4. *Journal of Neuroscience*, **22** (15), 6408–6414.

Guenther, S., Reeh, P.W., and Kress, M. Rises in [Ca^{2+}]i mediate capsaicin- and proton-induced heat sensitization of rat primary nociceptive neurons. 1999. *European Journal of Neuroscience*, **11** (9), 3143150.

Hägermark, O. 1992. Peripheral and central mediators of itch. *Skin Pharmacology*, **5** (1), 1–8.

Han, L., Ma, C., Liu, Q., *et al.* 2013. A subpopulation of nociceptors specifically linked to itch. *Nature Neuroscience*, **16** (2), 174–182.

Huang, S.M., Lee, H., Chung, M.K., *et al.* 2008. Overexpressed transient receptor potential vanilloid 3 ion channels in skin keratinocytes modulate pain sensitivity via prostaglandin E2. *Journal of Neuroscience*, **28** (51), 13727–13737.

Huang, S.M., Li, X., Yu, Y., Wang, J., and Caterina MJ. 2011. TRPV3 and TRPV4 ion channels are not major contributors to mouse heat sensation. *Molecular Pain*, **7**, 37.

Imamachi, N., Park, G.H., Lee, H., *et al.* 2009. TRPV1- expressing primary afferents generate behavioral responses to pruritogens via multiple mechanisms. *Proceedings of the National Academy of Sciences, USA*, **106** (27), 11330–11335.

Iwasaki, Y., Morita, A., Iwasawa, T., *et al.* 2006. A nonpungent component of steamed ginger–[10]-shogaol–increases adrenaline secretion via the activation of TRPV1. *Nutritional Neuroscience*, **9** (3–4), 169–178.

Jinks, S.L. and Carstens, E. 2002. Responses of superficial dorsal horn neurons to intradermal serotonin and other irritants: comparison with scratching behavior. *Journal of Physiology*, **87**, 1280–1289.

Johanek, L.M., Meyer, R.A., Hartke, T., *et al.* 2007. Psychophysical and physiological evidence for parallel afferent pathways mediating the sensation of itch. *Journal of Neuroscience*, **27** (28), 7490–7497.

Jordt, S.E., Bautista, D.M., Chuang, H.H., *et al.* 2004. Mustard oils and cannabinoids excite sensory nerve fibres through the TRP channel ANKTM1. *Nature*, **427** (6971), 260–265.

Karrer, T. and Bartoshuk, L. 1991. Capsaicin desensitization and recovery on the human tongue. *Physiology & Behavior*, **49**, 757–764.

Kim, S.E., Patapoutian, A., and Grandl J. 2013. Single residues in the outer pore of TRPV1 and TRPV3 have temperature-dependent conformations. *PLoS One*, **8** (3), e59593.

Klein, A.H., Sawyer, C.M., Carstens, M.I., *et al.* 2010. Topical application of L-menthol induces heat analgesia, mechanical allodynia, and a biphasic effect on cold sensitivity in rats. *Behavioral Brain Research*, **212** (2), 179–186.

Klein, A., Carstens, M.I., and Carstens, E. 2011a. Facial injections of pruritogens or algogens elicit distinct behavior responses in rats and excite overlapping populations of primary sensory and trigeminal subnucleus caudalis neurons. *Journal of Physiology*, **106** (3), 1078–1088.

Klein, A.H., Carstens, M.I., Zanotto, K.L., Sawyer, C.M., Ivanov, M., Cheung, S., and Carstens, E. 2011b. Self- and cross-desensitization of oral irritation by menthol and cinnamaldehyde (CA) via peripheral interactions at trigeminal sensory neurons. *Chemical Senses*, **36** (2), 199–208.

Klein, A.H., Iodi Carstens, M., McCluskey, T.S., *et al.* 2011c. Novel menthol-derived cooling compounds activate primary and second-order trigeminal sensory neurons and modulate lingual thermosensitivity. *Chemical Senses*, **36** (7), 649–658.

Klein, A.H., Sawyer, C.M., Zanotto, K.L., *et al.* 2011d. A tingling sanshool derivative excites primary sensory neurons and elicits nocifensive behavior in rats. *Journal of Physiology*, **105** (4), 1701–1710.

Klein, A.H., Carstens, M.I., and Carstens, E. 2013. Eugenol and carvacrol induce temporally desensitizing patterns of oral irritation and enhance innocuous warmth and noxious heat sensation on the tongue. *Pain*, **154** (10), 2078–2087.

Klein, A.H., Joe, C.L., Davoodi, A., *et al.* 2014. Eugenol and carvacrol excite first- and second-order trigeminal neurons and enhance their heat-evoked responses. *Neuroscience*, **271**, 45–55.
Knowlton, W.M., Palkar, R., Lippoldt, E.K., *et al.* 2013. A sensory-labeled line for cold: TRPM8-expressing sensory neurons define the cellular basis for cold, cold pain, and cooling-mediated analgesia. *Journal of Neuroscience*, **33** (7), 2837–2848.
Komai, M. and Bryant, B.P. 1993. Acetazolamide specifically inhibits lingual trigeminal nerve responses to carbon dioxide. *Brain Research*, **612**, 122–129.
Koo, J.Y., Jang, Y., Cho, H., *et al.* 2007. Hydroxy-alpha- sanshool activates TRPV1 and TRPA1 in sensory neurons. *European Journal of Neuroscience*, **26**, 1139–1147.
LaMotte, R.H., Shimada, S.G., Green, B.G., and Zelterman, D. 2009. Pruritic and nociceptive sensations and dysesthesias from a spicule of cowhage. *Journal of Physiology*, **101** (3), 1430–1443.
Le Pichon, C.E. and Chesler, A.T. 2014. The functional and anatomical dissection of somatosensory subpopulations using mouse genetics. *Frontiers of Neuroanatomy*, **8**, 21.
Lee, M.H., Yeon, K.Y., Park, C.K., *et al.* 2005. Eugenol inhibits calcium currents in dental afferent neurons. *Journal of Dental Research*, **84** (9), 848–851.
Lennertz, R.C., Tsunozaki, M., Bautista, D.M., and Stucky, C.L. 2010. Physiological basis of tingling paresthesia evoked by hydroxy-alpha-sanshool. *Journal of Neuroscience*, **30** (12), 4353–4361.
Liao, M., Cao, E., Julius, D., and Cheng, Y. 2013. Structure of the TRPV1 ion channel determined by electron cryo-microscopy. *Nature*, **504** (7478), 107–112.
Liu, L., Oortgiesen, M., Li, L., and Simon SA. 2001. Capsaicin inhibits activation of voltage-gated sodium currents in capsaicin-sensitive trigeminal ganglion neurons. *Journal of Physiology*, **85** (2), 745–758.
Liu, Q., Sikand, P., Ma, C., *et al.* 2012. Mechanisms of itch evoked by β-alanine. *Journal of Neuroscience*, **32** (42), 14532–14537.
Liu, L. and Simon, S.A. 1996. Similarities and differences in the currents activated by capsaicin, piperine, and zingerone in rat trigeminal ganglion cells. *Journal of Physiology*, **76** (3), 1858–1869.
Liu, Q., Tang, Z., Surdenikova L., *et al.* 2009. GPCR Mrgprs are itch receptors mediating chloroquine-induced pruritus. Cell, **139** (7), 1353–1365.
Macpherson, L.J., Dubin, A.E., Evans, M.J., *et al.* 2007. Noxious compounds activate TRPA1 ion channels through covalent modification of cysteines. *Nature*, **445** (7127), 541–545.
Macpherson, L.J., Geierstanger, B.H., Viswanath, V., *et al.* 2005. The pungency of garlic: activation of TRPA1 and TRPV1 in response to allicin. *Current Biology*, **15** (10), 929–934.
Mandadi, S., Sokabe, T., Shibasaki, K., *et al.* 2009. TRPV3 in keratinocytes transmits temperature information to sensory neurons via ATP. *Pflugers Archiv*, **458** (6), 1093–1102.
Marics, I., Malapert, P., Reynders, A., Gaillard, S., and Moqrich A. 2014. Acute heat-evoked temperature sensation is impaired but not abolished in mice lacking TRPV1 and TRPV3 channels. *PLoS One*, **9** (6), e99828.
Mason, J.R., Bean, N.J., Shah, P.S., and Clark, L. 1991. Taxon-specific differences in responsiveness to capsaicin and several analogues: Correlates between chemical structure and behavioral aversiveness. *Journal of Chemical Ecology*, **17** (12), 2539–2551.
McKemy D.D. 2013. The molecular and cellular basis of cold sensation. *ACS Chemical Neuroscience*, **4** (2), 238–247.
McKemy, D.D., Neuhausser, W.M., and Julius, D. 2002. Identification of a cold receptor reveals a general role for TRP channels in thermosensation. *Nature*, **416** (6876), 52–58.
Merrill, A.W., Cuellar, J.M., Judd, J.H., Carstens, M.I., and Carstens, E. 2008. Effects of TRPA1 agonists mustard oil and cinnamaldehyde on lumbar spinal wide-dynamic range neuronal responses to innocuous and noxious cutaneous stimuli in rats. *Journal of Physiology*, **99** (2), 415–425.
Moqrich, A., Hwang, S.W., Earley, T.J., *et al.* 2005. Impaired thermosensation in mice lacking TRPV3, a heat and camphor sensor in the skin. *Science*, **307** (5714), 1468–1472.
Namer, B., Carr, R., Johanek, L.M., *et al.* 2008. Separate peripheral pathways for pruritus in man. *Journal of Physiology*, **100** (4), 2062–2069.
Namer, B., Seifert, F., Handwerker, H.O., and Maihöfner C. 2005. TRPA1 and TRPM8 activation in humans: effects of cinnamaldehyde and menthol. *Neuroreport*, **16** (9), 955–959.
Nilius, B., Bíró, T., and Owsianik, G. 2014. TRPV3: time to decipher a poorly understood family member! *Journal of Physiology*, **592** (Pt 2), 295–304.
Nolano, M., Simone, D.A., Wendelschafer-Crabb, G., *et al.* 1999. Topical capsaicin in humans: parallel loss of epidermal nerve fibers and pain sensation. *Pain*, **81** (1–2), 135–145.

O'Neill, J., Brock, C., Olesen, A.E., Andresen, T., Nilsson, M., Dickenson, A.H. 2012. Unravelling the mystery of capsaicin: a tool to understand and treat pain. *Pharmacological Reviews*, **64** (4), 939–971.

Park, C.K., Li, H.Y., Yeo, K.Y., *et al.* 2006. Eugenol inhibits sodium currents in dental afferent neurons. *Journal of Dental Research*, **85**, 900–904.

Parker, G.H. 1912. The relation of smell, taste and the common chemical sense in vertebrates. *Journal of the Academy of Natural Science (Philadelphia)*, Ser. 2, **15**, 221–234.

Peier, A.M., Moqrich, A., Hergarden, A.C., *et al.* 2002a. A TRP channel that senses cold stimuli and menthol. *Cell*, **108** (5), 705–715.

Peier, A.M., Reeve, A.J., Andersson, D.A., *et al.* 2002b. A heat-sensitive TRP channel expressed in keratinocytes. *Science*, **296** (5575), 2046–2049.

Peyrot des Gachons, C., Uchida, K., Bryant, B., *et al.* 2011. Unusual pungency from extra-virgin olive oil is attributable to restricted spatial expression of the receptor of oleocanthal. *Journal of Neuroscience*, **31** (3), 999–1009.

Pogorzala, L.A., Mishra, S.K. and Hoon, M.A. 2013. The cellular code for mammalian thermosensation. *Journal of Neuroscience*, **33**, 5533–5541.

Prescott, J. and Swain-Campbell, N. 2000. Responses to repeated oral irritation by capsaicin, cinnamaldehyde and ethanol in PROP tasters and non-tasters. *Chemical Senses*, **25** (3), 239–246.

Reddy, V.B., Iuga, A.O., Shimada, S.G., LaMotte, R.H., and Lerner, E.A. 2008. Cowhage-evoked itch is mediated by a novel cysteine protease: a ligand of protease-activated receptors. *Journal of Neuroscience*, **28** (17), 4331–4335.

Reid, G., Babes, A., and Pluteanu, F. 2002. A cold- and menthol-activated current in rat dorsal root ganglion neurones: properties and role in cold transduction. *Journal of Physiology*, **545**, 595–614.

Riera, C.E., Menozzi-Smarrito, C., Affolter, M., *et al.* 2009. Compounds from Sichuan and Melegueta peppers activate, covalently and non-covalently, TRPA1 and TRPV1 channels. *British Journal of Pharmacology*, **157**, 1398–1409.

Riera, C.E., Menozzi-Smarrito, C, Affolter, M., *et al.* 2009. Compounds from Sichuan and Melegueta peppers activate, covalently and non-covalently, TRPA1 and TRPV1 channels. *British Journal of Pharmacology*, **157** (8), 1398–1409.

Ringkamp, M., Schepers, R.J., Shimada, S.G., *et al.* 2011. A role for nociceptive, myelinated nerve fibers in itch sensation. *Journal of Neuroscience*, **31** (42), 14841–14849.

Roberson, D.P., Gudes, S., Sprague, J.M., *et al.* 2013. Activity-dependent silencing reveals functionally distinct itch-generating sensory neurons. *Natural Neuroscience*, **16** (7), 910–918.

Rolls, E.T., Verhagen, J.V., and Kadohisa, M. 2003. Representations of the texture of food in the primate orbitofrontal cortex: neurons responding to viscosity, grittiness, and capsaicin. *Journal of Physiology*, **90** (6), 3711–3724.

Rosenthal, P. and Borsook, D. 2012. The corneal pain system. Part I: the missing piece of the dry eye puzzle. *Ocular Surface*, **10** (1), 2–14.

Sawyer, C.M., Carstens, M.I., and Carstens, E. 2009a. Mustard oil enhances spinal neuronal responses to noxious heat but not cooling. *Neuroscience Letters*, **461** (3), 271–274.

Sawyer, C.M., Carstens, M.I., Simons, C.T., *et al.* 2009b. Activation of lumbar spinal wide-dynamic range neurons by a sanshool derivative. *Journal of Physiology*, **101** (4), 1742–1748.

Schmelz, M., Schmidt, R., Bickel, A., Handwerker, H.O., and Torebjörk, H.E. 1997. Specific C-receptors for itch in human skin. *Journal of Neuroscience*, **17** (20), 8003–8008.

Schmidt, M., Dubin, A.E., Petrus, M.J., Earley, T.J., and Patapoutian, A. 2009. Nociceptive signals induce trafficking of TRPA1 to the plasma membrane. *Neuron*, **64**, 498–509.

Sikand, P., Shimada, S.G., Green, B.G., and LaMotte, R.H. 2009. Similar itch and nociceptive sensations evoked by punctate cutaneous application of capsaicin, histamine and cowhage. *Pain*, **144** (1–2), 66–75.

Sikand, P., Dong, X., and LaMotte, R.H. 2011a. BAM8-22 peptide produces itch and nociceptive sensations in humans independent of histamine release. *Journal of Neuroscience*, **31** (20), 7563–7567.

Sikand, P., Shimada, S.G., Green, B.G., and LaMotte, R.H. 2011b. Sensory responses to injection and punctate application of capsaicin and histamine to the skin. *Pain*, **152** (11), 2485–2894.

Simons, C.T., Boucher, Y., and Carstens, E. 2003a. Suppression of central taste transmission by oral capsaicin. *Journal of Neuroscience*, **23** (3), 978–985.

Simons, C.T., Carstens, M.I., and Carstens, E. 2003b. Oral irritation by mustard oil: self-desensitization and cross-desensitization with capsaicin. *Chemical Senses*, **28** (6), 459–465.

Simons, C.T., Boucher, Y., Carstens, M.I., and Carstens, E. 2006. Nicotine suppression of gustatory responses of neurons in the nucleus of the solitary tract. *Journal of Physiology,* **96** (4), 1877–1886.

Simons, C.T., Dessirier, J.M., Carstens, M.I., O'Mahony, M., and Carstens, E. 1999. Neurobiological and psychophysical mechanisms underlying the oral sensation produced by carbonated water. *Journal of Neuroscience,* **19** (18), 8134–8144.

Simons, C.T., O'Mahony, M., and Carstens, E. 2002. Taste suppression following lingual capsaicin pre-treatment in humans. *Chemical Senses,* **27** (4), 353–365.

Simons, C.T., Sudo, S., Sudo, M., and Carstens, E. 2004. Mustard oil has differential effects on the response of trigeminal caudalis neurons to heat and acidity. *Pain,* **110** (1–2), 64–71.

Smith, G.D., Gunthorpe, M.J., Kelsell, R.E., *et al.* 2002. TRPV3 is a temperature-sensitive vanilloid receptor-like protein. *Nature,* **418** (6894), 186–190.

Stevens, D.A. and Lawless, H.T. 1986. Putting out the fire: effects of tastants on oral chemical irritation. *Perception & Psychophysics,* **39** (5), 346–350.

Story, G.M., Peier, A.M., Reeve, A.J., *et al.* 2003. ANKTM1, a TRP-like channel expressed in nociceptive neurons, is activated by cold temperatures. *Cell,* **112** (6), 819–829.

Sudo, S., Sudo, M., Simons, C.T., Dessirier, J.M., and Carstens, E. 2002. Sensitization of trigeminal caudalis neuronal responses to intraoral acid and salt stimuli and desensitization by nicotine. *Pain,* **98** (3), 277–286.

Sugai, E., Morimitsu, Y., Iwasaki, Y., *et al.* 2005. Pungent qualities of sanshool-related compounds evaluated by a sensory test and activation of rat TRPV1. *Bioscience, Biotechnology, and Biochemistry,* **69** (10), 1951–1957.

Talavera, K., Gees, M., Karashima, Y., *et al.* 2009. Nicotine activates the chemosensory cation channel TRPA1. *Nature Neuroscience,* **12** (10), 1293–1299.

Tewksbury, J.J. and Nabhan, G.P. 2001. Seed dispersal. Directed deterrence by capsaicin in chilies. *Nature,* **412** (6845), 403–404.

Tsagareli, M.G., Tsiklauri, N., Zanotto, K.L., *et al.* 2010. Behavioral evidence of thermal hyperalgesia and mechanical allodynia induced by intradermal cinnamaldehyde in rats. *Neuroscience Letters,* **473** (3), 233–236.

Viana, F., de la Pena, E., and Belmonte, C. 2002. Specificity of cold thermotransduction is determined by differential ionic channel expression. *Nature Neuroscience,* **5**, 254–260.

Vincent, F. and Duncton, M.A. 2011. TRPV4 agonists and antagonists. *Current Topics in Medicinal Chemistry,* **11** (17), 2216–2226.

Wang, Y.Y., Chang, R.B., Waters, H.N., McKemy, D.D., and Liman, E.R. 2008. The nociceptor ion channel TRPA1 is potentiated and inactivated by permeating calcium ions. *Journal of Biological Chemistry,* **283** (47), 32691–32703.

Whitehead, M.C. and Kachele, D.L. 1994. Development of fungiform papillae, taste buds, and their innervation in the hamster. *Journal of Comparative Neurology,* **340** (4), 515–530.

Wilson, S.R., Gerhold, K.A., Bifolck-Fisher, A., *et al.* 2011. TRPA1 is required for histamine-independent, Mas-related G protein-coupled receptor- mediated itch. *Nature Neuroscience,* **14** (5), 595–602.

Wise, P.M., Wolf, M., Thom, S.R., and Bryant, B. 2013. The influence of bubbles on the perception carbonation bite. *PLoS One,* **8** (8), e71488.

Xing, H., Ling, J., Chen, M., and Gu, J.G. 2006. Chemical and cold sensitivity of two distinct populations of TRPM8-expressing somatosensory neurons. *Journal of Physiology,* **95**, 1221–1230.

Xu, H., Delling, M., Jun, J.C., and Clapham, D.E. 2006. Oregano, thyme and clove-derived flavors and skin sensitizers activate specific TRP channels. *Nature Neuroscience,* **9** (5), 628–635.

Xu, H., Ramsey, I.S., Kotecha, S.A., *et al.* 2002. TRPV3 is a calcium-permeable temperature-sensitive cation channel. *Nature,* **418** (6894), 181–186.

Zanotto, K.L., Iodi Carstens, M., and Carstens, E. 2008. Cross-desensitization of responses of rat trigeminal subnucleus caudalis neurons to cinnamaldehyde and menthol. *Neuroscience Letters,* **430** (1), 29–33.

Zanotto, K.L., Merrill, A.W., Carstens, M.I., and Carstens, E. 2007. Neurons in superficial trigeminal subnucleus caudalis responsive to oral cooling, menthol, and other irritant stimuli. *Journal of Physiology,* **97** (2), 966–978.

Zylka, M.J., Rice, F.L., and Anderson, D.J. 2005. Topographically distinct epidermal nociceptive circuits revealed by axonal tracers targeted to Mrgprd. *Neuron,* **45** (1), 17–25.

Index

Page numbers in *italics* refer to illustrations; those in **bold** refer to tables